GEODESY?WHAT'S THAT?

GEODESY?WHAT'S THAT?

✦

MY PERSONAL INVOLVEMENT IN THE AGE-OLD QUEST FOR THE SIZE AND SHAPE OF THE EARTH

WITH A RUNNING COMMENTARY ON LIFE IN A GOVERNMENT RESEARCH OFFICE

by Irene K. Fischer

iUniverse, Inc.

New York Lincoln Shanghai

GEODESY?WHAT'S THAT?
MY PERSONAL INVOLVEMENT IN THE AGE-OLD QUEST FOR THE SIZE AND SHAPE OF THE EARTH

iUniverse books may be ordered through booksellers or by contacting:

iUniverse
2021 Pine Lake Road, Suite 100
Lincoln, NE 68512
www.iuniverse.com
1-800-Authors (1-800-288-4677)

Portions of this text have been serialized in the American Congress on Surveying and Mapping's ACSM Bulletin Jan/February 2004-March/Arpil 2006 (Numbers 207 to 220)

The most complete version of this text is available at The Arthur & Elizabeth Schlesinger Library for the History of Women in America Radcliffe Institute for Advanced Study, Harvard University and at The American Institute of Physics Library, University of Maryland

ISBN-13: 978-0-595-36399-5 (pbk)
ISBN-13: 978-0-595-80834-2 (ebk)
ISBN-10: 0-595-36399-7 (pbk)
ISBN-10: 0-595-80834-4 (ebk)

Printed in the United States of America

We are grateful to Bernard Chovitz for his editing and supervision and making the ACSM publication possible, to Ed McKay for digitizing, and to Ilse Genovese, the Editor of the ACSM Bulletin for providing a venue in the Bulletin and for her fine layout and editing The present version was scanned at MIT from Irene Fischer's typescript and corrected and indexed by Michael Fischer. Thanks to friend Peter Ford of MIT's Center for Space Research for help in proofing-reading, and for interest in and enthusiasm for the memoirs.

Contents

FROM THE FEDERALESE ALPHABET

ACIC	= Aeronautical Charts and Information Center
ACSI	= Army Chief of Staff Intelligence
ACSM	= American Congress on Surveying and Mapping
AGU	= American Geophysical Union
AMS	= Army Map Service
AMSFE	= Army Map Service: Far East
AVN	= Allgemeine Vermessungs-Nachrichten
Baker-Nunn camera	= Satellite technique sponsored by SAO
BC-4	= Ballistic Camera-4, satellite technique sponsored by C & GS
C & GS	= U.S. Coast and Geodetic Survey
CO	= Commanding Officer
DESPA	= Department of Engineer Special Project Area
DIA	= Defense Intelligence Agency
DIA MC	= Defense Intelligence Agency: Mapping and Charting
DMA	= Defense Mapping Agency
DMATC	= DMA Topographic Center
DoD	= Department of Defense
EAPA	= Engineering Applications Project Assignment
ED	= European Datum
FBM	= Fleet Ballistic Missiles

FIG	= Fédération Internationale des Géomètres (International Federation of Surveyors)
FOUO	= For Official Use Only
FY	= Fiscal Year
GSEOPC-IV	= Geodesy/Solid Earth and Ocean Physics Conference IV
GEOS-C	= Geodetic Earth Orbiting Satellite (third in the series); or Geodynamic Experimental Oceanographic Satellite
GIMRADA	= (U.S.Army Engineer) Geodesy, Intelligence and Mapping Research and Development Agency
GS	= General Schedule (Federal Pay System)
GSA	= General Services Administration
HIRAN	= High Precision Radio Navigation System
IAG	= International Association of Geodesy
IAGS	= Inter American Geodetic Survey
IAU	= International Astronomical Union
IHR	= International Hydrographic Review
IUGG	= International Union of Geodesy and Geophysics
JGR	= Journal of Geophysical Research
MAR	= Mid-Atlantic Ridge
MATS	= Military Air Transport Service
MMD 68	= Modified Mercury Datum of 1968
NAD	= North American Datum
NASA	= National Aeronautics and Space Administration
NOAA	= National Oceanic and Atmospheric Administration
NOO	= Naval Oceanographic Office
NRL	= Naval Research Laboratory
OCE	= Office of the Chief of Engineers
OSU	= Ohio State University
PAI	= Project Assignment Instruction

PAIGH = Pan American Institute of Geography and History

PSAD 56 = Provisional South American Datum of 1956

R & A = Research and Analysis Branch (Division)

RIF = Reduction in Force (government mass firing of employees)

SAD 69 = South American Datum of 1969

SAO = Smithsonian Astrophysical Observatory (Cambridge, Mass.)

SECOR = Sequential Collation of Ranges, satellite technique sponsored by AMS

SNAFU = Situation normal: all fouled up

SSG = (International) 'Special Study Group (of the IAG)

TC = Topographic Center

TDY = Tour of Duty

TOPOCOM = Topographic Command

TR = Technical Report

UTM = Universal Transverse Mercator Projection

WGS = World Geodetic System

INTRODUCTION: JOYS AND WOES OF A GOVERNMENT RESEARCHER: MY PERSONAL INVOLVEMENT IN A GLORIOUS CHAPTER OF COSMOPOLITAN GEODESY

"And now, if you care to speak to us, Dr. Fischer—please take all the time you wish." It was my retirement party, complete with corsage, another medal, presents, speeches, and refreshments. At my request, it was held at the office, during office hours, so that it would be easier for everyone who cared to come. The speeches were over. I had intended to say just a few words of thanks and then say good-bye individually. But as I looked over the rows and rows of friendly and expectant faces, the realization hit me that I would be here just one more day, after having come here day after day after day for so many years. Was it really twenty-five years? It seemed like a dream gone by so fast. There were several people present who had left this agency years ago; several of my previous bosses had taken time out to come; some who could not make it that day, had called me before to say good-bye. And memories connected with these faces suddenly came alive and compressed those years into a rich kaleidoscope of personal encounters, insights, and changing times. I started telling about little events of the past, as they happened to come to mind, painting with these vignettes the atmosphere of happy work years—and I felt the nostalgic response from the audience as they remembered too.

The following day, my last day in the office, was taken up with administrative formalities and last good-byes. Many people told me again, how they had enjoyed my tales and how these had brought back their own memories of good times, along with mine. This gave me an idea which later grew into plans for this book.

The twenty-five years of my Government service (mid 1952–mid 1977) comprise not only an important part of my personal life (rather a distinct life out of my many different lives), but they coincide more or less with the existence of the "Geoid Branch" in the Department of Geodesy of the U.S. Defense Mapping Agency. More than that, they happen to span a glorious chapter in the history of geodesy when our scanty knowledge of the size and shape of the Earth developed from unconnected parts of the pre-satellite world to the know-it-all of the space age. And even more than that: the Geoid Branch took a leading role in international geodesy during that time, as is well known and acknowledged in the geodetic literature. It is not my intent here, however, to describe the technical development of geodesy as such. That has been done elsewhere by several authors, including my own long review article "The Figure of the Earth—Changes in Concepts' (*Geophysical Surveys*, vol.2, no.1, 1975, 3-54) which takes the story from ancient times to tomorrow. The present book, by contrast, is an entirely personal, subjective book about my personal experiences, trials and errors, goals and insights, while learning about and working in a fascinating field. It is based on my personal notes which I habitually took during those years, documents, diaries, correspondence of international scope, and the background of my many publications. As I sift through this material, memories come back to enliven the written evidence with the ambiance of the day-to-day life in the office. To keep this account authentically first-hand, it will of necessity be centered around my own experiences and omit those of others unless affecting mine.

Two topics of contemporary interest appear as a natural fall-out: my experience as a woman in a man's world, and my observations of the role of management in a research environment. These topics become clearer in hindsight. I will try to articulate my thoughts while I relive my story to tell it in these pages as it unfolds from the diaries. Especially the various management philosophies represented by inspiring leaders, cautious bureaucrats, and politicking underminers, will then appear in terms of their direct impact on daily life rather than as abstract generalities.

My story is meant also as a tribute to the many friendly souls at all levels who made my government service such an interesting and satisfying experience as I watched the "big wheels go around me in circles."

1

WHAT I REMEMBER ABOUT MY FIRST YEAR AT THE ARMY MAP SERVICE (mid 1952–mid 1953)

I. LOOKING FOR A JOB

Our daughter, Gay, wanted to go to college somewhere out of town. We had anticipated sending her to the University of Maryland as the most natural place to choose, since we lived in Takoma Park, Maryland, and there was a direct bus line to the college with a bus stop in front of the house. But we found out that we were naive, old-fashioned, definitely not with the times. "I wouldn't want to be caught dead there," she announced in fashionable teenager language, obviously influenced by her peers and the school counselor. So we tried to understand her world and to appreciate the importance of the four golden college years in the dreams, development, and eventual memories of an American youngster. As we could not afford an out-of-state college on one salary, I was going to look for a job. I had a very good formal education as a graduate of the University of Vienna, Austria, in mathematics, and extensive training in descriptive geometry at the Vienna Institute of Technology, and in geography and natural sciences at the Vienna Pedagogical Institute. I had several years' teaching experience and had worked for Professor John T. Rule at the Massachusetts Institute of Technology on U.S. Navy contracts for visual training aids, where I could apply my specialty of descriptive geometry to construct stereoscopic slides for courses in celestial mechanics, geometry, etc. For the benefit of American employers, I had my European documents officially evaluated by Professor Robert Ulich, Harvard University Graduate School of Education. Several years back, I had stopped working in order to stay home with our infant son until he would be old enough for full-day school. Michael would be in second grade next fall, and we could make arrangements with the Menard family up the street: only about an hour before and after school at the Menard home, and going to and from school with his classmate Dale Menard would solve the problem of my longer work day. With a Civil Service rating as a GS-9 (General Schedule, Grade 9), I applied at the David Taylor Model Basin, but when it came to an interview, Eric, my husband, could not see it: "You are not going to work for the Navy when I am working for the Army, at AMS (the Army Map Service)!" He took my curriculum vitae and background papers to Dr. John A. O'Keefe, Chief of the Research and Analysis Branch in Geodesy at AMS, who was interested in hiring me right away. The Personnel Office, however, was negatively impressed by my GS-9 rating and refused me the job they intended to fill at the GS-7 level, because I was overqualified. Obviously, they were not interested in quality. If I wanted the job, I had to bring them a GS-7 rating. So I had to ask the Civil Service Commission at the National Bureau of Standards for a GS-7 rating, which finally did get me a job in the "Long Lines Section" at a salary of $4,205. I was hired in August 1952, but started work in September after Michael was safely installed in his school routine.

This first experience with personnel policies was corroborated later when a well-meaning coworker advised me not to work so hard: a woman would never advance beyond a GS-7 at AMS anyway unless she was made a supervisor. Looking around, I saw that there were actually only GS-5s and GS-7s among the girls, except one, Pamela Anderson, GS-9 supervisor of the Occultation Section. Well, sometimes things turn out better than foretold. About a year and a half later, I got a GS-9 without being a supervisor, and I did quite well after that.

2. ENTERING A NEW WORLD

In my first interview with Dr. O'Keefe, Branch Chief, and Bernard H. Chovitz, Section Chief, they told me about some of the work that was going on, and pleasantly introduced me to everyone in the Branch. "You will need time to remember their names, but they know you now and will make you comfortable." And so it was. There was a very pleasant atmosphere of interested, capable people, whose strong individualities were described to me at a later time as "a group of handpicked characters."

There are two things that I still remember of that introductory conversation. One: that they tried to determine the size and shape of the Earth, and my nonplussed reaction: wasn't I taught that in grade school already? How come, they don't know? And two: a seemingly casual question by Dr. O'Keefe, whether I thought that the Mercator projection was a purely geometrical projection onto a tangent cylinder. Now, that one rang a bell! In anticipation of working for a mapping agency, I had scanned Deetz and Adams' *Elements of Map Projection* in the past week. I had been amused by the rather emotional remonstration against the error in many textbooks, which describe the Mercator projection as a cylindrical projection rather than a mathematical arrangement of the parallels to make the rhumb line intersect the meridians under a constant angle. So that question, a rather detailed one for an entrance interview at the GS-7 level, struck me right away as a shibboleth. And I watched with amusement the exchange of glances after my answer, and I knew I was accepted.

There was yet another shibboleth to pass in the first few days: a little informal mathematical quiz. Apparently, every newcomer was given such a quiz as a sort of placement aid. It consisted of reasonable problems for a recent college student, which required some specific college course formulas. Having been out of school for such a long time, I could not quite remember the exact coefficients, which one normally would look up anyway; but I had nothing in which to look them up. I would not want to give the impression of having any difficulty; so, rather than ask for such a formula book, which probably was on every desk, I decided to rederive that formula. It took a little longer but the hurdle was passed. In later weeks, I had the feeling that there were some more unobtrusive performance tests, but I did not mind since I was being introduced into an entirely new and fascinating field.

3. A BUSY PLACE FOR A HANDFUL OF PEOPLE

The first couple of weeks I spent struggling with and eventually subduing that unfamiliar creature, the desk calculator. Only then I could start to assist the other people of the Section in their varied tasks, and I learned as I went along. The "Long Line Section" had only four, a little later five, regular members at the time. Chovitz was Section Chief. (He later became Director of the Geodetic Research and Development Laboratory at NOAA, the National Oceanic and Atmospheric Administration; President of Section 2 of the IAG, the International Association of Geodesy, in 1975; and President of the Geodesy Section of the AGU, the American Geophysical Union, for 1978-1980.) Then there was Vic Lewicke, Rouleau, and me. A little later, Emanuel Sodano, who had been transferred from this Section to another one the year before, transferred back in. He was the author of Technical Report No. 7, an original contribution to the geodetic problem of very long lines, and I was impressed without yet understanding the subject.

As Chovitz put it: we were few in numbers, but "elite" in quality and that was what counted. The projects were extremely varied for such a small group, ranging from various map projection studies to the construction of geodetic tables and nomograms (diagrams allowing one to determine an unknown value from its graphically shown relationships to several known values) to Figure of the Earth studies (which later bloomed into a major undertaking), and weird topics thrown in, whose meaning I learned to grasp only gradually: geodetic datums (specifically defined coordinate systems for the triangulation in particular countries), datum transformations (mathematically changing one such coordinate system into that of a neighboring country), submarine gravity studies (gravity measurements made in submarines), deflections of the vertical (more about those later on), long lines (computation of length and direction of the shortest connection on the Earth's surface between any two given points; or finding the second point if the distance and direction from the known location of the first one are given), flare triangulation (using parachute flare signals as an intermediate observable object to lengthen the observable triangulation distances), guided missiles, and others. Also many diversified short-range requests from the outside needed attention and required much correspondence. And there were recurring administrative chores such as time sheets, reports of the work status, and estimates of future workloads. While Chovitz took the obvious lead in all these tasks, he was personally involved in a theoretical study of criteria to distinguish between various types of map projections. He had just published the first in a series of three papers on the subject ("Classification of map projections in terms of the metric tensor to the second order", *Boll. di Geodesia e Scienze Affini*, Anno XI, no. 4, 1952), and was working on a continuation which would be published in 1954 ("Some applications of the classification of map projections in terms of the metric tensor to the second order", *Boll. di Geodesia e Scienze Afini*, Anno XIII, no.1, with Italian translation added). I enjoyed helping with the long drawn-out equations. The third paper appeared another two years later ("A general formula for ellipsoid–to-ellipsoid mapping", *Boll. di Geodesia e Scienze Affini*, Anno XV, no.1, 1956).

I was particularly pleased and made to feel at home when I discovered at a later time that Chovitz knew of and appreciated the "Wiener Kreis" (Vienna Circle), the center of "logical positivism" in the 1920s and 1930s. I had been strongly influenced by the Wiener Kreis in my student days at the University of Vienna. Some of its members had been my revered teachers:

Moritz Schlick, Friedrich Waismann, Rudolf Carnap, Hans Hahn; and I have always been grateful to them for strengthening my ability to analyze concepts and statements for the scope of their meaning. Chovitz was specifically interested in Kurt Gödel, and I was happy to give him copies of Gödel's two famous theorems, which he had formulated as a young man in Vienna when I was a student there. ("Die Vollständigkeit der Axiome des logischen Funktionenkaküls" and "Über formal unentscheidbare Sätze der Principia Mathematica und verwandter Systeme," *Monatshefte für Math. und Phys.*, 1929 and 1930, resp.).

This unexpected interest brought back to me nostalgic memories of an important phase of my formative years. Even before my university years, friends who were already university students took me along to Fritz Waismann's evening classes on the foundations of mathematics and of scientific knowledge, which I greatly enjoyed and which led me to reading literature on these topics and to write a voluntary thesis on "The Development of the Number Concept" for the "Matura". This is the comprehensive graduation examination from the Austrian secondary school that is required for entry into the university. The certificate says that I passed "mit Auszeichnung" (*cum laude*) and that "the thesis indicates unusually highly developed mathematical reasoning." This official comment had no way of turning my head at that time, since all my friends were outstanding and several of my classmates had also written voluntary theses of high caliber in their field of interest. It only meant to me that I conformed to the level of my peers, that I had fulfilled my obligation so far, and that I was expected to do my best with the opportunities offered at the university. And these opportunities were many. Each year the catalogue of courses offered was eagerly awaited, since it differed each year for other than the required basic courses. You made your choice like choosing from a box of chocolate candy. You enrolled and paid for the required courses but you really went to a variety of other more interesting courses of your choice. Presence in lecture halls of one or two hundred students was, of course, not checked. So it was left to your supposedly mature judgment to make the most of these years, or to be a never finishing bum. The transition from the closely supervised high school to this university system was smoothed for me by my friends, a little older and therefore so much wiser, and by the fact that Professor Hans Hahn, mathematician, had evaluated my Matura thesis the year before because it had been beyond the scope of my high school mathematics teacher. Not only did I enroll in his courses, actually went there every day, took detailed notes, and took the voluntary semester exams, but I took pleasure in his clear, logical presentations not only in mathematics but also in his philosophical talks and papers as a member of the Vienna Circle.

This group of professors brought their background in mathematics and physics to bear on philosophy in contrast to the historical-linguistic approach of metaphysical systems. They saw it as their badly needed task to clarify what we think we know or don't know on the basis of scientific efforts. They tried to analyze statements about the world in a scientific fashion, distinguishing between logical, mathematical tautologies and statements which could or could not be tested in the physical world by empirical criteria; and they discussed the role of hypotheses in the construction of a scientific world view as a matrix from which testable specific statements could be derived. Their fight was directed mainly against confusing unprovable metaphysical notions with science, and thus provided us students not only with an invigorating atmosphere of probing discussions but also with a strict discipline in clear thinking. Some of their formulations were criticized as too narrow and extreme, such as e.g. that "the meaning of a statement is the method of its (at least theoretical) verification" and that sentences without such "meaning" are "meaningless" and thus inadmissible. It led to comments that the logical positivists were cutting off the limb they were sitting on because they admitted less and less as meaningful statements and had nothing much of interest left to talk about. Also the "verification" was attacked, because only "falsification" can give a definitive though negative test result. Some of such attacks seemed to (deliberately?) misinterpret the intent of the wording, such as the play on the two uses of the word "meaningless": a technical criterion for admitting or excluding a statement as scientific, and the non-technical usage of "nonsense". The sentence "the world is beautiful" would not pass the technical criterion and thus be meaningless as a scientific statement, but it would still have meaning as an expression of feeling. We thrived on such discussions and embroidered them on our Sunday hikes into the Vienna woods. The topics ranged from the concept of infinity in mathematics and why there are more decimal fractions than you can count even if you lived forever, to the question how do you know that you won't live forever or that the sun will rise again tomorrow. After all, you only know of past cases, not all future cases; so how can you predict? Well, these have been useful working hypotheses so far. If the sun does not rise tomorrow, the hypothesis is falsified and you'll have to scrap or revise it. And if you die, the hypothesis will be confirmed in your case, but you still cannot know for every case, should you then still care. On the impossibility to learn anything about the physical world from pure mathematics and logic, we loved to collect aphorisms such as Bertrand Russell's famous quip "Mathematics is the science in which we never know what we are talking about, nor whether what we say is true," which contrasts nicely with

Leopold Kronecker's view that "the counting numbers are the Lord's creation, the rest of arithmetic is man's work".

The Vienna Circle dispersed by emigration; the two main personalities, Professors Hahn and Schlick, had died before. While I did not keep up with the debates and developments in later years, the basic attitude of separating facts from fiction when trying to understand a scientific statement remained with me: to identify its factual content through rewording it as a method to be followed, which should lead to a predicted specific observable result. It means that if I were doing this, this, and that, I should find that this is the case (or not). In other words, I would understand the meaning of a scientific statement, if I could examine the conditions under which it would be either true or false. The analytical training I absorbed from the Vienna Circle stood me in good stead throughout my career.

Now, back to my environment so many years later: The Long Lines Section, as well as a Datum Section, an Occultation Section, and a Computation Section (the acorn which soon grew into a branch and later into a mighty oak, the current Department of Computer Science) belonged then to the Research and Analysis (R & A) Branch of about 30 people, all in one long room, served by one secretary and one telephone. Subdivisions between sections were accomplished by file cabinets. But dividing lines were crossed frequently to give amiable help in times of overload and pressing due dates, to assist by independent checking, or to engage in joint ventures. Thus Victor Montella helped out for a while in our Long Lines Section, Ted Gaskill and Sören W. Henriksen shared discussions with us, and our Rouleau and Lewicke helped out in their sections. J.W.H. Spencer and Don Stuart, Ed Early and Eileen Segal took care of our mass computations. To mention a few more names of these early days: Nora Moser, George Taylor, Jake Walkeau, Martha Carta, Jim Walker, Harry A. Lieberman, Isaac Lail, Howard S. Genatt.

The soul and driving force of this R & A Branch was its Chief, Dr. John A. O'Keefe (later at NASA, the National Aeronautics and Space Administration). He had participated as an astronomer in the 1948 Eclipse Expedition to northern Japan ("The National Geographic Society 1948 Eclipse Expedition to Rebun Jima", *Survey and Mapping*, Vol. IX, no.3, 1949) to contribute to an effort of determining the relative geodetic positions of the triangulation systems in India, Japan, and North America. The National Geographic Society had set up seven observation sites from Burma to Japan and the Aleutians. The U.S. Army helped with logistic support for the Rebun Jima site, and Lt. Cordova (who was to be our Director of the Topographic Center almost 30 years later) was involved in that, as he served on the Post Hostilities Mapping Programs in Japan and Korea at that time. The eventual goal of such intercontinental geodetic connections was a modern determination of the size and shape of the Earth.

For the same goal, occultation techniques were studied, and Dr. O'Keefe had just completed a detailed paper on four occultations observed at nine stations in the United States. The elaborate computations had been carried out in the Occultation Section under the guidance of Pam Anderson ("The Earth's Equatorial Radius and the Distance of the Moon", by J.A. O'Keefe and J.P. Anderson, *Astron. Journal*, Vol.57, no.4, 1952). Another paper was in the making: Technical Report No. 18: "A Rough Measurement of the Geodetic Relationship of Europe to North America, Using Visually Observed Occultations." Dr. O'Keefe's interest in these extraterrestrial methods for geodetic purposes made him naturally an early advocate of plans for a future artificial satellite whose geodetic possibilities he explored long before the actual event.

Dr. O'Keefe was instrumental in providing the U.S. Army with a new rectangular military grid, replacing the previously used polyconic projection with the Universal Transverse Mercator Projection (UTM). He wrote AMS Technical Manual No. 19 for explanation. A series of Technical Manuals were being prepared to provide tables for practical applications. At the same time, he explored, among other things, the theoretical basis of different procedures to transfer geodetic positions from one reference ellipsoid to another of different curvature via map projections, and had written several papers on the subject. (Specific ellipsoids are used as geometric earth models as part of the geodetic datum.) He kept close, personal contact with geodesists, astronomers, and geophysicists all over the world, responding to problems under discussion and formulating his own aspects and questions. His strong presence in the Branch was a constant delightful inspiration and stimulus, and his scientific curiosity and enthusiasm was contagious.

He was interested also in every one of his employees and tried to match project assignments and capabilities. He was a great, natural teacher. It was my personal luck that a man of his caliber would be interested in guiding my geodetic education from zero to something. I am grateful to him for introducing me to a happy second career for the next twenty-five years.

4. LEARNING ABOUT GEODESY

This education took the form of first letting me read and study the geodetic bible: the Jordan-Eggert *Handbuch der Vermessungskunde*, which was a treat for me and easily accessible through my familiarity with the German language. There was no English translation, and thus other novices in the field were given the English language primer of Hosmer *Geodesy* and never knew what they missed. To help matters in that respect, Dr. O'Keefe had asked Mrs. Martha Carta, a native German speaker, and a general assistant in his office, to translate some selected chapters of Jordan-Eggert into English for in-house use, and Chovitz, who read German fluently, was tasked to check that translation for geodetic correctness as it progressed.

Both, Dr. O'Keefe and Chovitz kept themselves available to answer my many questions while I was studying Jordan-Eggert. I remember one of Dr. O'Keefe's introductory explanations of the various different ellipsoids with different specific names. Knowing him by then as a sparkling entertainer, I thought that he was joking and pulling a string of such funny sounding creatures—Clarke ellipsoid, Hayford ellipsoid, Bessel ellipsoid, Everest ellipsoid, and several more ellipsoids—out of his hat for my amusement, and I laughed appreciatively. Little did I guess then that I would soon myself be an expert player with these flattened balls, knowing at the tip of my tongue their precise numerical size and shape and their use in the various parts of the world. Even less did I guess that some day there would be one carrying my own name (Fischer ellipsoid 1960, in three variations, and an update: Fischer ellipsoid 1968), and that a wide scope of fascinating related problems would engage my personal involvement.

At the same time I participated, of course, in the various tasks of the Section wherever I could be of use, be it by computing, checking, adapting formulas and procedures, resolving discrepancies and the like. This gave me a broad, informal introduction to the larger projects in work as well as to the type of outside short-range geodetic inquiries that we had to answer with technical expertise, in formal correspondence. There was Chovitz' map projection paper to be worked on as already mentioned. Then there was Lewicke's Laborde projection study (a special map projection using an ellipsoid instead of the simple sphere), where I helped off and on to check, correct, adapt, or derive long drawn-out formulas. There were discrepancies with formulas printed in respectable books; and the awe before a printed page had to be overcome, the printed errors to be identified, double-and triple-checked, and corrected. Sodano worked on flare triangulation, nomograms for the UTM scale factor, on UTM table constructions, and long line computations. S.W. Henriksen, in another Section, could use me for a while in some photogrammetric studies. Rouleau worked on datum problems, often on loan to the Datum Section, spent some time in the Library of Congress for a literature search, and was involved in several other projects. In general, everybody knew what was going on and helped when needed.

I was delighted, however, to be given my first independent assignment that no one else worked on: to find out whether it was possible to sight a signal on Taiwan from mainland China, considering the size and curvature of the Earth and the high mountains on Taiwan. Dr. O'Keefe personally took me around to the "area desks" in the Foreign Control Branch, specifically to meet Mary Lou, Chief of the Asia Pacific Desk, who was always very helpful and informative in many ways. He took me to the map library, the book library, the geodetic reference library, and introduced me to the people in charge, and showed me how to find things and whom to ask for assistance if needed. The geodetic reference library was organized and headed by Mr. Manfred Sobernheim. It included translating and abstracting services provided by Mrs. Alma Plachte (Russian), Mr. Milos Achin (almost any European language) and others. Sobernheim proved to be of invaluable help throughout the years, very knowledgeable, always ready to produce, procure, or draw attention to a pertinent item, in almost no time. He had his own, efficient, but apparently unorthodox library system, a thorn in the side of the main library, which for years tried to destroy his little kingdom and, alas, eventually succeeded. A uniform, orthodox library system is more important, of course, in some people's eyes, than quick, knowledgeable service to the researchers!

Dr. O'Keefe also showed me with pride the big electronic computer, UNIVAC–I, a unique prize possession at that time, which placed the Army Map Service in the forefront of computing agencies. I was duly impressed although it would be some time before I took advantage of this marvel. For the tasks at hand, my desk calculator was quite sufficient. While working on the Taiwan-China question, I was wondering about the great amount of close supervision I received, in comparison to my experiences in my student years so long ago when writing my two graduation theses at the University of Vienna and the Vienna Institute of Technology, respectively. Those theses, one in mathematics ("Die dichteste gitterförmige Lagerung kongruenter konvexer Körper") and one in descriptive geometry ("Modifikationen und Verallgemeinerungen der Zyklographie"), were of so much larger scope, yet nobody bothered looking at what I did between sending me a slip of paper with the title of the respective thesis and receiving the finished product. Well, this is a different life and a different country, I told myself. Later when I noticed that this little project had no repercussions or further extensions as other projects in the Section used to have, I wondered whether

I may have just passed another of these little informal tests to let my supervisors know how I went about doing things.

Particularly intriguing were the problems Dr. O'Keefe and Chovitz were involved in together, and I learned by listening to their discussions, reading up on the subject, and helping with computations. One topic dealt with the geodetic aspects of errors in guided missile computations, specifically whether and how much the irregular shape of the Earth (the geoid) would deflect the missile. Chovitz' Technical Report No. 9, "The Effect of the Undulations of the Geoid on Guided Missile Paths," had just been published, following Technical Report No.6 on the same general subject. Now, Dr. O'Keefe and Chovitz were interested in formulating a statistical expression for the disturbance of missiles by these undulations as described in terms of known deflections of the vertical, that is, the slopes of the geoid along a path overflown by the missile. This study led eventually, in 1954, to Chovitz' TR-16: "A Statistical Analysis of the Effect of Deflections of the Vertical on Inertial Guidance Systems". Now and then I helped, for instance in computing a numerical example of following a hypothetical path across the United States. Such deflections, known from the small differences between astronomically and geodetically determined positions of a place, are little telltales of significant geodetic relationships, and as such became one of my favorite tools in later years.

Another of Dr. O'Keefe and Chovitz' long-standing studies concerned submarine pendulum gravity measurements and a doubt about their reliability, since the vertical vibrations of the submarine might affect the pendulum period. Various theoretical approaches were pursued to study vibrations and refine formulas, but had not yet produced a clear-cut answer. Now it was suggested to try also a numerical example of using submarine gravity readings together with the more conventional land gravity readings to see whether they would blend into smooth patterns. Mr. Donald A. Rice, Chief of the Astronomy and Gravity Section of the U.S. Coast and Geodetic Survey (C & GS), an authority on gravity work, had written a very readable article in the *Bulletin Géodésique* (No.25, 1952), also published as AMS Technical Report No.2, on "Deflections of the Vertical From Gravity Anomalies." He suggested that one compute the deflections of the vertical at two Spanish stations on the Mediterranean coast from the gravity anomalies (these are deviations from a specific model of global gravity values) of the surrounding area out to a radius of about 450 kilometers, and to compare the results with deflection values of these stations as known from the astrogeodetic network (that is, the differences between their position as astronomically observed and geodetically computed from the triangulation). If such deflection values, computed by two so entirely different methods and source data, agreed, everything would be considered all right. If they disagreed, the submarine gravity data might be the culprit, and one would have to see whether the disagreement could in fact be traced to them.

I was allowed to do this work and became more and more involved in it. The two types of deflections did not agree at either of the two stations in Spain and they did not agree at a third station on the North African coast either. The gravimetric deflections were much more positive than their astrogeodetic counterparts on the European Datum and one had to investigate all possible reasons for the disagreement, including a possible misalignment of the European Datum. I experimented with variations of the computing procedures, compared the approach by "free air" gravity anomalies with the one by isostatic reductions, and added more test stations which eventually numbered twenty-three, surrounding the western part of the Mediterranean Sea; but that took another year. The choice of the Mediterranean Sea was incidental for a reliability test in principle; it was due to the availability of sufficient data in that area. For me as a one-time European, however, this choice added a particular, happy flavor and I imagined, as so often later in global geodesy, that I was going on a trip to the blue skies and waters of Spain, Sicily, and North Africa.

Then there was the new Section assignment called Figure-of-the-Earth studies, which aimed at a modern determination of the size and shape of the Earth from all available and obtainable geodetic data. It was in its preparatory stages then, starting with schemes to select, collect, and organize these data; and with plans and studies of how to process them later. It was slated to become a major effort of international scope and importance. I was fascinated by the prospect.

5. THE "FIGURE OF THE EARTH" PROJECT—

This grandiose project was envisioned and guided by the Division Chief, Mr. Floyd W. Hough. During World War II, Col. Hough served with the Corps of Engineers and formed the famous "Hough Team" which followed closely behind the advancing American front lines in Germany, and managed to collect enormous amounts of geodetic data including German-captured data of the U.S.S.R. This unique collection later formed the nucleus for the Army Map Service library and the basis for the research activities in the Geodetic Division of which he was the first Chief. In the post-war period, he played a decisive role in the readjustment of the European triangulation network on the International (Hayford) Ellipsoid, which was later formalized as the European Datum of 1950. His

Report on the recommendations of the "Commission for the Adjustment of the European Triangulation" for the adjustment procedures to be followed (*AGU Transactions*, Dec. 1948) was later included in the documentation collection of the "Commission Permanente Internationale des Triangulations Européenne, recommended by the 1960 Symposium in Lisbon. He also succeeded in getting the UTM grid system adopted by NATO as a uniform military system.

Under Mr. Hough's guidance, the geodetic arc along the 30th meridian through Africa was scheduled to be completed in 1954, with the Assistant Division Chief, Mr. David L. Mills, directly leading the fieldwork on the last missing link of about 600 miles in the Sudan and Uganda. This arc would be a crucial new part of the Figure-of-the-Earth project in our Section.

Several people in Mr. Hough's front office were involved in this project in one or another way, and I gradually learned to identify and to appreciate them: Mr. W. Mussetter (astronomer, with long and interesting field stories to tell), Mr. John S. McCall (geodesist), Mr. F. Plachte (who had been part of the "Hough Team"), Mr. Bill Lane, Mr. Ted Skowron, Mr. Frank Fleischman. Miss L. Heiser was the Division secretary. Later, also Miss Alice Mason (later in DIA Headquarters) was there.

Mr. Hough maintained personal contact with the geodetic authorities in foreign countries and encouraged international cooperation and professional friendship. He was awarded the Legion of Merit and the Exceptional Civilian Service Awards by the U.S. Army. When the Defense Mapping Agency Topographic Center (DMATC), the successor of the Army Map Service, installed a Gallery of Distinguished Civilian Employees in 1975, Mr. Hough was one of the first to be chosen. At that celebration, the reunion with his former subordinates and his former superior, Col. John G. Ladd, brought out their respect and affection for their prestigious leader. He died half a year later. The immediate activity on the Figure-of-the-Earth assignment, that I noticed around me in that first year, was to collect pertinent data: mean elevations the world over as read from appropriate maps, and lists of astronomic stations with their corresponding geodetic positions (deflections of the vertical). The elevations were to be read under two different schemes: a global collection was to consist of mean elevations for one degree square area units, with modifications in the higher latitudes (Phase I); and a collection of much denser regional readings within a radius of about 450 km around a selected number of astrogeodetic stations (Phase II). For the second scheme, transparent templates had to be designed to subdivide the region into area units decreasing in size toward the respective station, so that the greater effect of nearby topography would be balanced by the smaller area for which the mean elevation was estimated. These templates had to be designed and constructed to fit the various maps of different scales as overlays. These were just the groundwork preparations and they seemed already like a very big order. It was decided to utilize services from other units, even by outside contracts. "Map Intelligence" (Mr. M. Folstein) was to choose the best maps, Field Offices were to read the elevations, Mr. Plachte of the front office would collect lists of existing astrogeodetic deflections for Eurasia, the Americas Branch (Mr. H.C. Fuller) would do he same for the Americas, the "Cartometric" Division was to preplot selected astrogeodetic stations on the chosen maps, and UNIVAC (J.W.H. Spencer and D. Stuart) would do the mass computations. Eventually, students at the University of Maryland were hired under contract to expedite the reading of the elevations in Phase II. Chovitz had the responsibility to organize all these efforts and, together with our little group, to write the specifications, check and analyze the returned elevation sheets and to correct errors. For instance, we noticed in the first returns a frequent tendency towards overestimations, due to an inexperienced psychological inclination to confuse printed elevation values for extreme points with average elevations; and we had to alert the readers against such misinterpretations. To reduce errors, two students would produce independent elevation sheets and we would compare them, judge the discrepancies with respect to allowable limits, spot-check them, and "adopt" a final sheet. To further avoid errors, I devised a "foolproof" color code for the map units (feet, meters, fathoms), so that no computation was to be made by the students to convert one into another.

Elevations for Phase II in Central Europe would be read by Dr. E. Cigas and his group in Frankfurt am Main, who had made the readjustment of the triangulation there after the war for the Army Map Service; and correspondence and maps went back and forth. UNIVAC would produce a listing of the global collection of Phase I from the Field Offices, so that we could check for gaps and fill these in. Then we had to produce formulas and instructions or UNIVAC to turn these data, including the appropriately sized polar caps, into topographic deflections at specific points. We determined that the simplification was acceptable to consider the topographic mass of a one-degree square as if concentrated at its midpoint; and this would significantly shorten the computations. The instructions included isostatic corrections and evaluation of the deflection values at hypothetical points at 100 km intervals. For each evaluation, only the topography beyond a 450 km radius from that station was used (the "distant" topography), because the regional topography was handled in Phase II. This scheme permitted us to draw graphs of isolines, from which the effect of the distant topography could be extracted for any station by simple interpolation.

These plans were established now, but it would take a long time to carry them out for the whole world. I did request, however, to have the Mediterranean area of the gravity study processed first. The main part of the project, procedures to turn this material into a new Figure of the Earth, was way in the future and mysterious to me at the time. Obviously, it would be a tremendous job and, just as obviously, I wanted to be in the midst of it.

The scope of the new project in our Section caused its name to be changed from "'Long Line Section" to "Geoid Section" in spring 1953, which gave me a first glimpse into the administrative machinery, and let me meet Miss Mary F. Crabb, job analyzer, for rewriting our job descriptions. The name "Geoid Section" was to reflect the ultimate goal of deriving the form of the geoid (that is, the mathematical figure of the Earth at sea level height) together with a reference ellipsoid most closely approaching it. After having recovered from the surprise of my first AMS days, that one did not know the Earth's size sufficiently well, I plunged into this challenge and started to study in due time chapters of other texts besides Jordan-Eggert, such as Hayford, Hayford and Bowie, Rice, Tanni, Heiskanen, Vening-Meinesz, Helmert, Bomford, Hopfner, Bäschlin, etc. I started my little hand written library of notes, excerpts, definitions, and formulas, as I had done in student times.

In a happy mood of anticipation, I felt reminded of the story of the two brick layers: one unhappy about the heavy work of laying bricks day in and day out to earn a living, and the other doing the same work but singing happily all day long about his good fortunes. Asked about the cause of his happiness, he said: I am doing glorious work, I am laying bricks for a great temple, I am building a Temple for the Glory of the Almighty!

This first year at the Army Map Service was a very busy year for me indeed. These vibrating activities, questions, searches, and new vistas were an extraordinary experience of entering an entirely new field about God's own Earth, under the guidance of these exceptional men. How lucky can one be, to have a second career opened to you under such circumstances. I thanked Dr. O'Keefe for a fascinating year of study, for which one normally would have to pay tuition. But here I had all this and a paycheck too!

2

MY GROWING INVOLVEMENT AND THE FIRST PUBLICATIONS (mid 1953–mid 1956)

1. COMING TO GRIPS WITH THE NEW PROFESSION

As if designed for me personally at the right time, a formal graduate course in geodesy was offered by Dr. A.D. Sollins, of the U.S. Coast and Geodetic Survey, at the Graduate School of the Department of Agriculture, in fall 1953 and spring 1954, and I grabbed that opportunity happily. Dr. O'Keefe had made propaganda for this course in his branch, but there were not many takers. The Government did not pay the tuition then as it does now, which made it a different ball game then versus now. It also meant going downtown after working hours and coming home late to my little family. But I thought it was an opportunity not to be missed, and I took both semesters. I remember Mr. J. Walkeau, Chief of the Datum Section, and Mr. E. Sodano from our Section going with me, and maybe one or another more who then dropped out. At the beginning of the second semester, Walkeau left the Army Map Service (replaced by Mr. C.W. Kroll), but he continued the course with me.

It was an excellent course, which nicely dovetailed with my work in the Office, providing theoretical depth and perspective to it and also opening other chapters of geodesy for me. The basic concepts of geodesy became clearer. While field geodesists observe astronomic positions and measure distances and directions on the Earth's surface, geographers combine these data into coherent geodetic surveys leading to maps, and theoretical geodesists miraculously deduce from all that the shape and size of our planet Earth. They distinguish between the confusingly detailed topography of our daily world of mountains and valleys, and the wonderfully smooth surface of the oceans (Fig. 1); and they dream of continuing that watery smoothness under the mountains—chopping them off so to speak, since we know their measured elevations—to arrive at a smooth Earth, the "geoid." "Gea" is Greek and means earth, the Greek goddess of the Earth, and "geoid" means then "something like the Earth." So that is what they are after! But how to describe the shape of this fancy Earth? Well, the trick is to compare it with something we know from geometry such as a sphere or, if more sophisticated, an ellipsoid and find the differences from such a reference figure from place to place. The smaller the differences, called "geoidal deviations" or "geoidal heights," the better the fit, and the closer is the claim to have deduced the shape and size of the geoid, that is, "the figure of the Earth." To make things more complicated and thereby more interesting, there is also another way to determine those differences: gravimetric measurements give a clue of where there is more mass underfoot than elsewhere and that would indicate that the geoid is higher here above the reference figure than elsewhere. These ideas linked the course to my two major interests in the office: the Figure of the Earth Project and the gravimetric surveys in the Mediterranean. But there is a long arduous way from visualizing a goal and divising procedures to actually getting there.

THREE SURFACES

Ch. 2, Fig. 1: "Three Sufaces"

In the second semester, we were asked to give a seminar presentation, for which I chose A. Prey's spherical harmonic representation of the Earth's topography. This is a mathematical technique of describing first the large features and gradually adding smaller and smaller features in an orderly fashion. I had great fun making a rather dramatic presentation of how the topography changed with increasing detail and the familiar geographic features gradually appeared, as degrees and orders of the spherical harmonic expression were added—only to have another course participant object violently: "That is not what the Bible says how the Earth was created."

Work in the Office was always varied and, in such a small group, one naturally helped with each other's projects; but I gravitated more and more towards volunteering for those jobs that were related to the size and shape of the Earth and gravity reductions, both theoretically and in specific areas of the world.

Mr. Harry Lieberman, Administrative Assistant to the Branch Chief at the time, was working on an extension of the geoid chart for Central Europe to all of Europe and to parts of North Africa and the U.S.S.R., using all available deflection data, including the so-called "deflections of the vertical," that is the differences between astronomic and geodetic positions of specific stations (Fig. 2).

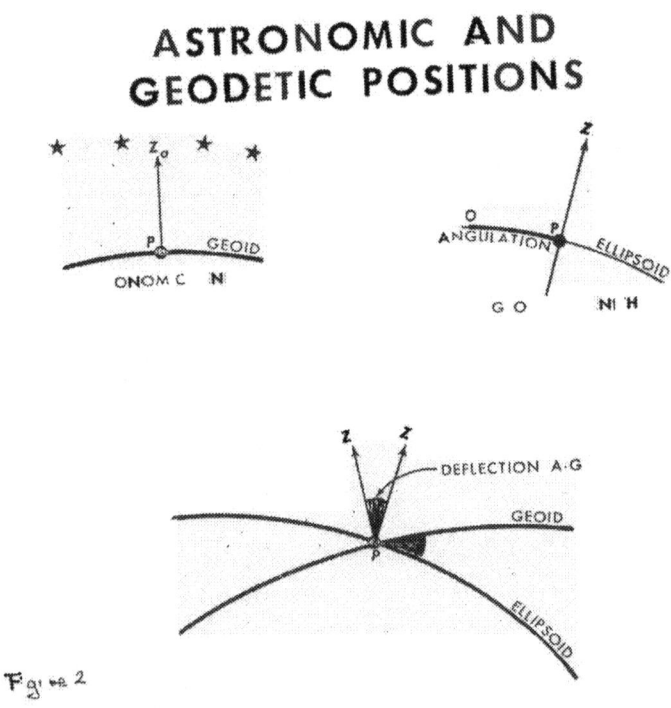

Ch. 2, Fig. 2: Astronomic and Geodetic Positions

The astronomic position of a point P is the direction of the vertical (the plumb line) towards the observed place among the stars, the astronomic zenith Z_a. The geodetic position of point P is the perpendicular to the chosen reference ellipsoid serving the triangulation survey, the geodetic zenith Z_g. These two zenith directions at point P are usually not identical. The angle between them is "the astrogeodetic deflection of the vertical," a basic value in geoid studies. It equals the angle of intersection between the geoidal and ellipsoidal surfaces at P, since each zenith direction is by definition perpendicular to its respective surface. This relationship is also a basic fact in geoid studies (Fig. 3).

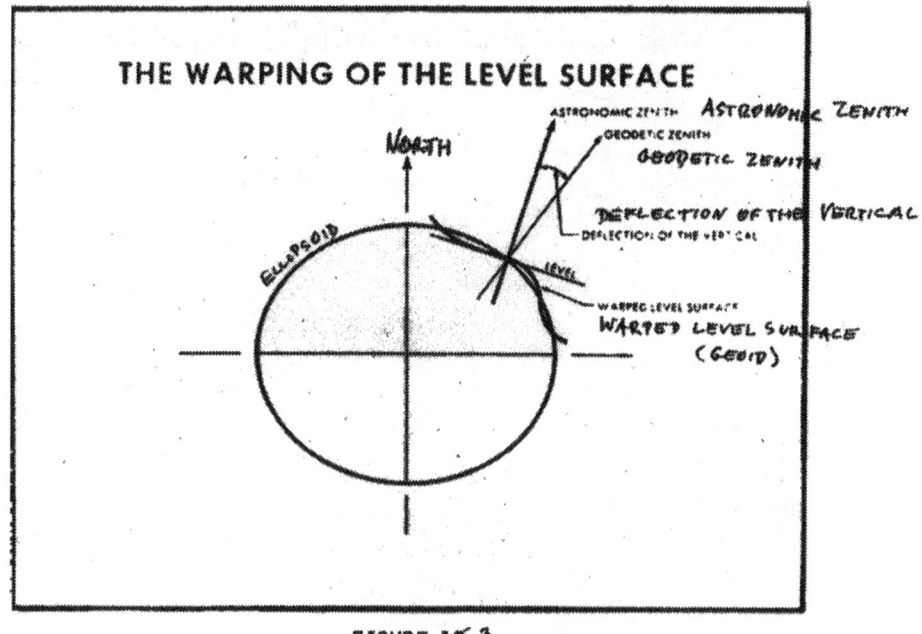

Ch. 2 Fig. 3: The Warping of the Level Surface.

Lieberman's study was a first attempt to give a consistent picture of the geoid for this whole region, as referred to a consistent geodetic survey. In this case it was the European geodetic network referred to the so-called "International Ellipsoid" which is defined by a specific shape and size; both together are called the "European Datum." I helped Lieberman draw the contours of the geoid charts, studied his procedure of modifying the position of the reference ellipsoid for a closer fit, and how to change the ellipsoid itself to a better fitting one for making the geoidal heights (the distances of the geoid from the ellipsoid) even smaller.

The submarine gravity study in the Mediterranean region was considered more and more my special job. I was permitted to extend the investigation to 23 coastal stations all around the Western and Central Mediterranean Sea, and collected gravity data from the various submarine expeditions and from land surveys in the surrounding countries. I experimented with modifying the assumptions about rock density in computing topographic–isostatic reductions, and with different procedures using isostasy or not, drawing isolines one way or another, and watching the effect of such variations—and I gained insight by doing and speculating. After having done the laborious chore of map readings for elevations and gravity values for the regions around the first three stations, I was permitted to farm out these chores for the other 20 stations to the good souls who supplied such services for our larger Figure-of-the-Earth project. I drew up specifications and color schemes which I had found useful in my own experience with the first three stations and passed them on to the Field Offices through staff channels (Mr. C.T. Williford). I expanded my study to include also the effect of the "distant" topography, that is, beyond the radius of 450 km around the station, which had not been done for the first three stations. I knew from my reading that gravity data covering the whole globe were needed for accurately computing a deflection value at any place, and I did not want to take anybody's word for it that an area within 450 km would suffice. I computed the effect of the "distant" gravity information taken from a world chart by L. Tanni (incomplete but the best there was) and combined it with the effect of the "distant" topography computed as part of our global study, and found what I had been told: that the effect was negligible. Now I believed it; the fact was worth demonstrating, at least to me. Checking and double-checking was not necessarily dull; it occasionally assumed the flavor of a "who done it" story. So, for instance, I worried about an odd-looking gravity value in Spain until I managed to trace the derivation to a map where the first digit of a spot elevation of over 1000 m. had been cut off by the left margin of the map, so that the mountain had computationally disappeared. Another worry concerned a station in Sicily named Monte Gemini, where a deflection value did not seem to conform to pattern, did not seem to make sense. It kept me puzzled for quite a while, until persistent checking of source material eventually revealed a confusion of two similar sounding station names: Monte Gemini and San Giovanni Gemini, where the geodetic position of the first had been paired erroneously with the astronomic position of the second. Straightening out "The Mess of Monte Gemini," as my Memorandum for the Record was entitled, was my first small contribution to the International Association of Geodesy, which maintained a listing of all deflection values in Europe, and where my correction of an incorrect value was welcome. Catching little gremlins like these helped me overcome a sense of uneasiness in the face of those heaps of little numbers on my desk which were supposedly telling me something about the great big world outside. How could they? Yes, they were, or came from, observations; but these admittedly had errors, and then they had been cooked in a big, black pot of computational witchery which gave them an eerie tinge, even though I had brewed that witchery myself. But now there were these two experiences: one, that such a little number, apparently not much different from the next one, did reveal under close scrutiny in Washington, D.C., that someone in Sicily had goofed; and two, that I had begun to develop a sense of consistency between patterns of little numbers and underlying realities or hypotheses, which made me expect certain behavior and scrutinize offenders.

The Mediterranean study clearly resolved the doubt about the validity of submarine gravity measurements. There was no difficulty in combining them with the conventional land measurements. The almost constant discrepancy of about six seconds of arc between astrogeodetic and gravimetrically computed deflections would have to be explained by a misfit of the European Datum in this area. Later studies of mine and others confirmed that the European Datum, at first considered to be good for the whole world, would not be good for Africa and Asia.

While I was still working on this project, I had the good fortune to meet Professor F.A. Venig Meinesz, the famous Dutch geophysicist, in person at a conference at the U.S. Coast and Geodetic Survey in November 1953, arranged in his honor. I remember well the white-haired gentleman who also visited the Army Map Service on the following day and showed kind interest in my Mediterranean study. Actually, his submarine gravity values were on trial in this study, but came out with flying colors.

2. THE LONG ROAD FROM MANUSCRIPT TO REPRINTS

The gravity study was completed in February 1954 and first made into AMS Technical Report No. 13, entitled "The Deflections of the Vertical in the Western and Central Mediterranean Area." It was promptly classified, which means restricted circulation within Government only, with a few special exceptions, and cumbersome handling each time the document is used. Dr. O'Keefe decided, however, that the study was scientifically important enough for wider circulation and he recommended publication in the *Bulletin Géodésique* in a letter to Col. Georges Laclavère, International Association of Geodesy (IAG) in Paris. It was published in the December issue of 1954, and I subscribed to the *Bulletin* since then.

The letters surrounding government release and publication were again "firsts" in my education and experience, to be repeated and suffered through so many more times throughout the years, but interesting in a nostalgic way. After I had removed the classified decimals of my numerical results, the request for clearance was signed by the Assistant Commanding Officer, Lt. Col. Elmore G. Lawton, on 12 March 1954, and addressed to the Chief of Engineers, Attention of the Chief, Technical Liaison Division. It came back with a stamp on it saying: Approved for Publication, dated 25 March 1954; that is less than two weeks round trip. Today, that round trip has been lengthened artificially (not required by Defense Department Regulations for purely technical papers) by many government miles and reams of covering letters at intermediate stops of unpredictable duration. It takes now ingenuity and alertness to find the hiding places and to engineer the next step, all the time watching the calendar date advance toward Doomsday. My last technical study in government, which I had worked very hard to complete before my retirement in 1977, I submitted for clearance eight weeks before that date, ample time, I thought. But it almost did not make it; it needed machine guns to move it from roadblock to roadblock, and in the end it took the personal intervention of a friendly soul in the highest echelons of the Defense Mapping Agency's headquarters to cut further artificial corners and let me send the paper out to the journal of my choice. Ah, the abandoned simple efficiency of bygone days.

The galley proofs from Mr. J.J. Levallois, Secrétaire Adjoint de l'Association Internationale de Géodésie, came addressed to me with a form letter to Monsieur Irène Fischèr. In today's time of Women's Lib they could not have gotten away with such a form letter which clearly indicated that women were not expected to be authors. But my reaction was one of pure fun with that title and the French pronunciation of my name. Our order for two hundred reprints had to go through the American Embassy (then as also now), but despite a request for a rush order (there is a customary short deadline for ordering reprints with the return of the corrected galley proofs), it took the Embassy from November until February before they contacted the Bulletin Géodésique at a time when the printer's proof had already been destroyed. Levallois was very understanding and had the article printed again; but he commented in a letter to me that "next time it would be better to ask directly for the number of reprints needed, which will avoid what we call in French 'les lenteurs administratives'." He was so right, but the advice did not do us any good. The ruling still persists and I had repeated trouble of the same kind or even with embarrassing delays of payment also in later years. Ah, the preserved inefficiency of bygone days!

This first publication received international attention in the form of a letter from Dr. Pierre Lejay, President of the Bureau Gravimétrique International in Paris, expressing great interest and asking specific questions; detailed favorable reviews in foreign geodetic journals such as the *Zeitscrift für Vermessungswesen* (Oct. 1955, by Martha Näbauer) and *Geofizikai Közlemények* (vol.6, no.1-2, 1957, by J. Renner); and eventual inclusion into the archive of the "Commission Permanente Internationale des Triangulations Européenne." I have an impressive-looking form letter from the Institut de France, Académie des Sciences, acknowledging in beautiful calligraphy the receipt of the article which would be placed into their library and mentioned in their bibliographic bulletin.

While I was finding my way to the right people in the right offices to get my manuscript and my drawings for illustrations transformed into the approved format of an Army Map Service (AMS) Technical Report (TR) No.13, I was asked to apply my newly acquired expertise also to two other Technical Reports, No. 14 and No. 15. The latter was a delightful explanation of the occultation method, written by Dr. O'Keefe, and there were a lot of illustrations available from which to make an appropriate selection. I enjoyed helping with this lively picture booklet, wrote for copyright waivers where needed, and saw it through the publication office. TR-14, entitled "An Investigation of the Geoid in Europe and Asia" was Harry Lieberman's work with which I was familiar already. Lieberman had left in September 1963 for an extended tour of duty at the Army Map Service Far East (AMSFE). I was assigned to see his paper through to publication as a Technical Report. The remarkable thing about this study was that its single-handed successful achievement anteceded a group attempt by the Study Commission No.14 of the International Association of Geodesy (IAG) under the chairmanship of Brigadier Guy Bomford in Oxford, which was presented to the IAG Assembly in Rome in 1954. Lieberman lost out in scientific visibility, because TR-14 had been classified due to its U.S.S.R. part. But Chovitz succeeded to persuade the authorities, to salvage the unclassified part of the work so that Mr. Hough could take it to Rome, and to have it published eventually. I was happy to shorten this interesting paper to meet clearance regulations and to edit it for a belated publication in the *Bulletin Géodésique*, September 1955. A footnote gives credit to the author's independence and the early time of the study in

1953. Chovitz also suggested to Mr. Hough to take the unclassified version of my Western Mediterranean study along to Rome because of its pertinence. Sodano's work on long lines was already prominently mentioned in the summary of U.S. activities to be presented there.

3. THE OFFICE ENVIRONMENT

All these publication activities required attention to illustrations including designing them, inking them, and lettering them; and of course getting the manuscript typed, proofread, corrected, and maybe typed again. Also work memos, instructions, and letters needed to be typed, approved, corrected, retyped, and retyped, then as well as now. A change of command from Col. J. G. Ladd to Col. J. D. Abell added some more paper work. In February 1954, Chovitz had to write a detailed memorandum for the new Colonel explaining the significance of the section projects for the next few years, in preparation for the Colonel's visit to the Branch in March. And there were the periodic status reports to be typed, of course. But there was only one branch secretary for all this, Margy, and she was also the time keeper, and she answered the telephone (one telephone for the whole Branch business!). No wonder that her desk was piled high with work waiting. Too much paper work piled high before you may eventually be counterproductive and produce instead a sense of futility or rebellion. One day when Rouleau brought her his report to type too, on top of it all, it was apparently the last straw for the poor thing; and she greeted him with friendly exasperation: "I wish you would drop dead". To which the perfect gentleman replied with a deep formal courtesy bow, "After you, madam."

And there was the bell, remember? It rang in the morning to mark the beginning of the workday and it rang at its end eight hours later. Thirty minutes for lunch in the cafeteria were your own time and could include a short stroll in the sun. No eating at your desk allowed anytime. If you reached your desk after the morning bell, it may cost you an hour's annual leave; so there was a stampede in the lobby every morning to catch the elevator. And there was a stampede out of the offices at the sound of the evening bell. Some people tried to get their coats to their desks beforehand so that they would be first when dashing out. I remember a supervisor dashing out right after such a guy, bringing him back in and letting him wait for quite a while, because he was not supposed to get his coat ahead of time. I also remember that a colonel came out of his office at stampede time and was almost thrown over. Then we had big signs in the halls: "Don't Run, Walk."

It made you feel back in grade school, and I am sure it fostered a grade school mentality.

Dr. O'Keefe came around quite often, always with new ideas, challenges, interesting aspects, valuable information, always encouraging and ready to help. He was often seen using an inverted wastepaper basket as a stool next to the person's desk he wanted to talk to; for it was the first duty of a supervisor—so he explained for the benefit of us future potential supervisors—to take your time with a subordinate employee, tell him exactly and patiently what you want done and how. Don't talk down to him physically or otherwise, but sit down with him on the same physical level so that a clarification of misunderstandings or of any problems becomes easier. Dr. O'Keefe tried to ease my transition from school mathematics (where problems have clear-cut solutions and assumptions are spelled out or posited) to applied mathematics in geodesy where solutions are requested when not nearly enough information is available, and where answers are required when the assumptions are not even clear. Your job as a technical expert, he used to say, is to give technically sound information but also to give that information in a form that is useful and meaningful to the colonels and generals who need it. So if he asks you for a number, don't lecture him that there is not yet enough information around to produce that number. Since he needs, or thinks he needs, that number now and can't wait for it to the end of days, so give him a number that is better than the one a non-technical aide might pull out of his pocket. From technical background you might easily know whether such a number would be positive or negative, whether it is a thousand rather than five, whether it is between 20 and 30 or between 200 and 300. In other words, you might be able to give a range, and that might be all that he really needs. But be sure that your answer or advice is sound to the best of your knowledge so that he can rely on it. Then he will respect you and also listen if you suggest that certain surveys or studies should be made to serve him better in the future. It was good sound advice, and I thought of it often throughout the years. Dr. O'Keefe also took pains to separate the joy of research from the dilemma of its purpose in the Defense Department. "Look at it this way", he said, "if we did not prepare the capability of geodetically pinpointing a target in a war, the war would not stop, but the inaccurate bombing would endanger and destroy whole areas instead of military targets. We have an obligation to prevent that by reducing error from geodetic ignorance. Your obligation is to pursue research into the Figure of the Earth the best you can and to participate in the efforts of the geodetic community. We will back you up. Our obligation is to use what we need, which is the reason why government is involved in research." It was a masterpiece in the art of building positive employee relations: it established all-important trust, challenged the employee's potential, and elicited commitment.

Dr. O'Keefe was a demanding teacher and supervisor. He would come around with numerous short requests to find out for him one thing or another on short notice. If he sensed some perplexity on my part of how to go about that task, he might say: "Why don't you use your head. Well, can do?" And what else is there left to answer than: "Yes, can do". At one of several dinner parties at his house, which my husband and I were privileged to attend, one of the guests asked me whether I also worked at the Army Map Service where Dr. 'Keefe worked. "Yes," I answered, "I am one of Dr. O'Keefe's little slaves." After a couple of seconds when one could have heard a pin drop, Dr. O'Keefe shot back: "Now wait a minute, that shoe is on the other foot." Both comments were correct in a way, I guess. Dr. 'Keefe was definitely the prime mover in his little kingdom which he had created with his brilliance and enthusiasm. And I, among others, was happy to do his bidding, and felt enriched in doing so. On the other hand, Dr. O'Keefe went out of his way to educate, encourage and protect his employees, straightening the way for each one where he could.

There were only a few women in the whole branch then (none other in my section) and we used to go together to the cafeteria for the coffee break, standing around the tables. No sitting down was permitted; in fact, the chairs had been removed to another closed section. Pam Anderson would tell us of her four puppies, which she lovingly raised, of their individual behavior and their occasionally vivid dreams which she could watch from their eye movements in their sleep. When she married Henriksen who was not a dog lover, at least not to that extent, she compromised to keeping only two puppies. I am sure, puppies never had it so good as with Pam.

For lunch, my husband, whose office was in the Intelligence Division on the second floor, always came up for me so that we could share the morning experiences, leaving the afternoon events for entertainment on the ride back home. Much later I was told that the exclusiveness of the two of us at lunch time was widely noticed with benevolent amusement, as was our walking down the hall arm-in-arm, European fashion, which must have looked funny to American eyes, without our being aware of that. For special occasions, however, the women used to go out for lunch together, taking an hour annual leave beyond the statutory thirty minutes lunch period. Such occasions were salary raises that naturally called for celebration. The pay system provided for automatic within-grade step increases by law every year for the lower grades, and every year and a half for the higher grades starting with GS-11. There were periodic job audits by Personnel

Management, which possibly could result in a grade promotion. When I noticed that, of the small group of women, the black members were not invited along with the others, I was in for another kind of education in American ways, namely that they would not be admitted to Washington's restaurants. For me as a European refugee from the Nazis and with a Jewish historic memory of more than a millennium of persecution of a systematically humiliated minority, this reality of American life was incompatible with the dream of the Land of the Free to which I had fled. I refused to go along under these circumstances and was thus instrumental in breaking up the habit of these excursions. When my time came to celebrate my promotion to GS-9 in April 1954, I invited the women for dinner to my own home instead. It was arranged for a Monday evening directly after work so that I had the Sunday before to prepare and set the table. My guest book shows that there were Martha Carta, Nora Moser (later Taylor), Pam Anderson Henriksen, Thelma Cooly (later Robinson), Naomi Millet, Eileen Segal (later Kantwell), and Gwelda. One more black girl whom I liked, was invited and expected but, to my regret and mystification, did not show although she had been at work. She had an embarrassed, flimsy excuse the next morning. Otherwise, it was a pleasant evening. The promotion itself was unexpected after the pessimistic indoctrination at the start of my career, but naturally very much appreciated. It gave me the good feeling that my supervisors looked out for me enough to break the Army's known anti-feminine bias. The incident with the black girl was disturbing, but I was advised not to say anything; that I did not know enough about this side of America. There were a very few more bewildering experiences such as no response to warm letters of congratulations, but in general I had no trouble, I believe, in reaching a person to person link without regard to color. I had felt uncomfortable about Mr. Andrew Glucic's low pay at that time compared with my raise when I discovered that this fellow refugee from Nazism had been the head of a geodetic organization in his native Yugoslavia. Dr. O'Keefe tried to utilize this man's knowledge in a better position and therefore had rescued him recently from a low-paying dull routine job in the Foreign Control Branch. He explained to me the workings of Army pay regulations and that he intended to promote Mr. Glusic as fast as legally possible. I had many interesting conversations with Andy Glusic, this fellow European, in the years to come. When I became Branch Chief, I nominated him an Honorary Member of my branch. He remained a good friend until his death in January 1970.

4. THE PITFALLS OF GEODETIC DATUMS AND DATUM EXTENSIONS

In the spring of 1954, there came a request, among others, to investigate the difference between coordinates on British datum and those obtained from a SHO-RAN tie (short range electronic navigation system) between Norway and Scotland. It gave me a chance to become familiar with the intricacies of datum conversions, connecting the British Isles to European Datum both ways, via Norway and across the Straits of Dover to France; and speculating about the discrepancy in the loop closure when reaching a specific point via both routes.

More intriguing because more speculative was a subsequent request to investigate the relationship between the European and the Tokyo Datum. There was H. Lieberman's TR-14, carrying the European Datum tentatively along the 52nd parallel into Manchuria. But how could one get from there to Tokyo? No satellites were as yet in the geodetic tool kit, mind you. But there was a little geodetic Memorandum No. 660 which established conversion formulas between the Shinkyo Datum (Manchurian Principal System) and the Tokyo Datum on the basis of an overlap at several stations near the 38th parallel in Korea. Combining this with Lieberma's information, I came up with a tentative answer to satisfy the customer.

But then I started wondering and speculating about these two geodetic systems which employed the same Bessel Ellipsoid of 1841 as their reference figure, yet led to different coordinates for points on their common border. The geodetic surveys had started routinely in each country with the establishment of the respective datum points, that is, Shinkyo and Tokyo respectively, and were extended routinely until they overlapped in Korea, displaying discrepancies there in terms of latitude and longitude. Memo 660 was a mathematical formulation derived from these discrepancies for the purpose of computing differences at other points and thus switching from one system to the other if needed. Specifically, one could use Memo 660 to compute the latitude and longitude of Tokyo in the Manchurian system, and vice versa, Shinkyo in the Tokyo system. Trying to visualize what was going on here, it became obvious to me that there must be a spatial relationship between the two reference ellipsoids, the Bessel as used in the Shinkyo Datum and the Bessel as used in the Tokyo Datum, whose geometrical identity was just incidental. That identity was a historical remainder of past expectations that the Bessel Ellipsoid was the world ellipsoid, disproved by the very differences between these two systems as well as by the 1924 election of the Hayford Ellipsoid to the honor of being _the_ world ellipsoid. If one could establish that spatial relationship, for which Memo 660 apparently was a two-dimensional aspect, then one could at the same time establish the geoidal height of each of the two datum points with respect to the ellipsoid of the other system, while each had a zero geoidal height by definition in its own system. With my background in descriptive geometry, the three-dimensional viewpoint was much more meaningful and even imperative to me than the geodetically conventional two-dimensional shifting and rotating of maps until they fit into a neighboring survey when stretched to scale (the so-called "Anfelderung" method). Eventually, I came up with a possible spatial configuration that would satisfy the two-dimensional relationship in Memo 660. And proudly, I showed my handiwork to Chovitz and Dr. O' Keefe.

Chovitz was mildly interested, but Dr. O'Keefe's rejection was total and devastating. He tried to explain to me that conventional geodesy used the ellipsoid just as a mathematical computation device, a set of tables to be consulted during processing, without the slightest thought of a third dimension. Therefore, my beautiful construction of a three-dimensional picture from conventional surveys was inadmissible. The horizontal control network, carried out from point to point on the ground, was routinely reduced to sea level (the geoid) and then treated as if that surface were identical with the computational reference, the ellipsoid; in technical language, it was "developed" onto the ellipsoidal surface with the horizontal geoidal distances held fixed in length. I found discussions of the two viewpoints in the literature: the "development" method versus the "projection" method where the horizontal geoidal distances were not held fixed in length, but are further "projected" onto the ellipsoid, with their length changing accordingly. To my mind, the "development method" was just plainly wrong, because it treated as identical what was different. If the surveys had been computed that way, then one had to make some allowances for a cumulative error over long distances, but on principle the three-dimensional concept of a projection from the geoid to the ellipsoid had to be maintained.

Dr. O'Keefe, kindly and patiently, let the discussion go on for weeks and weeks; and I was not going to give up without understanding these mysteries. The motto which Ludwig Boltzmann, famous Austrian physicist, had placed in front of his "'Mechanik", would have described my fighting spirit: "Bring vor, was wahr ist.

> Schreib so, dass es klar ist
> Und verficht's, bis es mit Dir gar ist!'

(Propose what is true. Write so that it is clear; and fight for it till it kills you.) In December 1954, Dr. O'Keefe went on a trip or several weeks, but on the way he still made another patient try to make me see the light. I have and cherish a ten page letter from him, closely hand-scribbled on American Airline stationary, "In Flight With American", delightful as usual. It says in part: "…The hydraulics people who usually are the best customers for height data, do not want the heights above the ellipsoid; they want the heights above the geoid, since water will run on the ellipsoid, but not on the geoid. Of course, there are also occasions when the thing required is the height above the ellipsoid. This is true in astronomical observations of the moon or in certain other problems….It would then be logical to measure simply in x, y, z…Reference to an ellipsoidal surface under these circumstances is mere obscurantism—the use of a notation hallowed by time and endeared by familiarity, but most awkward for any practical purpose—like English orthography, or the foot-yard-furlong-chain-rod-mile-inch system (grain-ounce-pound-stone-ton, or teaspoon-tablespoon-cup-pint-quart-gallon-hogshead-acre foot). These are echoes of Sumerian scribes, with their sexagesimal systems printing on bricks with a stylus, of Greek armillary spheres and of…Like all such systems, they are well adapted to keeping out the uninitiated. Conversely, they are very poorly adapted to training new people, or to developing new ideas…"

Then the letter discusses practical examples of a floor plan, a cadastral survey of your property, and a city survey, where a plane rectangular coordinate system is adequate because the effect of the ignored curvature of the earth on horizontal angles and ground distances is almost zero. Use of latitude and longitude instead would make things immediately more complicated, more expensive, and less accurate. In practice, one would make a compromise by determining only one or two astronomical positions in the area and find others by calculation from measured distances. His system would be just as good as the first one, only more awkward and complicated. The letter continues: "The real damage would come when somebody said, Look here, the astronomical latitude and longitude of a point represents the direction of the normal at that point to the geoid. Now, how about your geodetic latitude and longitude—what do they represent the normal to?'—Up to now, all was well. 'The surveyor had been happily taking his lengths and distances and applying the spheroid tables and getting latitude and longitude. When he wanted a distance, he reversed the process, and everybody was happy. So now he goes and looks in a fat book about surveying and finds out about geodetic latitudes and longitudes representing the normals to the ellipsoid, so he answers—they are the normals to the ellipsoid—figuring that now the philosophical pester will go away and fall into a well or something.".…(When the surveyor fits a small net into a larger one,) "the philosopher says, 'I assume that you have some notion of height above the ellipsoid, to go along with your latitude and longitude.' 'Well—' says S. 'Let's not quibble', says P. 'In any case, I can see clearly now how you must have proceeded. I will convert your latitude, longitude and height into x, y, z and I will make a simple translation of them, and then I will turn them back into latitude, longitude and we shall see.' So he does this. The surveyor goes on figuring λ, φ while the philosopher is working, hoping that all will be well. Suddenly the philosopher leaps up from his desk. 'My friend', he says, 'I find that these formulae of yours clearly imply that the difference in geoidal height between the zero milestone and the old oak tree on the Dogan plan is 25.3 millimeter. I can prove this, says he, because otherwise the shift formulae that you gave me cannot be interpreted as the translator of a part of the ellipsoid.' 'Well', says S., 'I never said they were'. 'You should have' says P.; 'when you used the language of λ and φ, you certainly implied space directions to a definite surface'. 'Did I?' says S., 'I never meant to.' 'The discussion now becomes heated, and the philosopher insists that the surveyor is just plain wrong. Finally, a mathematician appears. He listens to the argument and says: 'You are both wrong. You, S, should not have used latitudes and longitudes. They were not needed in your problem. If your surveys had become so big that you were hurt by the effects of spherical excess, you could have modified them slightly by assuming that they were on a surface of a certain Gaussian curvature. Since the corrections would have been pretty small, you could have got by with a very poor value of the Gaussian curvature. Then you could have carried out all your manipulations almost as if in a plane. And you, P, should have taken a more operational view of S's work. What S meant by latitudes and longitudes and ellipsoids never was more than a way—not a very handy way either—of marking some reference lines on the earth's surface. All of this business about the center of the ellipsoid and the translation in x, y was something that got dragged in when you started to use his latitudes and longitudes for purposes for which they were not intended. Immediately both P and S leap on M, and beat him to death with a blunt instrument."

In a P.S., Dr. O'Keefe mentioned the background of Memo 660, saying that he had given instructions to derive it in rectangular coordinates, and not in λ and φ. If that order had been obeyed, there would have been no temptation to bring in questions about the axes of the ellipsoids and about relative geoidal heights…"It would have been obvious that a shift could be found which would reconcile the x, y coordinates at the boundary, but that this need have nothing to do with the space relationship of Tokyo and Shinkyo. And finally, that the x, y relationship could be turned into a λ, φ relationship if anyone were so perverse as to insist on it."

The picture drawn of the pre-satellite surveyor contrasts interestingly with today's surveyor who is probably a lot more knowledgeable and sophisticated, or at least is guided more decisively into doing what and how the philosopher (rather the modern geodesist) wants things done, and by the mathematician—engineer into using a little black box which will give him the results he needs at the push of a button.

Dr. O'Keefe was right, of course, in his description of local surveys that had no need for a third dimension, but I did not think that it applied to my interest in continental and global surveys, where a three-dimensional approach seemed mandatory to me. I knew his paper "The Isoparametric Method of Mapping One Ellipsoid on Another," written two years earlier (*Trans. AGU*, Vol.34, no.6, 1953), where he explained a cartographic routine procedure for an ellipsoid change by going from one set of geodetic positions to the coordinates of some standard map projection via the appropriate ellipsoid tables; then converting these map coordinates to the second set of geodetic positions via the corresponding ellipsoid tables. The incurred inaccuracy would be less than the acceptable cartographic map distortion. While this paper seemed straightforward for the purpose of the mapmaker and a limited map area, the intermediate step of the map projection seemed extraneous to my search for a three-dimensional picture. A later paper "'The Development Method as Projection of a Deformed Ellipsoid" articulated Dr. O'Keefe's preference for mapping within the geoidal surface and using the Gaussian curvature as a parameter for its difference from the ellipsoid (Professor Marussi's theory); it was interesting as the mathematical basis for the working habits of his "Surveyor", but it did not appeal to me for the same reason. True, for his "Surveyor", height was irrelevant and the survey example contained no or at most two astronomic positions. But there were a lot between Tokyo and Shinkyo, and these could be paired with geodetic positions to produce the so-called deflections of the vertical, which in turn could be made to reveal the hidden spatial relationship of the geoid to the ellipsoid. So I went to work on the list of Manchurian deflections which we had from a 1942 report by Omori.

I studied Hayford's graphical procedure of turning an area-wide distribution of deflections into geoidal contours and applied it to Manchuria. The resulting geoid chart on Shinkyo Datum looked nice and smooth. It reached the U.S.S.R. border point to where Lieberman had carried the European Datum, which I now could extend to all of Manchuria. Then I studied Japanese work on the geoid in Japan, but that could not be extended to the Manchurian border for lack of deflection data outside of Japan. Some time back, we had made a request for astrogeodetic deflections in Korea, in the context of the comprehensive Figure-of-the-Earth studies, but that would still take a long time. Now, I studied Tanni's gravimetric world geoid as a possible means of speculative connection. Lieberman's TR-14 had worked out its relationship to European Datum and various reference ellipsoids, which I could adapt to get some approximate geoidal height of Tokyo on the European Datum. How valid would that number be, considering that Tanni's data were scanty and averaged, and that Lieberman's derivation had been made in far-away Europe? What could I compare it with as a check?

It contradicted the tentative result from that ill-famed Memo 660, even if plausible uncertainties were included. In a 10-page paper with several computational enclosures, dated March 1955, I reviewed for my own check and benefit all steps of my reasoning from the beginning, all analyses, checks, alternatives, and speculations, and what needed to be done when more information became available. Here are two long notes from Dr. O'Keefe and a short concluding memorandum by Chovitz summarizing that my reasoning was correct for the projection method but that not sufficient information was available to carry it through, as Dr. O'Keefe had tried to tell me to begin with. An uneasiness remained, however, until a break came for me when I found some correspondence in AMS files about some more of the background of that mysterious Memo 660. It indicated that it had been an interim solution needed for mapping under pressure of time and circumstances, and based on poor geodetic material; and that the mathematics had been set up to impose the condition that the distance between Tokyo and Shinkyo must not change in the conversion. This imposed condition had made me see the light at last: it enforced a mathematical incompatibility with the three-dimensional approach, where distances in general do change with a datum change. While I knew about holding the distances fixed in the development method, I had felt that this could be treated as an "uncertainty" in a limited area. But mathematically enforcing that condition at the fringes of an area spanning about 2,000 km was a different thing. I finally realized that it was not a matter of neglecting something insignificantly small, but of using mathematics suited to the purposes of two-dimensional geodesy, yet inherently contradictory in a three-dimensional context.

I had learned much more geodesy from this bizarre case than originally expected. I was grateful for the many patient discussions and for the opportunity to work things out in depth in a practical application. I was more convinced than ever that the future lay with the three-dimensional approach, and I concentrated on computing geoid charts or profiles to collect three-dimensional data. At about that time, we received deflection data for a near-meridional arc in Great Britain and I turned these into a geoidal profile. I did the same for the four meridional arcs used in our Figure-of-the-Earth study.

By the turn of the year 1955/56, a request for consultation came from Venezuela via the Inter American Geodetic Survey (IAGS), concerning the derivation of a gravimetric deflection at La Canoa, to be used to define there a datum point for the South American continent. When Mr. J. Koslozki (IAGS) and Dr. Romero (Venezuela) came with the material several months later, it turned out that the consultation call was just pro forma: the computations had already been done and the plans were fixed. Chovitz and I noticed that the gravity data around La Canoa covered only an area within a 75 km radius (later enlarged to 150 km), quite insufficient according to our experience with gravimetric deflections in the Western Mediterranean study, where we had used a 450 km radius. Lack of money precluded an extension. Also, we wondered about the location of a continental datum point at one end of the continent, and in a geologically rough region, in the mountains near the Caribbean Sea and the Puerto Rican trench, which would have an effect on the actual deflection, but not be covered by the limited gravity survey.

This problem gave me some more insight into the role of datum points, datum definitions and their consequences, and I became increasingly involved in these questions. I did not foresee then, however, that the rectification of this particular choice for South America would become one of my major tasks some day, requested as a resolution to a protracted controversy between the South American states. When the "La Canoa Datum" with its incompletely derived deflection of the vertical was nonetheless called the South American Datum, Mr. Hough insisted on the name "Provisional South American Datum of 1956", to remind cartographers (and geodesists) that it was only meant to give a start to triangulation computations, but not to be extended over the continent. His misgivings were not heeded, yet were well founded as it turned out later.

5. A NEW FIGURE OF THE EARTH

The Figure-of-the-Earth project was still the major project in our section. I took an increasingly greater part in it as my widening background studies enabled me to grapple with, analyze, and formulate its many-sided aspects. In February 1954, Professor W.A. Heiskanen, internationally renowned proponent of the isostatic and gravimetric approaches in geodesy, visited us and showed great interest in our plans. He was Director of the Finnish Geodetic Institute in Helsinki and also of the Department of Geodetic Science which he had established at Ohio State University, and he and the Finnish members of his faculty seemed to commute between the two institutions. He requested and received a print-out of our listing of 1° by 1° area mean elevations, which was part of our Phase I. This phase comprised the calculation of the effect of the "distant" topography on the deflections of the vertical at hypothetical stations at 100 km intervals. I plotted and contoured the information in two graphs, one for the meridional and one for the prime vertical component. These graphs confirmed globally what had been seen in the advance computations for the Mediterranean area (and already published in that article), namely, that this effect was on the order of a tenth of a second of arc and thus negligible. This was good to know and it simplified procedures.

Phase II, the regional collection of mean elevations in specific area schemes within 450 km of selected astronomic stations, was much more time consuming due to the sheer multitudes of numbers collected, double-checked, corrected, and processed. While parts of this phase dragged on till fall 1955, Chovitz and I prepared for the next and major step: how to extract from these heaps of numbers a modern determination of the size and shape of the Earth.

The crucial new items prompting such a determination were the two new geodetic arcs, completed in 1954, spanning more than 90° in latitude; the longest arcs ever measured in geodetic history (Fig. 4). They reached from Canada into Chile, and from Scandinavia via the historic 30th meridian all the way through Africa. The northern part of the African arc had already made history in ancient times when Eratosthenes (3rd century B.C.) calculated the size of the Earth from two types of information: the distance from Alexandria to Syene (Aswan), and the angle between the sun's rays and the vertical pointer of a sundial in Alexandria on solstice day, when in Syene the sun's rays could reach the bottom of a deep well. In the late 19th century, Sir David Gill, Her Majesty's Astronomer Royal at the Cape of Good Hope, started a survey along the 30th meridian from its southern end, in the hope of carrying it north to Egypt for an eventual connection with the observatory at Greenwich, to facilitate a better longitude determination for South Africa. His unfinished dream was made generally famous by Jules Verne in his science fiction novel *Meridiana*. It was finally fulfilled by our Army Map Service team under Mr. D. L. Mills. What a historic perspective! It filled me with a sense of awesome history and privilege of participation.

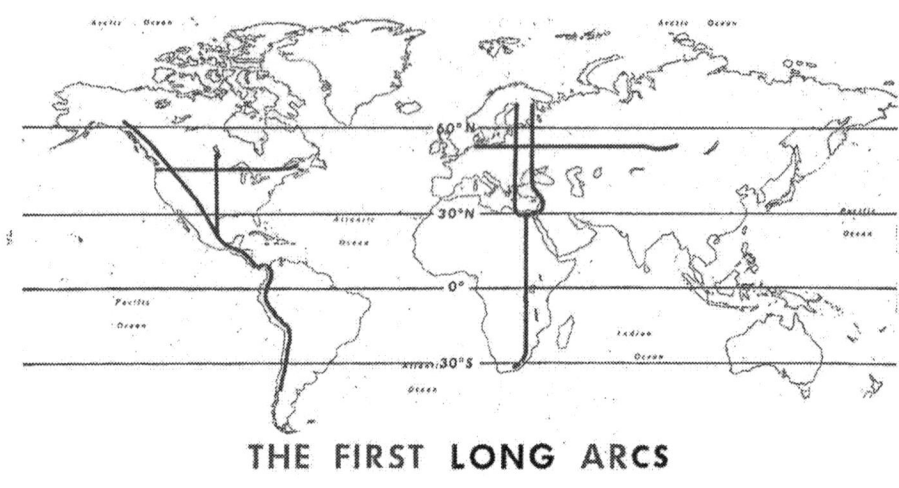

THE FIRST LONG ARCS

Figure 4

Ch. 2, Fig. 4: The First Long Arcs

Curiosity about what these unique overlong meridional arcs would reveal won the day. It was decided to make a preliminary arc solution first and to delay the plan of a comprehensive solution including all other geodetic networks. The idea to be formulated mathematically was this: Ideally, the geographic position of a place, that is, its latitude and longitude, should be the same by whatever method they were established. But the astronomic position (from direct astronomic observations) and the geodetic position (from measuring distances and angles through a geodetic network starting at another already established place) are in general not the same. Why not? Because the geodetic computation is tied to an a priori assumption about the size and shape of the Earth which is not necessarily correct. It takes the form of a geodetic datum established at one time from too limited information, and retained for the sake of consistency as a reference framework. If that reference datum is changed to another one, the geodetically computed positions will change, making their differences from the observed astronomic positions smaller or larger depending on the change. Devising a datum change, which would make those differences vanish or at least make them as small as possible, would result in a "best fitting" datum. Because a geodetic datum is defined in terms of a specific ellipsoid in a specific position at the datum point, such a solution or a "best fitting ellipsoid" would give us a good approximation for the size and shape of the Earth. But since the irregular Earth (or rather the geoid) is not a smooth mathematical ellipsoid, differences between astronomic and geodetic positions (deflections of the vertical) will still exist even if minimized in a best fit. These will give us more details of the geoidal shape, as shown on geoid charts. It is obvious that a solution for a best fitting ellipsoid may vary with the extent of the data to be fitted. That is why the unsurpassed length of the new arcs made them so interesting and us so expectant.

But where does this plan leave all the hustle and bustle of Phase II, which kept us busy for so long? The purpose of this activity was to meet a string of worries about the other part of the deflections, namely the astronomic positions to which the geodetic positions were to be fitted. We were worried that the direction of the plumb line, the vertical, which is the reference direction for the astronomic observations, could be affected, "falsified", by the attraction of the irregular topographic masses around it. If there is a mountain on one side of the station, could it not pull that plumb line out of direction? So we were going to try and estimate by how much all the topographic masses (out to about 450 km distance in Phase II) would deflect the vertical from its "normal" direction at the specific stations, and correct for this "falsification" before entering the main computation or a best fit. Then you start worrying whether there are unknown dense masses underground, which also would deflect that vertical direction; what about the unknown densities and unknown distribution of these unknown masses? This type of questions and attempted answers is technically called "topographic-isostatic corrections", but a name does not solve any mystery, it only conceals the complexities of a scientific puzzle. One learns to appreciate Albert Einstein's famous comment: "Der Herrgott ist mutwillig, aber nicht böswillig." (The Almighty is mischievous but not malicious.) In that hopeful spirit we tried our best to estimate the topographic effect from the map readings in Phase II, and we examined isostatic theories about the unknown underground masses to add an estimated isostatic effect.

Thus we made plans for two sets of computations: with and without topographic-isostatic corrections. The first one would supposedly give us a world ellipsoid after correcting the input data or all the irregularities as best as we could. The second (technically called a "free air" solution) would give us a world ellipsoid fitted to the Earth "as is". The original assignment had only called for a topographic-isostatic solution, but we had asked for permission to do a "free air" solution alongside for comparison and received authorization in fall 1954.

By the end of 1954, the new arc in Africa had been adjusted on the extended European Datum, and the American arc on the extended North American Datum. It was decided to add variations (alternative meridians) for both arcs in their northern parts where so much more data existed, and also to add the two longest available parallels, to utilize at least some representation of the existing data to the east and west. Thus we used in the Western hemisphere the coastal arc to Alaska, in addition to the 98th meridian, and also the 44th parallel. In the Eastern hemisphere, the African arc was continued through the 24th meridian in Europe, with a variant through the triangulation along the Eastern Mediterranean to the 32nd meridian; and the 52nd parallel was added. We planned a number of partial solutions or analysis, from single arcs and some combinations of arcs; our prize possession, the electronic computer UNIVAC, would make that possible.

In summer 1955 we persuaded the UNIVAC Branch to let us do our own programming in order to expedite matters and facilitate analyzing. I greatly enjoyed taking their in-house course and happily started programming my own work; but when I sent it over for processing, it turned out that they had deliberately taught us an obsolete compiler that they did not use anymore.

It was obvious that they did not want us to learn their trade secrets; they wanted to defend their territory and keep us dependent. That fight was to go on for years with much frustration on both sides, before the researchers were allowed to do their own programming.

For us, the refusal of direct access to UNIVAC operations was a blow in view of the due date to have computations finished by the end of the year. We could have achieved a quicker turn around for experimental solutions and saved all that time of writing detailed requests for every step. Well, it was going to get much worse later on, before it got better in that respect.

The last batches of Phase II map readings had to be expedited, but we were short on people. Chovitz wrote a request at least for summer help and there came two college students and three soldiers. Lewicke had been transferred into the Datum Section to replace Montella who was about to leave. Rouleau had transferred to Math Computation Branch already the year before. That left only Chovitz, Sodano, and myself; and I planned to go on an extended vacation with my husband to Paris, Rome, Israel, and Athens. Sodano was very busy with his own projects: UTM zone to zone tables, UTM extension beyond 80° latitude, long lines inquiries, and several other things. Chovitz also had several things going besides pushing the Figure-of-the-Earth work. And we all shared in giving help to each other's specials and to coping with outside requests. I tried to get done as much as possible before leaving on vacation, wrote a summary of the plans for the various partial arc solutions for UNIVAC, and a list of "Things to Do" for the time when I was gone and when I would be back.

It was going to be our first overseas flight and I was sort of scared. "Flying puts one a little too much in the Good Lord's hand" was a comment I had once heard and liked. Dr. O'Keefe offered an observation from his perspective: Catholics have it easier. They know there is an 'upstairs' where they would go." But I thought of Bernard Shaw's vision of the "upstairs" (uninteresting, boring, singing hymns all day) and I was not eager. There is also a Jewish tradition of the "Academy on High" where one can hear the sages of all times discourse on God's word all day long. But I did not think they admitted women there; and even if they did, I was just not yet interested. I wanted to be alive, be with my family, enjoy this colorful world, and compute the size and shape of the Earth. The airplane seats in 1955 were roomier and much more comfortable than today. One could tip the back almost horizontal and thus really sleep through the night, which saved that many hours of anxiety. Also, your patronage of the airline was much more appreciated than today. Each of us received a colorful certificate (which I still have): "From the Ancient and Mythical Realm of Neptune, Rex, Court of the Dawn, the Sun and the Moon," testifying with a golden stamp that we made one trip (there was space for golden stamps up to ten trips) over "Oceanus Atlanticus from New York to Paris on this 18th day of August 1955, via Trans World Airlines," and covenanting us to "use, recommend and support Air Transportation to help foster amity between nations and good will amongst the peoples of the Earth," signed by the President of TWA, Inc. It reminded us of another awesome crossing several years earlier, the crossing of the equator by ship, when the crew mumbled some dark predictions of what Neptune would do to first-time passengers crossing "The Line". Suddenly they yelled: "See the line?!" and passengers leaned over the railing trying to see a line, and then there was a big splash dousing a screaming crew member as a vicarious sacrifice for all of us.

I awoke at daybreak to a fantastically beautiful sight. We were way above the clouds; there was brilliant sunshine and blue sky; and white clouds enveloped the Earth, which looked like a precious jewel wrapped in fine white cotton balls for protection. Here and there a slight parting between the clouds permitted a glimpse of the waters beneath, reflecting the rays of the sun like the fire of the jewel. "The heavens declare the Glory of the Lord, The firmament shows His handiwork", it sang in me with the psalmist and Beethoven, and Isaiah's "The heaven is My throne and the earth is My footstool" seemed written all over. It was an awesome and unforgettable experience, strengthened but never paralleled by future flights. When we neared the European coast, one could catch glimpses of land from between the clouds. The people down there would have a cloudy day while we had bright sunshine. They were invisible for their smallness, crawling like ants on the ground. Yet, the tiny creatures all over the lands had been able to collect millions of purposeful measurements, processed them, stored them in office safes and libraries and put them at the disposal of the Army Map Service for a new determination of the size and shape of this wondrous Earth, my jewel Earth. And I was privileged to be at the center of this activity, if this man-made vehicle did not fall down now or on the way back. P-L-E-A-S-E don't let it fall! It was with great relief that I felt the touchdown on terra firma and, after the ensuing maddening drive on the runway, the final stop at the prescribed spot at the airport. I offered a brief yet heart-felt "thank you" prayer that "He had kept us in life, and sustained us, and let us reach this time." This blessing of gratitude has become a standing and meaningful ritual at the landing of all our future flights.

When I returned to the office, refreshed and eager, I immersed myself in analyzing the numerical returns from the UNIVAC as they came in, trying to visualize their significance. I plotted each solution to see graphically the change from the previously plotted geoid profiles along the meridianal arcs on the original reference datum, the North American (using the "Clarke Ellipsoid") and the European Datum (using the "International Ellipsoid), respectively. The summer help had left or would soon leave. But Thelma Cooley transferred from the Occultations Branch into our branch and was a great help to all of us. Pam Henriksen, the Chief of that branch, left AMS to raise her own, human family (alongside the puppies) and her branch was going to be reoriented towards future satellite work under Walt Henriksen. Chovitz and Henriksen had attended, several months earlier, a satellite conference where Wernher von Braun had been speaking; and Dr. O'Keefe tried to develop some satellite capability in the branch. He had written a paper on "The Geodetic Significance of an Artificial Satellite' and another one on orbit theory, and he was working with Charles Batchlor of the Occultations Section on a study to determine the effect of the Earth's potential field on the position of a satellite. I too tried to tool up and started to study textbooks on astronomy and celestial mechanics. But the priority job at this point was to finish the arc solution for the new Figure of the Earth and to write a Technical Report about it (TR-19). Chovitz wrote part and I wrote part. Chovitz added a detailed account of formulas and uncertainties in long appendices, and I added all the illustrations including many deflection and geoid profiles, and I pointed out their significance. I also included a discussion against the notion of a triaxial Earth. The final solution of this Report showed that the earth model should be somewhat smaller than the International Ellipsoid chosen in 1924. Its value of the flattening, 1/297, (that is the measure of its difference from a sphere) was adopted in this result as the one most plausible. I had derived and wanted to include also a solution for a better fitting value of the flattening, 1/298.1 (a little less flat at the poles). I was not permitted to do so, however, and had to take it out of the illustration summarizing the "Comparisons". "You would make a fool of yourself and of us,' I was told, "because the flattening is very well known by now from several authoritative sources"; and these were enumerated to me. As a relative newcomer still to the field, I was not sufficiently familiar with these proofs or evidences and their definitive, last-word quality, and so I had put my foot in blissful ignorance where angels fear to tread. To general chagrin and upset, the satellite of a couple of years later had also not studied the literature and thus was not at all impressed by these authoritative sources, and insisted on a flattening value (298.3) that had been considered disproved long ago. With a beginner's luck, my value from the arc solution had been near that.

My wish to include my fascinating geoid profiles was humored, however. To me, they represented the result of an important learning experience. Plotting the deflections along the American meridianal arc, for instance, showed them as random little values (a few seconds of arc), but computing from them the geoidal heights (distances of the geoid from the reference ellipsoid) brought out their three-dimensional significance as slopes of the geoidal surface (Fig 5). Imagine the profile as a vertical cut into the Earth and add one little height increment (slope times the distance for which it is valid) to the next along the profile, then the shape of the geoid will be traced out, showing graphically its discrepancy from the reference ellipsoid (Fig 6). Going south along the American arc showed the geoid going up and up and up as compared to the North American reference ellipsoid (the Clarke Ellipsoid of 1866) to an incredible height of more than 200 meters in Chile. This showed not only that the Clarke Ellipsoid of the North American Datum did not fit in Chile, but also that a better fitting ellipsoid would have to be larger (less curved) in order to agree with the geoidal trend. This surprising result was paralleled, though in reverse, along the African arc (Fig 7). Going south from the Mediterranean Sea, the plot of deflection values looked innocuously random, but the geoid derived from them dived down and down below the International Ellipsoid of the European Datum, an unexpected 200 meters of negative geoidal height in South Africa. The European Datum was thus shown to be a poor choice for Africa. A better fitting ellipsoid would have to be smaller (more curved).

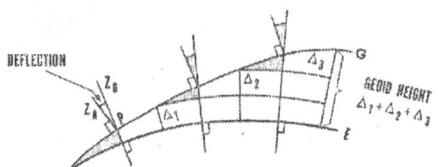

THE GEOID
from
ASTROGEODETIC DEFLECTIONS

○ THE DEFLECTION OF THE VERTICAL
IS THE ANGLE BETWEEN ASTRO ZENITH Z_A AND GEODETIC ZENITH Z_G;
EQUALS THE SLOPE OF THE GEOID FROM THE ELLIPSOID

○ HEIGHT INCREMENTS Δ = SLOPE × DISTANCE

Figure 5

Ch. 2, Fig 5: The Geoid from Astrogeodetic Deflections.

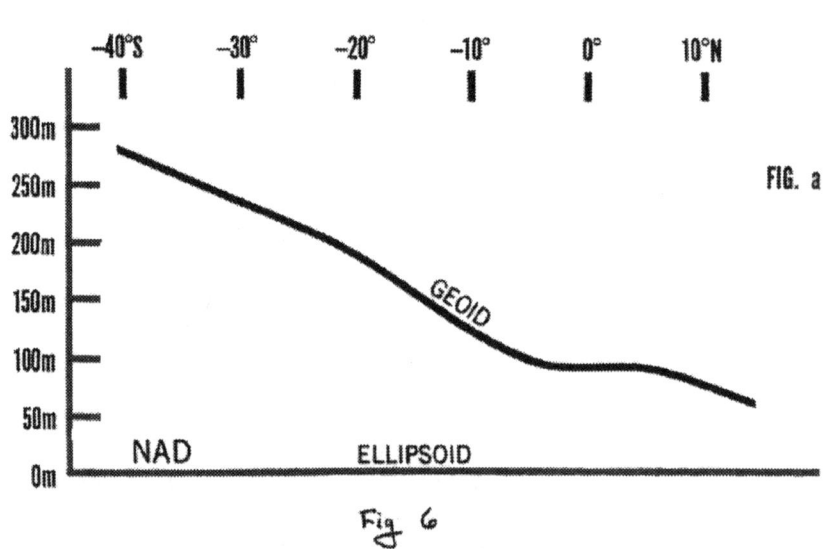

Fig 6

Ch. 2, Fig 6: The American Arc

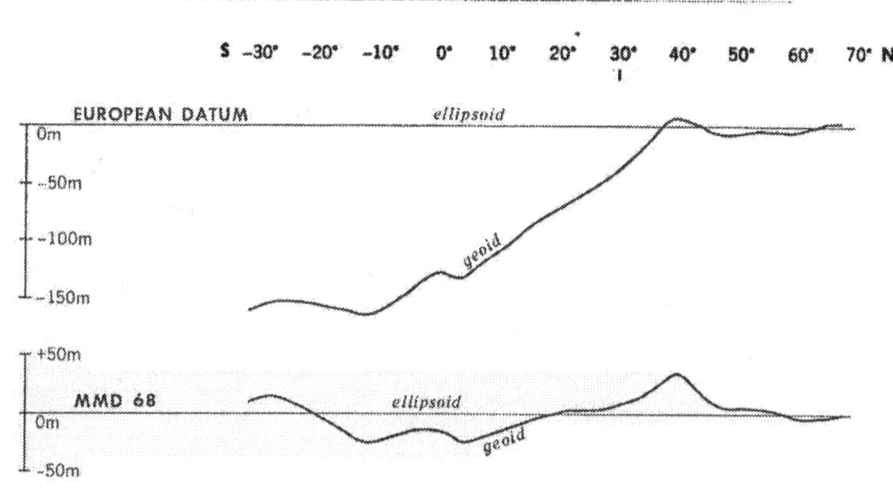

Fig 7   Poor fit of European Datum in Africa

Ch. 2, Fig 7: The African Arc

Figure 8 depicts these results in a global cut through an irregular, potato-shaped Earth, the geoid. The European ellipsoid E_1 had been fitted fairly well within Europe without knowledge of the rest of the world. Figure 8 suggests that a better fitting world model would have to be smaller than E_1 and larger as well as rounder than E_2. I thought that this three-dimensional way of looking at geoidal heights as the criterion for a good or poor fit and for deciding on a remedy, was much clearer and made more sense than the two-dimensional interpretation of deflections as mysterious little differences between astronomic and geodetic positions, and I decided to work with geoid heights as the basic values from now on.

THE GEOID AND TWO ELLIPSOIDS

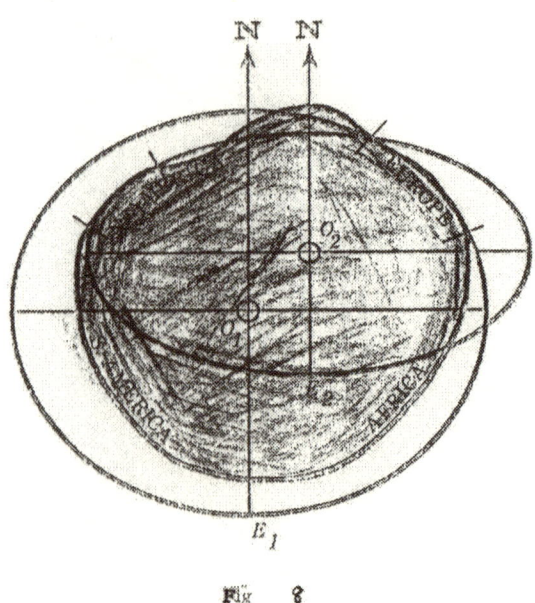

Ch. 2, Fig 8: The Geoid and Two Ellipsoids

For publications in the *Transactions of the American Geophysical Union* (AGU) we made a shorter version under the same title, "A New Determination of the Figure of the Earth from Arcs" (Vol.37, no.5, Oct.1956, 534-545I later reprinted in Buenos Aires, *Revista Cartogrfica*, no.6,1957). We also prepared a presentation for the Annual Meeting of the AGU in May 1956. I joined the AGU as a member that year. The presentation at the meeting (given by Chovitz) was very well received. Professor Heiskanen, recipient of the William Bowie Medal at the same Annual Meeting, called it "the most important piece of geodesic information presented in the last two decades," and he congratulated Mr. Hough and his coworkers on "the splendid task they had just accomplished."

Publicity about our new Figure of the Earth began with a press release before the AGU Meeting and a press conference at the Meeting. Notices appeared in the press the following day. *Time* magazine devoted two full columns to our paper and included a sketch explaining Eratosthenes' derivation in relation to ours.

Another paper headed its account of our work with "New Report Shrinks Earth by Half a Mile." *The Daily News* brought two columns under the title "It May Be a Smaller World Than You Think." *The New York Times Magazine* placed us under the "Oddities of the Year" in their end-of-year review in December. We even made "Strange As It Seems" by Elsie Hix, who brought a cartoon. What more is there to wish for?

3

TOWARDS INTERNATIONAL PARTICIPATION (mid 1956–end 1958)

1. ON MY OWN.

In July 1956, Chovitz transferred into the Technical Developments Staff, first on a ninety-day detail, but practically for good. It was a minor earthquake for the Branch and even more so for our Section. Before he left, he explained the contents of his files and turned parts of them over to Sodano and parts to me according to our preponderant interests. Sodano's interests were flare triangulation, geodetic table construction, nomograms, and long lines where he had already made a name for himself. He had written TR-7: "Inverse computation for long lines; a non-iterative method based on the true geodesic", 1950. I was interested in expanding the geoid studies which had prompted the change in the section's name from "Long Lines" to "Geoid". The suggestion was made to create two sections. This happened *de facto* but not *de jure*. When Chovitz did not return from his detail and Sodano was asked to take over, I was so involved in my work that I rather welcomed the prospect of not having to spend time on the perennial administrative trivialities, and that was the only difference I could see. Sodano was easy to talk to. I admired his capabilities in analyzing geodetic requirements and finding a practical solution. I liked his common-sense directness and down-to-earth judgments. He impressed me with his work on his house, which he single handedly enlarged for his growing family, even digging a basement underneath. And we had common interests in raising figs. We exchanged plants. I still have a mock-orange shrub blooming every year that comes from his gardens and he got one of my fig trees, which he dug up with his impressive strength while his small son watched in admiration and tried to help. Sodano devised ingenious ways to keep his fig trees alive during the winter in this, for figs, marginal Washington climate.

There were a lot of other things but figs to talk about, however. There were the frequent outside requests for information on map projections, deflection values, long line computations and the line that needed researching, writing letters and proofreading the typed outgoing version. There were relatively short assignments that took priority over everything else and occupied us all. Of course, help in checking and double-checking was exchanged all around. Fortunately, we had again some temporary student help for the summer months, but that was not always an unmixed blessing considering the time spent in teaching them at least some minimum rationale for our mysterious activities.

In my own one-person section within the section, I was very busy indeed. Dr. O'Keefe tried to fill the vacuum left by Chovitz' departure by coming around more often with advice, encouragement, and new ideas, always responsive to questions and the need for discussions. He believed in encouraging all of his employees to read freely around their subject to widen the scope of their understanding and to take advantage of our great AMS library including the opportunity to borrow from the Library of Congress. He also maintained a shelf of periodicals in the office to make us aware of current literature and discussions. He was always full of work suggestions for anyone of us to pick up, and he encouraged everyone to grow according to ability; the sky was the limit. He permitted freedom of action to see things through on our own, as long as we stayed within the general policies and kept him informed. It seems that I had then more freedom of action than at the end of my career as a GS-15 when the fumbling management tried to restrict people's normal speech and actions to the scope of kindergarten permissions, subject to elusive hierarchical decision-making. In hindsight, it shows the interesting contrast between supportive and restrictive management, with the ensuing psychological impact on the work force.

Among several things, Dr. O'Keefe suggested that I try to explore the effect of terrestrial mean gravity anomalies on a satellite orbit; and he wanted me to help in his own orbit investigations for 'Project Vanguard' by taking computation work off his hand and shepherding the trial solutions through UNIVAC. So I needed time to study some astronomy and celestial mechanics. I also needed time to follow up another of Dr. O'Keefe's suggestions, namely to study the European geoid for indications of a possibly still extant effects of the ice-age. Maybe one could discern a bulge in the geoid in some distance from the area where the ice load had pressed down on the Earth's crust, and forced the material to move sideways. Such a bulge might be masked when plotted against the backdrop of an ill-fitting reference ellipsoid, but now that we had derived a new and supposedly better-fitting ellipsoid than one had used before, even if it was only preliminary, there might be a chance to consider that question. So I went and found literature on the postglacial uplift in Fennoscandia, and while I was at it, I looked at the story of the ice age in America and the USSR, and the living example of the Greenland ice still pressing down. There were the French and British expeditions across the Greenland ice, complete with observed gravity profiles. Would these tell anything? The other part of this project was to establish the best and most comprehensive representation of the European geoid or better of the Eurasian geoid. There was Lieberman's TR-14 where the Eurasian geoid was tentatively referred to a variety of ellipsoids and where the mechanics was demonstrated of how to change from one to another. And there also was an interesting TR-12, a translation of a Russian document on geoid work across the USSR, by B.V. Dubovskoy, 1939. It contained a geoid chart up to 130° east longitude that showed increasingly crowded contour lines towards east when referred to the Bessel ellipsoid as used in the 1932 triangulation; this indicated an unusually steep rise in the geoid. I started to pay more attention to Russian geodetic literature with the help of friendly and knowledgeable Mr. Colonna from the "Foreign Control Branch". Mr. Colonna knew not only foreign geodesy but also languages including Russian. He also knew and loved old clocks. He took our two 150-year-old clocks into his loving care and restored them to life. He was a very pleasant and competent colleague, loved his work, was esteemed by everyone, and was sent away on his seventieth birthday as an unwilling victim of the government's wasteful age restriction rule.

The Russian triangulation of 1932 on the Bessel ellipsoid had been changed to the current triangulation on the Krasovskiy ellipsoid in 1942; the steep, about 400 meter rise of the geoid towards east, revealing a misfit of the old datum, must have played a role in the decision to change. In Molodenskiy's papers I found a discussion of my old problem of "development" versus "projection" method and, much more important, an elegant and practical way of changing the first into the second. The basic idea is simple: a scale change is needed for every distance measured on the geoid, but treated as if it were on the ellipsoid in the adjustment. In practice, however, it is rather complicated because the neglected scale changes had been lumped together with other kinds of errors and distributed throughout the network in the conventional adjustment procedure. The resulting incorrectness is negligible in the ordinary practical case in a limited area, but it might show up in long distances, as it did at the eastern end of the long Russian geoid profile. And it might show up in our two exceptionally long meridianal arcs too.

Molodenskiy's rectifying formulas were thus very useful and welcome. I adapted them to our case and gratefully called the improvements the "Molodenskiy Corrections', which term caught on in the literature.

It was obvious by now that my efforts for the ice-age project" dovetailed with those for the major 'Figure-of-the-Earth project' which I had inherited from Chovitz. The "arc solution' had been done as a preliminary shortcut because of the curiosity about the effect of the new long arcs. The original commitment was a comprehensive solution utilizing all networks. But I intended to do things differently; I wanted to stress a three-dimensional approach. While the "arc solution" was based on minimizing deflections, and my geoid profiles had been an interesting side line, I wanted to make geoidal heights the basic input this time, and minimize them to derive a most closely fitting ellipsoidal reference surface. So I needed and was looking for geoid charts of any area. I also wanted to apply the "Molodenskiy Corrections" in order to conform to the "projection method" and to explore at the same time the practical allowable limits for their neglect.

When I heard Dr. O'Keefe mention that the Astronomy Branch could well use up-to-date geoidal profiles across the United States, similar to the strip which Hayford had computed almost half a century ago, I jumped at the chance for permission to do this myself. I would do the whole North American continent, not just a strip, the same way as I had done the Manchurian geoid. Dr. O'Keefe thought that I underestimated the work involved, that it was impossible for one person to do all that, and that he could not give me enough people to help. So I scouted around the plant to borrow people and eventually found a friendly ear. I managed to persuade Mr. Homer C. Fuller, Chief of the Americas Branch, that my project was the most interesting, most important, and most fascinating project there was under the sun, and who else but the Americas Branch should have a chance and privilege to work on an American geoid chart. Mr. Fuller was intrigued; while he could not spare his own people, he arranged for the Field Office in San Antonio, Texas, to work on my project, provided I could furnish clear and detailed directions, which I promised, of course. In September 1956, we had a conference with all the people along the chain of authorities connecting me with the Field Office, and I was in business. I wrote "foolproof" instructions and explanations for the Field Office and requested that interim phases would be sent up to me for inspection and correction or additional explanations as needed. In the meantime, I made computations by another method for selected profiles crisscrossing the continent, in order to have something for a comparison with the expected results from the Field Office. All this would take a while, of course.

The new Figure of the Earth, although preliminary, was of great interest to the geodetic community because it was significantly smaller (by more than 100 m in its equatorial radius) than the International Ellipsoid, and people wanted to use "the latest" in their projects. We called it the "Hough Ellipsoid" in honor of our Department Chief whose vision and leadership had conceived and guided this work. I received several requests in connection with it such as computing geocentric distances at specific places (Commander J. C. Tison, then U.S. Coast & Geodetic Survey; Col. Clair E. Ewing, then Patrick Air Force Base). Dr. Samuel Herrick, California, and Dr. J. W. Siry (then Naval Research Laboratory) visited AMS and recommended adoption of the Hough Ellipsoid for satellite work. This was endorsed by Col. Robert Miller, OCE. So I established a "Hough Datum" or "Vanguard Datum" for the satellite project Vanguard and computed "Hough coordinates" for a list of Vanguard tracking stations for Col. Miller, Dr. Siry, and Col. Ewing. This was the first operational world datum. While it was replaced as a world datum in later years by updated versions, the Hough ellipsoid is still being used in some places (Samoa, Wake, Eniwetok, Kwajalein). Visitors then wanted to be briefed on it and on related topics.

There were many other shorter assignments to attend to, such as contoured deflection charts for some countries; memoranda to be composed such as persuading the Commanding Officer that gravity observations were needed and where and why; manuscripts to be evaluated and the like. Thus, I had always several things going at the same time, but I thrived on it and was happy. When I was a very little girl and wanted more time for whatever I was doing, my father used to tease me: If you get up one hour before day, then you will have 25 instead of 24 hours in the day". I remember I was puzzled and it took me a little while to figure that one out. Now, I was trying to do the most in those eight office hours and took my speculations home with me in order to streamline what was to be done the next day and the next and get at least ten hours work into the eight. There was a certain urgency about it, because the freedom given me to play with these precious toys implied for me the obligation to come up with something useful, to integrate the bits and pieces of information lying around for the asking, into better insight into the grand question of the size and shape of our Earth. When a colleague told me: "I could never be interested in the Figure of the Earth", it struck me with amazement and disbelief that someone could fail to see this question as the core and pinnacle of all of geodesy to which all geodetic activities funneled their contributions.

I suppose I was in a similar frame of mind as my daughter when she had been a member of the junior high school band. Mr. Hoose, the band director, had an enormous influence upon the children. He asked them to come an hour early to school in the morning to practice and they came. He asked them to stay late after school or return in the evening and they stayed and returned. He asked them to spend their weekends practicing, in order to be able sooner to climb the next rung in the ladder of achievements and they did. He succeeded to have an excellent, motivated, and happy band and the band was the central, absorbing interest in the school life of the members. My daughter described the phenomenon thus: "And if Mr. Hoose would ask us to jump in the lake, we would jump in the lake." And how do the other children who are not in the band, react and fill their time? "Oh, there is nothing for those poor kids. They are left out, and they lead dull and miserable lives".

_My_ Mr. Hoose was the Figure of the Earth. While I would not be ready to jump into the lake but would surely find ways around the lake instead, I did feel sorry for those who were unresponsive to the supremacy and grandeur of the ancient quest that had fired the imagination of mankind since millennia—and that might get a unique contribution from our holdings and talents at the Army Map Service.

In February 1957, Dr. O'Keefe hired an assistant for me. Mary Slutsky had worked for the government already several years, but was bored by now by the routine work and looked for something more interesting. At her first interview with Dr. O'Keefe, she was taken around to talk to some of us, and I painted for her a colorful picture of what I was doing and she accepted the job. At her first workday, she found herself assigned, by the routine of the Personnel Office, to the Datum Section, which also needed people. But they had not counted with Mary. She told them that this was not the job why she had wanted to make a change and come here. If she could not work on the Figure of the Earth, she would go back where she had come from. Good for her and good for me! She worked with me for the next ten years. She was a very conscientious worker, tracking down every potentially noxious detail. We routinely checked each other's computations; while I never found an error in hers, there was no chance that one of mine could slip by her.

In March 1957, I had a big surprise of something that I never knew existed and thus saw as an event in a strange world: I received, it seemed out of the blue, a medal. I was told that it was the Meritorious Civilian Service Medal of the Army (signed by Major General E. C. Itschner, Chief of Engineers) and that it was the first time that the Army Map Service had ever qualified for such a distinction. I was told about it the day before, so as to be sure to be here and bring my husband. Col. Abell presented the award, I said a few words of appreciation, and there were refreshments. There are some photos showing me with Col. Abell, Dr.O'Keefe, Mr. Mills, Mr.Culley, Mr. Chovitz, and a proud husband. All looked a lot younger then.

Mr. Hough had retired from government at the end of January and Mr. Mills was his successor. Mr. Frank L. Culley had come to us from the U.S. Coast and Geodetic Survey to be the Assistant Chief. He also was the Secretary of the Geodesy Section of the AGU. The following July, Mr. Hough received the Legion of Merit award and the Exceptional Civilian Service award in a ceremony in the open court of the Pentagon. I had a personal invitation to attend. It was a beautiful sunny day and I felt very proud of my former Chief. In August, Chovitz too received the Meritorious Civilian Service award. I was the only one who had trouble with the lapel pin: it was made for men's lapels. This anti-feminine bias was corrected several years later, however, when new rosettes were issued and one was sent to me (in December 1961).

2. IUGG, TORONTO 1957

The International Union of Geodesy and Geophysics (IUGG), of which the International Association of Geodesy (IAG) is one of seven constituent associations, used to hold a General Assembly every three years (now every four years). At the Rome Assembly in 1954, Mr. Hough had been the only delegate from the Army Map Service, but now in 1957, six people were allowed to go to Toronto: Mr. Mills, Mr. Culley, Dr.O'Keefe, Mr. Chovitz, Mr.Sodano, and myself. Mr. Hough was there too; and so were some family members: Mrs. Hough, Mrs. Mills, Mrs. Culley and daughter, Mr. Sodano's father, and of course my husband.

In November 1955, Chovitz had been nominated by Mr. Hough to be a member of the Special Study Group No. 10 (SSG-No.10) led by Brigadier Guy Bomford of Oxford University, England. The purpose of the Study Group, which belonged to Section V of the IAG, was the "Determination of the European Geoid from Astrogeodetic Deflections of the Vertical", continuing the work of the Study Commission No. 14 at the Rome Assembly. When Chovitz left the Geodetic Division in July 1956, he suggested to Bomford with Hough's concurrence that I take his place there, whereupon I had received a very kind letter of invitation, the first of an abundant correspondence with Bomford over all these years. Bomford also invited Chovitz to stay on.

Within SSG-No.10, I found myself suddenly in the awe-inspiring international company of great geodesists, whom I had only known by name so far but would meet in person in the course of time. The members under Bomford's chairmanship were: Mr. J. J. Levallois (Paris), Hofrat Professor Dr. K. Mader (Vienna), Dr. H. Wolf (Frankfurt am Main), Professor A. Marussi (Trieste), Professor V. R. Ölander (Helsinki), Professor Dr. C. F. Baeschlin (Switzerland), Mr. C. A. Whitten (U. S. Coast and Geodetic Survey, Washington D.C.), Professor Dr. K. Ledersteger (Vienna), General I. Seref Dura (Ankara), Dr. Djordje Nicoloch (Yugoslavia), General G. Spiliotopoulos (Athens), Mr. Don A. Rice (U.S. Coast and Geodetic Survey, Washington D.C.), Mr. B. H. Chovitz (AMS), and myself.

In spring 1957 when the first circulars for the Toronto Assembly came around, Chovitz suggested that we submit a paper together on "The Influence of the Distant Topography", that we had worked on in connection with Phase I of the Figure of the Earth project. He also would offer our listing of global elevation readings and talk about our already published long arc solution. I had several things in work and wondered whether there was time enough to get something finished and turned into a paper for Toronto. Time was at a premium, but with Mary Slutsky on board and Archie Carlson helping out for several weeks, and the Field Office coming through on schedule, I managed to write and submit for clearance a paper on the North American geoid chart at the end of June, finish the chart itself and another paper on the Hough Ellipsoid at the end of July, and got them all cleared and 200 handout copies each produced before leaving for Toronto end of August. Also Sodano had his three papers (on flare triangulation, long lines, and U.T.M.), with the assistance of Cooley-Robinson, ready on time for Toronto. All this reads like a fairy tale now compared with today's obstacle feats of the modern bureaucracy. An added pleasure then was the concluding letter from Mr. Richardson, San Antonio Field Office, thanking for the opportunity of doing this interesting work and wanting to be considered again for similar jobs.

The international atmosphere at the Toronto Assembly was exhilarating. I have a photo of the impressively multinational plenary session being addressed by the Canadian Prime Minister Diefenbaker, and photos from the reception by Toronto's Mayor Phillips at the Hotel Royal, and photos from a marvelous Sunday boat trip on the Georgian Bay of Lake Huron. There was an excursion to Stratford for a great performance of Twelfth Night. I remember glimpses of people clusters standing around on the university grounds, chatting in the dining and social rooms, lecture halls, garden paths, and on the broad steps of the buildings, to get acquainted and discuss problems. There was the curious spell of experiencing books and letters, even formulas, turn into live people who mostly looked so very different from what you might have thought. The formula that I had used to minimize the geoidal heights turned into friendly Dr. J. de Graaff-Hunter, President of the AG. When I delivered my handout package, I found busy Mr. J. J. Levallois smiling at me in a cordial welcome. (Remember the one with the understanding observation of "les lenteurs administratives," and with that first French form letter addressing me as "Monsieur"). Big books turned into Sir Harold Jeffreys, past IAG President, F. Baeschlin, and Brigadier Guy Bomford. And there were many more such magic wand transformations.

Then we overheard conversations in some oddly familiar yet oddly unaccustomed language, German studded with strong Viennese dialect in between, rising out of a life of long ago. But that brought back poignant memories and doubts: could these German-speaking people have been Nazis at that long-ago time? You would never get a plausible answer to such a question. So I decided to avoid any awkwardness and not say any German word in their hearing; my accent would have identified my former home town right away. I let them talk English to me, waiting courteously till they found the right word in English, and I kept that up for years at future meetings. Obviously, they did not recognize my accent in English.

Sodano made a big splash with his papers which were discussed at length. His method of long-line calculations (an extension and improvement of his TR-7) was acknowledged as the most precise among competitors. He then reported on field tests, using the light on an airplane or parachuted chemical flare as visible intermediaries between non-intervisible distant stations to determine the direction between them. These were newer developments, an outgrowth of his TR-10 of 1952.

Special Study Group No.9 reported on various methods of using the Moon or a future artificial satellite for geodetic purposes, thus drawing attention of this international forum towards an impending, yet unknown jump in the future developments. The individual contributions by Dr. O'Keefe, Dr. W. Markowitz (chairman), Dr. F. L. Whipple, and others, were published in the *Bulletin Géodésique* (No.49, September 1958). Chovitz presented our common paper, "The Influence of the Distant Topography on the Deflection of the Vertical' and offered a printout of our global collection of mean elevations, upon request. He also reported on our new Figure of the Earth, published the year before, and mentioned briefly the significance of my papers to be presented later in the Geoid Section, that is, Section V.

Brig. Bomford, Chairman of SSG-No.10, and also President of IAG Section V, had a pleasant way of putting people at ease, which I had felt already in the previous correspondence. With his marvelous sense of humor he explained to us the didactical usefulness of his bulging middle to quickly impress on his students in the classroom the notion of the equatorial bulge of the Earth. He had accommodated my papers in his program and had some kind words to say about them. He tried to impress on me the need for establishing a connection between the hemispheres so that world geoid charts on one and the same datum could be drawn; and he appealed to me for American enthusiasm to get such work funded. I could easily assure him of my personal enthusiasm, if that would help.

My North American geoid chart presented here, was significant in three ways (1) it was the first chart ever to cover the whole continent (up to 68° North), where before only a narrow east-west strip across the U.S.A. had been computed by J. F. Hayford in 1909, and the southeastern part of Canada by C. H. Ney in 1952; (2) it was part of the world scheme that Bomford had mentioned and (3) it had a fascinating feature that had unexpectedly appeared while I worked on it, like a magic picture coming up when children rub a seemingly empty piece of paper, it showed a big depression in the geoid around the Hudson's Bay, which could be related to Dr. O'Keefe's idea that the ice age should have left some still noticeable traces connected with the expected postglacial uplift. Mr. J. E. R. Ross, Dominion Geodesist, Canada, to whom I had written for permission to publish this chart, as it was based partly on Canadian data, had expressed his amazement at the consistency of this chart with geological literature and congratulated us for this important and excellent work. I had put aside the ice-age study in Europe, due to the other pressing work, but I would surely follow it up again. It was a pleasure to meet Mr. Ross here in person, and also Dr. J. E. Lilly (and Mrs. Lilly), Ross' successor who from now on would be my source and correspondent for Canadian deflection values.

The paper on the Hough Ellipsoid was also important for three reasons: (1) I had changed the method from the customary two-dimensional "development" to the three-dimensional "projection" method, discussed and applied the so-called Molodenskiy Correction, and demonstrated the difference it made for our long arcs; (2) I had used all available networks as had been our original commitment; and (3) I represented these networks in terms of geoidal heights rather than deflections, and derived a world solution by minimizing them. There was a fourth reason for presenting this paper at Toronto: There would be a delegation from the U.S.S.R. and I had a specific question, preferably to Dubovskoy or Molodenskiy, in connection with the long east west geoid profile on the Bessel ellipsoid. Since no new numerical information was let out of the U.S.S.R., this 1939 profile was very important and I wanted to be sure to apply it correctly. My question was: was this profile computed according to the "projection" or the "development" method? In the latter case, I would have to apply that Molodenskiy Correction to it; this would make a non-negligible difference of about twenty-five meters in geoid height at the east end, where I would have to connect it to the Manchurian geoid. The time of the Dubovskoy article, 1939, was the time when such questions were discussed in the Russian literature, so it could have been done either way. I could not expect an answer to a letter or an informal question to an individual so I tried a public question by presentation and publication in addition to an attempt at personal contact. For that purpose, I had computed my paper for both cases with alternative solutions. Page 6 gave the one case, and quoted results separately for the Western and Eastern hemisphere. Page 6a contained the other case for the Eastern hemisphere, should the Russian material need the correction which I had applied. Page 6a was inserted so that it could be ignored within the narrative, if that assumption was not true. After presenting and distributing my paper with the stress on this open question, I sought to make contact with a Russian geodesist. One could not do that casually, because the Russian delegation was obviously not allowed to mingle and converse with us freely. Although some of them wanted to be friendly and sat down to chat, they might suddenly jump up and leave in the middle of a sentence when a colleague of theirs passed by and whistled or gestured ever so slightly. They all lived at the Hotel Royal and not as we did, individually housed at the university's dormitories or at various hotels. The two people I had hoped to meet, Dubovskoy and Molodenskiy, had not come. I was told that both were ill and also that Dubovskoy had changed to a very different field after the war. When I put my question to Professor Y. Boulanger, he was very friendly but claimed that he did not know or had forgotten since it was so long ago. Then I tried Mr. A. A. Izotov who also at first said that he did not know or remember but I did not buy that, since I knew Izotov from the literature as being in that very field. Eventually, a meeting was arranged, with Dr. F. Bros from Czechoslovakia to be Izotov's interpreter. It seemed prudent at this point to have my own interpreter and I found the Israeli geodesist, Mr. B. Goussinsky, whose mother tongue had been Russian, willing to help me. I also brought the English translation of Dubovskoy's article, to be as technically specific as possible. When Izotov saw this report he was clearly upset and wanted to know how I got it and where the Russian original was. I said: "On my desk in my office." Then he said that if he could have seen my original he would have known where it came from and could have answered my question. That did not make much sense to me with respect to my question but it illuminated the circumstances. The impression that he was afraid to say anything, was underlined when later on he did not even want to answer questions about his own paper, presented and distributed at the meeting.

A stalemate! But not quite. Several months later I received a letter from Goussinsky with a translation of Isotov's report on the Toronto Assembly, mentioning both of my papers in detail. He reported that my Russian source material was taken from Dubovskoy and he gave my numerical results, quoting the numbers for the Western hemisphere from my page 6 and those for the Eastern hemisphere from page 6a, not even mentioning that two sets of solutions had been computed and published. So here was my answer, loud and clear: Dubovskoy's work was done with the development method and needed the correction that I had inserted on page 6a. I reported this to Brig. Bomford. Just to be quite sure, I asked him to use his good offices as President of an IAG Study Group to double-check with Dr. I. D. Zhongolovich, who was now Secretary of the IAG Section V (Geoid Section). The answer was a clear confirmation. Zhongolovich reported back that the Dubovskoy document had been lost during the war! it was exactly known, however, that the computations in it had been done on the Bessel ellipsoid with reductions only to sea level and not to the ellipsoid.

3. CONSOLIDATION BEFORE THE SPACE AGE

Coming home from Toronto in the middle of September 1957, I considered two jobs as high priorities: to pursue the ice-age project and to round out the paper on the Hough Ellipsoid into a comprehensive Technical Report. There were several items to consider: the African profile along the 30th meridian looked a little different from Bomford's, and the cause of the discrepancy must be found new deflection data brought back from Toronto should be incorporated; ways of connecting the hemispheres must be devised and tried out; new deflections in Korea were available at last and a three-dimensional tie between Manchuria and Japan should be constructed now so that this whole area could be incorporated into the world datum for that purpose, also a geoid chart of Japan based on the latest collection of Japanese deflections must be constructed and extended through Korea to a common point with the Manchurian Principal System. It meant a lot of work, but Mary Slutsky was there to help, and I persuaded Mrs. Carta who welcomed an occasional change from her administrative duties, to do some computing and plotting. Mr. Fred Wilson ("Freddie the Artist", then in the Americas Branch under Mr. Fuller) promised to ink six of the illustrations. Others would be done by Dr. N. Moon's section. Mr. Ray Lorah, then in the Foreign Control Branch, would ink the large Japan and Manchuria charts and photograph them down to size. I enjoyed inking my own numerous illustrations (I had learned to do that to exacting standards in my student days), but I had been warned to be careful and not let the job analyzer catch me at it, which might cost me my grade as a researcher.

Some foreign visitors came through AMS on their way home from Toronto: Dr. Erik Tengström from Uppsala who had reported there on his interesting work on the geoid in Sweden, and Brig. A. H. Dawson, Director of the Military Survey, London. There also came Dr. R. K. C. Johns who had been with the Canadian Geodetic Survey under Dominion Geodesist J. E. R. Ross, and had originally studied geodesy under Baeschlin in Switzerland, the past IAG president whom I had met in Toronto.

Suddenly, on 4 October 1957, we were aroused by the successful launching of the U.S.S.R. Sputnik. It was quite a shock for Western space scientists who for some reason had not counted with the possibility of a Russian "first." Some took it almost as a personal affront. The face-saving reaction was, "Ah, it is just a crude stunt; lifting a big and heavy chunk into space to get attention; it has no scientific value; it cannot be compared with our forthcoming small and sophisticated Vanguard." Yet it electrified the scientific community to press harder for preparedness when the American satellite would come true. Dr. O'Keefe had been very busy and productive already, as mentioned earlier. His new paper (together with C. D. Batchlor of the Astronomy Section) called "Perturbations of a Close Satellite by the Equatorial Ellipticity of the Earth" (*Astronomical Journal*, 1957, vol.62, 183-185) was ready with mathematical wherewithals, and a mood-setter called "Geodesy comes of age with Vanguard" (*Astronautics*, August 1957) claimed that "geodesy will make greater advances in the next 18 months, through the use of the artificial satellite, than in the last 50 years." He arranged for us a course in celestial mechanics taught by Professor E. Dwyer, Georgetown University, after working hours at the plant. Participants were, I believe, S. W. Henriksen, Werner Kahn, Marvin Marchant, Howie S. Genatt, C. D. Batchlor, myself, and Don Stuart and Ed Early from the Computations Branch. Werner Kahn, who did not live far from us, used to take me home in the evening, which gave a chance to talk things over. Dr. O'Keefe also arranged to get the ephemeris of the Sputnik's rocket for us to learn. He continued his own work on the Vanguard orbit where I had helped a little. It was now adapted to an AMS project called "Betty" for which I computed the coordinates of AMS tracking stations in South America on the Hough-Vanguard Datum. A 1977 issue of the office newspaper TOPOCOMENT shows a picture of Col. F. O. Diercks (who was Commanding

Officer of AMS, 1957-61) and Mr. Charles Andregg (who was Chief of AMS Operations and Planning Staff in 1957, and DMA Deputy Director in 1977) at their 1977 visit of the NASA satellite tracking station outside of Quito, Ecuador, reminiscing about the events twenty years earlier when they had been instrumental in acquiring the tracking system for this AMS site.

By January 1958, I had my marbles more or less lined up and began to write Technical Report No. 22: "A Tentative World Datum From Geoidal Heights." I had several new features to report besides the three-dimensional approach in using geoidal heights and the projection method, which had been the centerpiece of the Toronto presentation. (i) I had constructed a new, up-to-date geoid chart of Japan on Tokyo Datum. (2) Tokyo Datum and Shinkyo Datum had been connected through Korea and a three-dimensional transformation formula had been derived replacing the old Memo 660, which had had me, going in circles some ages ago. (3) The Japanese geoid looked very different on Tokyo Datum (very steep) and on Shinkyo Datum (very smooth, like the Manchurian geoid), although both datums used the same Bessel ellipsoid. This exhibition became a school example to explain the idea of a well-or ill-fitting geodetic datum, which becomes rather obvious from a three-dimensional viewpoint. (4) The Manchuria-Japan region was now incorporated into the scope of the world datum, and other areas had been updated by new data brought back from Toronto. (5) The Krasovskiy ellipsoid, used in the U.S.S.R., was investigated as to its overall fit for the Eastern hemisphere and found not to fit in Africa nor in the Far East. (6) I had connected the two hemispheres by means of W.A. Heiskanen's gravimetrically derived "Columbus Geoid", which he had presented at the AGU meeting in May 1957 and also at Toronto. My idea had been this: each hemisphere was on its own reference system and the gravimetric Columbus geoid was on a third system. If the latter was centered at the center of the Earth as it was claimed to be, and if one could devise a way to relate it to the North American Datum on its one side and to the European on the other, then it would serve as a link between the two. To derive the required relationship between my North American geoid chart, e.g., and Heiskanen's, I would match the geoidal heights at corresponding points in the two systems and stretch the gravimetric system to fit the established scale of the North American geodetic network (like stretching an elastic wristwatch band to the right size) from such a match one could derive transformation formulas to carry one set into the other. After doing the same for the European side, one could combine the two transformations into one between the North American and the European systems. To be sure, gravimetric geoid heights were not as reliable as my astrogeodetic ones because they presupposed global gravity coverage which did not (and still does not) exist. But Heiskanen's new chart was an improvement over Tanni's, and the best there was. (7) I also connected the hemispheres by another method, utilizing a survey tie across Greenland, which the Air Force had just completed. This tie was classified and its discussion was thus confined to a classified report. (8) The ice-age study was extended to include besides the Hudson Bay also Fennoscandia, Siberia, and Alaska. The surprising thing here was the striking correlation between my geoid contour charts and ice age hinge lines, both established by the different methods and purposes of their specialized fields.

Dr. O'Keefe liked what I had assembled, particularly the attempts to connect the two separate geodetic networks of the hemispheres in several ways. Because of the importance of such connection and the urgency at this time of the dawning satellite age to establish a geodetic world datum as best as one could, he pushed the completion of my work for an AMS staff briefing and for a presentation to the AGU meeting on May 5, 1958. Although it was already past closing time for the printed program, he managed to get me included as the last speaker in the morning session.

I was quite nervous all morning. Two years earlier, Chovitz' and my "New Figure of the Earth' had been presented at this forum with great success, but Chovitz was the speaker then, and proper delivery and possible nervousness had been his and not my problem. I don't remember any particular nervousness in presenting my papers in Toronto, probably because I was so busy there that I did not have tine to think about it. There was the major excitement of the international setting for two weeks with a crowded schedule of several purposeful conversations. Since one almost never got out of the limelight, this became the normal state. But I do remember my nervousness on the singly morning of May 5, 1958; it became a sort of prototype of the butterflies before my many future presentations and of a way to cope with them. I was bothered by the big names in the audience: there was C. A. Whitten, C. I. Aslakson, W. A. Heiskanen, U. A. Uotila, J. L. Orzel, my former chief F. W. Hough, and many others who certainly knew more geodesy than I did, since I had not come up through their ranks of practical field experience. I still saw myself as a relative newcomer from another field, approaching geodetic problems with my training in geometry, logical analysis, and uninhibited common sense. Was it not impertinent for me to speak to them about their technical problems? Now, wait a minute, why not use some of that common sense on yourself: These people may know a lot more than you do, but they do not know what you worked on this year and they want to know so why not tell them? What will I say? I haven't a prepared speech to read, because I found such readings awfully boring and had proudly decided to speak freely. All I had were eight slides and slivers of paper the size of my palm, with a few keywords of what to mention with each slide. What will I say after "Thank you, Mr. Chairman?:—You'll say "Ladies and Gentlemen. It is my pleasure and privilege…" Don't be silly; you have talked to classes for years with no trouble at all.—But those were kids, that's different. No, it isn't. You have something to tell them, that they don't know yet; the same with kids. So pretend to yourself, that you are in front of a classroom telling kids what they ought to know. O.k., I'll pretend they are kids. But I never addressed kids as "Ladies and Gentlemen". I might go blank after that. In fact, I feel blank right now, as if I had forgotten all I ever knew. Now stop that nonsense. Don't you remember having felt like that before? When? Well, just before those oral and written exams you took in your student days, which you then passed very well. What helped you then? Yes, I remember a large, ugly stair well in the Vienna Institute of Technology, where I had to climb several floors to be tortured up there. I wished the building burnt down or any other emergency would prevent that examination that day. Did I really wish that? If that happened, I would have to come back another day. All my preparations would be for nothing. I wanted to take that exam, did I not? And that as soon as possible, so I could go on with my plans for tonight and tomorrow. Then I told myself: When I see this stair well again, I will have passed the exam and be free to follow my plans for tonight and tomorrow. Concentrate on those future plans. And with that, the apparent paralysis was broken. With the sound of the first exam question, the blank had vanished, and I enjoyed the free use of the treasure of a thorough exam preparation.

And it worked the same way at that AGU presentation. As soon as I heard myself say "Thank you, Mr. Chairman," I enjoyed myself tremendously, telling the kids out there about the new things that I had worked on. I had a private good omen too: it was my late father's birthday, and I knew he wanted me to succeed.

"A Tentative World Datum' was submitted to the *AGU Transactions* for publication. Because of the wider interdisciplinary interest of the ice age part, this topic was formulated as a separate article, "The Impact of the Ice Age on the Present Form of the Geoid," in the same issue (*Journal of Geophysical Research*, Jan. 1959). The *AGU Transactions* had just changed their publication policy by placing scientific articles into the monthly series of the *J. G. R.* and retaining the *Transactions* as a quarterly for general AGU affairs. And this paper did attract attention from outside of geodesy. I received an invitation from the British Glaciological Society to join. I had a commending letter from the Directorate of Physical Research at the Department of National Defense in Ottawa, Canada, who also sent me the latest glacial map of North America (published by the Canadian Geological Association in Toronto), obviously as an encouragement to continue in more detail. There was a good letter from W. L. Donn, co-author with M. Ewing of "A Theory of Ice Ages" which I had quoted. And there were quite a number of requests for reprints. Professor R. J. Lougee, Clarke University at Worcester, Mass., whose work in the Algonquin hinge line I had used, wrote me about his new discovery that there had been really only one glacial stage in America, not four as generally assumed, and he invited me to listen to his announcement at the AAAS Convention in December 1958. His new theory did not convince his colleagues, however. His book was published after his premature death; no time to defend it.

Many years later, I still found my ice age paper quoted in the literature, e.g. by W. Kich, Regensburg, in "Das Eis der Erde und die Geodäsie" (*Zeitschrift für Vermessungswesen*, No.11, 1971). In summer 1972, my husband and I participated in a geographical field trip to the Maritime Provinces in connection with the Geographers' Congress at Montreal we were shown the still persisting evidence of the ice age in Nova Scotia under the guidance of expert Professor J. Brian Bird, McGill University. He had reproduced my geoid contours around the Hudson Bay depression in his book *The Physiography of Arctic Canada* of 1967.

The "Tentative World Datum", published in the same issue, received also much attention, with a lengthy abstract by the U.S.S.R. Academy of Sciences and elsewhere. The International Dictionary of Geophysics, in its review article on geodesy (written by H. Wolf, Bonn) said, "…Of the more recent research employing astrogeodetical methods of determining the Earth's dimensions, the works of Krasovskiy and Irene Fischer count among the most important. The Krasovskiy ellipsoid forms the basis of the State triangulation in the U.S.S.R., cf. Izotov (1959), and Irene Fischer (1959) calculated the well-known ellipsoid dimensions for the 'Hough ellipsoid'…. The Central European Geoid was calculated by Wolf (1949b) on the basis of the Central European network. It was later on extended over all Europe and Asia by Brig. Bomford (1960) and Lieberman (1955) respectively, a geoid calculation in the global sense being made finally by Irene Fischer (1959). We are indebted to K. Ledersteger, too, for further research and results".

The excitement of consolidation before the approaching space age gave rise to an article in *Life* magazine (May 12, 1958) about AMS activities, called "The Missile-era to Chart the Earth", written by John Dille, obviously inspired and dramatized by Dr. O'Keefe. It showed among other things also a picture of Mr. Hough explaining on a blackboard the idea of the ellipsoid named after him. Another picture shows Dr. O'Keefe with a relief model of my North American geoid chart, made by Sylvia Palmer, then of the Relief Map Branch. It is built in layers representing 5m geoid height increments, indicating clearly the depression in the Hudson Bay region. This heavy exhibition piece has been one of several eye-catching show pieces in my office until I left.

A pleasant interruption was the request for a review of George Strasser's excellent publication *Ellipsoidische Parameter der Erdfigur 1800-1950*. It is a very informative collection of the numerous attempts during more than 150 years to determine the size and shape of the Earth, all with more or less different results. One can see clearly the dependency of a specific result on the specific set of data used in its derivation, which explains its success as a representation of this data set and its failure to be the world model, especially in those days of very limited data coverage.

In March 1958, Dr. Walter D. Lambert (retired from the Coast and Geodetic Survey) visited our Branch, and his 1928 paper on "The Figure of the Earth and the Parallax of the Moon" was mentioned in connection with the fact that we had now a better Figure of the Earth than was available then. Dr. O'Keefe suggested to see what difference our new geodetic knowledge would make, if inserted into that paper, and whether it would clear up the discrepancy between observed and computed parallax which Lambert had determined.

Around that time, Dr. Ben Yaplee (Naval Research Laboratory) tried to measure the distance to the Moon by radar and discussed with Dr. O'Keefe the implications for the size of the Earth. Naturally, I wanted to understand these interrelations, and I went to the Naval Observatory library to find literature and observational data for the observatories at Greenwich and the Cape of Good Hope, where the classical determination of the Moon's parallax had been made in 1911. This grew into a complex and fascinating study which kept me busy off and on for the next several years.

Part of the consolidation effort before going on to new things, was Miss Slutsky's TR-30. Mary had shown much interest in the Figure of the Earth project including the work we had done before she came, and I persuaded her to familiarize herself with the material stored in our files on Phase I with a view towards formalizing it in a Technical Report. We thought that this work should be made available to the geodetic community for use in global studies. Mary began to collect and organize the topographic-isostatic deflections in the files and then proceeded to augment the material for a better global coverage. Occasionally, she could employ summer help for the tedious map readings. Because of the relatively low priority, her numerous additional contributions to the collection of stations, and her other responsibilities, it took a while before TR-30 was completed. It was a very useful collection and copies were requested still many years later.

In October 1958, Rev. Pierre Lejay, the founder and Director of the Bureau Gravimetrique International in Paris, and his assistant, Dr. Suzanne Coron, both of whom I had met in Toronto, visited Georgetown College Observatory; and its Director, Rev. Francis J. Heyden, arranged in their honor a Scientific Colloquium to discuss informally problems and projects in the gravimetric field. We heard Dr. George P. Woollard tell us about progress in establishing a consistent global gravity network; and Rev. Lejay about new gravity data received or expected at the Bureau. Dr. J. L. Worzel reported on new data in the Pacific and across the Atlantic ridge observed on cruises by the Lamont Observatory (which would bear on my study of the Mid-Atlantic Ridge in distant 1977). Col. George Woodring of the Air Force told about work on the ice cap stations in Greenland; Mr. Donald A. Rice and others discussed priorities in survey planning, Dr. Lloyd Thompson talked on instrument testing and airborne gravity measurements, and Dr. Coron on her isostatic studies in the French Alps. On his way home, Rev. Lejay suddenly died on shipboard, three days after we had said good-bye.

I have a photo of the colloquium participants. Besides the people already mentioned, there were A. Baldini, Lt. B. Murray, A. Cosler, W. M. Kaula, Bela Szabo, Gunnar Leifson, B. Lane, R. McShane, Urho Uotila. Kaula had joined AMS by the end of 1957, after he had received the first M.A. degree in geodesy that Ohio State University conferred. I had known of Captain Kaula's interests already from his two manuscripts which several months earlier I had been asked to evaluate before publication when he was at the Army Engineer School at Fort Belvoir: one on the accuracy of gravimetric deflections and another on deflections in New Britain. He now worked on a global representation of gravity anomalies by $1° \times 1°$ square mean values, extending the coverage to unobserved areas by statistical theories. Eventually, he would produce from that a gravimetric geoid, in competition to Heiskanen's Columbus Geoid.

Around this time, an attempt was made to upgrade the image of the geodesist, and the Civil Service Commission announced a new examination to create a prestigious profession. I was a "mathematician" up to then and was not inclined to change that label. But I was asked to set an example for a high caliber geodesist and thus help the upgrading effort. I was made a "geodesist" and so were others. I was a GS-12 then. The grade scale was much lower then than now, mostly because the names of the administrative layers were upgraded since (without a corresponding rise in authority or leadership, however). Mr. Hough and his successor Mr. Mills were then division chiefs at GS-14, but had much more independence than the department chief has now at GS-15. Dr. O'Keefe as Branch Chief was unfortunately limited to GS-13 (maybe one reason why AMS lost him to NASA in 1958). Although there was a Section Chief layer between him and me, he made me a GS-12 to acknowledge my contributions. He and Mrs. Iola B. Rust, Personnel Office, championed the idea that a capable employee can raise the value of a job to the company by giving it a superior status not foreseen in the original more modest job descriptions. They reformulated my job duties at the higher grade to include my activities on the international scene and my internationally acknowledged geodetic contributions. This new principle of job evaluation, that the incumbent can raise the grade of the position by superior contributions, that the incumbent can rank the position and not only the other way around as usual, led a few years later to the special pay category of "Research Geodesist". Job descriptions pertaining to such research positions cannot meaningfully center on a fixed position with predetermined duties, but rather give representative examples of the activities of this particular incumbent. My job descriptions read like that.

4. A LITTLE NUMBER CAME DOWN FROM THE SKY

Suddenly, the other shoe came down with a thump. True, it had been expected since the uproar about Sputnik, and one could hear the time ticking, but it was still sudden when the first geodetic result was extracted from the observations of the Vanguard satellites: a new value for the flattening of the Earth (a measure by how much the ellipsoidal earth model differed from a sphere). With Dr. O'Keefe's foresighted preparations, it was only natural that he would be the first to announce such a result. The satellite had been launched on March 17, 1958, and already on June 24, 1958, the Harvard College Observatory sent out Announcement Card 1408:

> Oblateness of the Earth by Artificial Satellites.—The Commanding Officer of the U.S. Army Map Service and Dr. John A. O'Keefe report the following results from artificial satellites "A preliminary determination of the oblateness of the earth has been made at the Army Map Service. It has not been possible to process the observational material; so far it has only been possible to analyze

the elements published by NRL. A preliminary study by H.G. Hertz and M. Marchant, based on the NRL orbits of 1958 β_2 for 26 March to 6 June 1958 indicate that the oblateness is near the value of $1/298.38 \pm 0.07$. This value is considerably smaller than the international value of $1/297.0$. In view of the general interest in this constant, it has been decided to put out this value in spite of its very preliminary nature. The determinations of $1/f$ with estimated standard deviations, when available, are:

Inverse Oblateness

Method used	Node	Perigee
Satellite		
1958 α	298.0 ± 0.3	297.8
1958 β_2	298.38 ± 0.07	298.3

June 24, 1958	Fred L. Whipple

HARVARD COLLEGE OBSERVATORY
ANNOUNCEMENT CARD 1408

Oblateness of the Earth by Artifical Satellites. The Commanding Officer of the U. S. Army Map Service and Dr. John A. O'Keefe report the following results from artificial satellites: "A preliminary determination of the oblateness of the earth has been made at the Army Map Service. It has not been possible to process the observational material; so far it has only been possible to analyze the elements published by NRL. A preliminary study by H. G. Hertz and M. Marchant, based on the NRL orbits of 1958 β2 for 26 March to 6 June 1958, indicate that the oblateness is near the value of 1/298.38 ± 0.07. This value is considerably smaller than the international value of 1/297.0. In view of the general interest in this constant, it has been decided to put out this value in spite of its very preliminary nature. The determinations of 1/f with estimated standard deviations, when available, are:"

Inverse Oblateness

Method used:	Node	Perigee
Satellite		
1958α	298.0 ± 0.3	297.8
1958β2	298.38 ± 0.07	298.3

June 24, 1958   FRED L. WHIPPLE

Ch. 3, Fig. 1: Announcement Card1408

After the happy excitement that AMS had been "first" with this announcement subsided a little, we looked at what we had and we were dismayed. Some thought it was not right that the satellite had again taken sides in the east-west conflict in favor of the wrong side. The Russians with their Krasovskiy ellipsoid (flattening 1/298.3) had said all along that the International flattening 1/297.0 used in the West was wrong and that they were right. But we had all the proofs in our pocket; did the satellite not read the literature?

The adopted value of the Earth's flattening (the fraction by which the polar axis is foreshortened in terms of the equatorial radius) changed from $1/297 = 0.003367$ to $1/298.3 = 0.003352$, that is, by a difference of 0.000015. Such a small number and such a big fuss! It means that the earth model is a wee little bit less flattened at the poles. Yet it changed our working plans. I would have to revise my new "Tentative World Datum" that I had just presented at the AGU meeting and sent out for publication. True, most of the work was not affected, that is, the work within each hemisphere. Only the connection of the two hemispheres contained the assumption of the old, more flattened form of the earth ellipsoid and would have to be replaced.

Instead of the Hough ellipsoid I would have to decide on another, less flat ellipsoid. From the correlation between flattening and earth radius, discussed in the earlier paper, "A New Figure of the Earth from Arcs", one could expect that the radius of a "rounder" model would be smaller when fitted to basically the same data. Also Kaula had to modify plans for his new gravimetric geoid to accommodate the new value; if he finished in a reasonable time, I could use it instead of the Columbus geoid.

Some geodesists of note could not bring themselves to accept the satellite verdict and insisted for a long time that there must be a serious error. By contrast, Dr.O'Keefe explored the geophysical implications of the new fact of life and pointed out ("Zonal Harmonics of the Earth's Gravitational Field and the Basic Hypothesis of Geodesy", *J.G.R.,* vol.64, 2389-2392) that it contradicted the previous hypothesis that the Earth was essentially in hydrostatic equilibrium. If that was not the case, then the actual flattening (derived by satellite) and the hydrostatic flattening (of a theoretical fluid Earth in equilibrium) would be expected to be different, and therefore the latter needed to be rederived without using that hypothesis. Henriksen started to analyze the proofs for the old value to see where they went wrong, and came up in due time with a paper: "The Hydrostatic Flattening of the Earth" (*Annals IGY,* vol.12, 197-198), where he found a new hydrostatic value of about 1/300. It was later slightly refined by Dr. O'Keefe, and again by Sir Harold Jeffreys, England, as 1/299.67. This was suggested as the appropriate theoretical starting value of interest to geophysicists when studying the forces that shape the Earth as it actually is.

Dr. O'Keefe's analysis of the observational material collected by the satellite, also led to (among other things) the now famous pear-shape of the Earth ("Vanguard Measurenents Give Pear-Shaped Component of Earth's Figure", with co-authors A. Eckels and R. K. Squires, *Science*, vol.129, 565-566, 1959). With his flare for dramatics and his great talent to drive the significance of a difficult concept home for the grasp of the uninitiated, he managed to transform a drab mathematical coefficient in an abstract theory into a colorful household picture that even made the comics in the newspaper. A Peanuts' cartoon in 1959 shows Charlie Brown happy with a new globe, something he had always wanted to have. Then Linus tells him that according to the latest scientific discovery, the world is not round, it is pear-shaped; upon which Charlie Brown is so disgusted that he throws the new globe away. This was cartoonist Schultz's glorious contribution to popularize new science to the very young generation. He must have been very successful because one day I received a telephone call from some government office downtown, requesting authoritative advice: It seemed—and I could just picture it—that an elementary school teacher had called for help about what she should tell the kids. She had taught them that the Earth was round, but one tyke apparently would take no nonsense from the establishment and insisted that he knew that the Earth was shaped like a pear. What should the teachers be told to tell the kids?

About 12 years later, I wrote a booklet called *Basic Geodesy, An Initiation into the Mysteries of Geodetic Concepts*, which was published by the U.S. Army Engineer School, Fort Belvoir, Va., and widely distributed. I had wanted to include this cartoon, but quite unexpectedly and contrary to customary experience, I was refused copyright permission and thus could not rescue this old forgotten masterpiece from total oblivion. I did get permission, graciously, to use it in my own briefings, which I did to the delight of many audiences. The Army editors, though personally sympathetic, did not dare to allow me to include even a paraphrased version of the cartoon nor the real-life story of the schoolteacher. Even the subtitle of the pamphlet was too much for the Army and was at first dropped but then cautiously tucked away into the Preface.

PEANUTS: (c) United Feature Syndicate, Inc.

4

THE MERCURY DATUM
(1958–winter 1960/1)

1. FROM BEDFORD, DECEMBER 1958, TO HELSINKI, 1960 IUGG

In October, 1958, Project Vanguard was transferred from the Naval Research Laboratory (NRL) to the new National Aeronautics and Space Administration (NASA), and most of the people connected with that project (e.g. Dr. J. Siry) went along. At the AGU meeting, May 1958, Dr. O'Keefe had reported the recommendation of the "AGU Committee on Geodetic Applications of Artificial Satellites" that in view of the critical need for new studies of satellite orbital theory, experts in celestial mechanics should be relieved from other duties as far as possible and allowed to pursue these studies." Dr. O'Keefe himself transferred to NASA by the end of November 1958, and several others from our capable group of "handpicked characters," who were involved in satellite work, followed soon afterwards: Dr. Hans Hertz, Mr. Werner Kahn, and Mr. Howard S. Genatt. Mr. William Kaula who had been asked to take over Dr. O'Keefe's job at AMS, made no secret of his plans to stay with AMS only as long as he needed to finish his gravity analysis and get a few papers published; then he would move on too. There was a good-bye picture taken of the whole Research and Analysis Branch, reprinted recently in the *Topocomment* of 8 July 1977. Mrs. Dorothy Davis was Branch Secretary then. The place did not look the same after Dr. O'Keefe left. If Chovitz's leaving had been like an earthquake upheaval, Dr.O'Keefe's leaving and the subsequent exodus of these other friends seemed ominous. There were still Marvin Marchant and Henriksen left of that group, and there were Sodano, Glusic, and Marian Hardy to talk things over with, and Kaula although for a limited time. I kept in touch, of course, with the émigrés. I myself did not even consider joining them then, because I had my work cut out here for a long time to come and was too involved to leave it. And busy I was.

Kaula, Henriksen (with Pam, his wife) and I attended the Conference on Contemporary Geodesy at the Smithsonian Astrophysical Observatory, Cambridge, Mass., December 1-2, 1958, where I heard Dr. Maurice Ewing (Lamont Geological Observatory, N.Y.) describe his fascinating vision of a network of bench marks on the ocean bottom in support of an urgently needed geodetic world datum; and I also talked to him about my ice-age work where I had quoted him. I listened to discussions between experts about current methods in geodesy and where one could or should go from there.

December 3-4, Kaula and I attended the "Seminar on Military Geodesy" at the Geophysics Research Directorate, Air Force Cambridge Research Center in Bedford, Mass., where I presented a paper on my plans for updating my Tentative World Datum to incorporate the satellite-derived flattening value. Since a competing proposal developed at Ohio State University (OSU) under Heiskanen's guidance was presented by ACIC (Aeronautical Charts and Information Center), I pointed out the difference in approach to the crucial problem of combining the separate geodetic networks of the two hemispheres to a uniform global systems my use of 202 geoidal heights evenly distributed over the whole surveyed world in about 5° intervals versus the OSU use of only one gravimetrically derived deflection value per entire hemisphere, even though computationally checked (yet not quite corroborated) by a cluster of several auxiliary points in their neighborhood. The inaccuracies in individual geoid height values would be constrained by the global coverage, while the inaccuracy in the one deflection value would result in a precarious alignment of the whole hemisphere as if balanced on the tip of one pencil, with a systematic error increasing with distance from the pencil point. One could see that error already now in the discrepancy of the OSU results with values derived in their own "Columbus Geoid". Kaula reported on his statistical approach to extrapolate the scanty gravity information into unobserved areas, which differed from the OSU approach, and on his plans for a new gravimetric geoid chart based on the new flattening and thus replacing the Columbus Geoid. I intended to use Kaula's results.

Our military bosses at the Seminar refrained from voicing an opinion on the relative merits of the competing proposals and gave their approval for future work in a Solomonic fashion. These were not the exact words, but I remember their meaning as something like this: Both AMS and ACIC are commended for excellent work. Now, be good children, keep on working on what you think is best, and then add up your final results and divide by two; for we can use only one system. There was a very nice "thank you" letter from the Seminar's organizer, Mr. Owen (Obie) Williams, and a letter of commendation for our "superb presentations" from the Air Force Geophysics Research Directorate to our Commanding Officer, Col. F. O. Diercks, who added a delighted cover letter of appreciation in turn. Col. Diercks was Commanding Officer at AMS from 1957 to 1961, known for his care for the needs of the AMS people and most appreciated by these in turn.

In a professional sense, the Bedford decision was a letdown, although one might concede that there was not much else the military administrators could be expected to say. For my own technical integrity, however, I pursued my plans for an unclassified updating of my Tentative World Datum with undiminished scrupulousness, intent to produce a technically sound product.

Some newly accumulated information needed to be incorporated. The new geoidal arc through Iran made it possible to include all of India and surrounding countries. The requested information about the Dubovskoy profile in the U.S.S.R., discussed earlier, had been received from Zhongolovich through Bomford; and it changed that area somewhat, and as a consequence also the Far East. Some minor improvements were made here and there. My widely distributed geoid chart of the North American continent was considered important enough by itself so that I was asked to write a brief article about it for *The Military Engineer*, a magazine to which the mapping agencies regularly sent some (usually unsigned) information about their activities.

Something odd, however, happened on the way from me through the liaison office at AMS to the magazines: a paragraph had been changed to read as if Mr. Fuller had made the chart. Two more names were mentioned but not mine. Mr. Fuller felt embarrassed; I knew he would never claim someone else's credit. I went to see the liaison people to find out why my manuscript had been changed without consulting me, but, of course, nobody remembered doing it. Someone in my office surmised that people "downstairs" could not accept the idea that a woman could do something important and therefore inserted the names of these three well-known men. This was a new angle of government life for me.

The combination of the two hemispherical geodetic systems would be straightforward if there were a good geodetic connection across Iceland and Greenland. Even if the new tie made by the Air Force were not classified, however, it would not have served my purpose adequately, because it was made with the old two-dimensional philosophy, that is, with horizontal coordinates only. The tie was comparable to a rope hanging across Iceland-Greenland, whose length was measured without regard to the slack the rope might assume in a three-dimensional view. Thus the actual distance could be much smaller. How much? I requested that astronomic observations be made along the path to determine the geoidal profile and thus the slack of the rope. And P-L-E-A-S-E do them quickly before the ice around the stations melts or shifts. But it takes a long time for such a request to be heeded, processed, and carried out, particularly because an expedition to the Greenland ice cap was possible only during the summer season. Since I could not wait for years, I tried several reasonable assumptions and studied their effects on the resulting versions of a new world datum. In an additional version I used also Kaula's new gravimetric geoid in place of Heiskanen's. Eventually, three variations were published which happened to yield only slightly different earth radii 6378,166m, 6378,160m, and 6378,155m.

These numbers make more sense when seen in the context of geodetic histories. In 1924, the International Association of Geodesy (IAG) declared the Hayford Ellipsoid (equatorial radius 6378,388m, flattening 1/297) as _the_ International Ellipsoid, _the_ model for the Earth as a whole. Relying on this judgment, it was adopted for the European Datum in 1950, also in some other countries, in astronomy, and in theoretical studies in general. This prestige was shattered by our new Figure of the Earth in 1956, based on the evidence of the two long historical arcs; the radius was unexpectedly reduced by more than a hundred meters, which is a lot in geodesy and therefore attracted so much attention. The application of the new satellite-derived flattening (instead of that of the International Ellipsoid) dropped the length of the radius by another hundred meters because of the mathematical correlation between radius and flattening. You may visualize the meaning of such a correlation, if you think of a piece of a broken vase which had an approximately ellipsoidal shape and a certain volume. Attempting to re-determine the original size, you might try several models of about the same volume. If you squash it slightly at the poles (greater flattening), the equatorial radius gets larger. If the flattening is less (the shape is nearer to a sphere which has a zero flattening), then the radius is smaller. And the flattening _was_ smaller than the old one. The two effects together resulted in the tremendous and important drop in size of more than 200m. For perspective, I may add here a hindsight observation of later years: the later developed satellite capability of measuring distances all around the globe, reduced these early results by only another twenty or so meters.

This study was published in July 1960 in the _Journal of Geophysical Research_ under the title "An Astrogeodetic World Datum Based on the Flattening of 1/298.3," later presented to the international community at the IUGG Assembly in Helsinki in 1960, and picked up for use in various endeavors. Brig. Guy Bomford, England, picked the version with the smallest radius for a uniform readjustment of the whole Southeast Asian region as the new "South Asian Datum". Geonautics, Inc. (contractors of NASA) picked the version with the largest radius for use in NASA's manned space flight project "Mercury" and coined the terms "Mercury Datum" and "Mercury coordinates." In documents, e.g. of the Jet Propulsion Laboratory or of the Marshall Space Flight Center, the world datum was identified as "known as the Astrogeodetic World Datum of Irene Fischer, 1960," which glade me feel awkward and I pushed instead the catchy title of Mercury Datum which caught on. NASA retained the datum for the subsequent manned space flight programs Gemini and Apollo. It was used also in NASA's space trajectory programs, by the U.S. Navy in their electronic navigation systems, and the U.S. Air Force for their test ranges. It was the Defense Department's only authorized unclassified geodetic world datum for more than the next ten years. The detailed presentation at Helsinki was published in the *Bulletin Géodésique* ("The present extent of the astro-geodetic geoid and the geodetic world datum derived from it," no.61, Sept.1961) and translated from there into German by the Geodetic Commission of the Bavarian Academy of Sciences. Like my previous papers, it was reviewed in detail in foreign journals. I used it, of course, in my own work which veered for a while away from the Earth to the mysteries of the lunar parallax.

But this gets me ahead of my story. Let us return to the everyday coloring of office life between Bedford and Helsinki. We had a new addition to the Geoid Section: Rosemary B. Phillips, a very intelligent, very pretty and pleasant young woman. I believe she told me that her family called her "pixy" and I could see why. She and Marvin Marchant, who joined us for a few months, worked on Kaula's statistical extrapolation of gravity anomalies into unobserved areas in order to achieve global coverage from relatively scanty observational data and from that the gravimetric geoid which I used in my world datum. Kaula in turn used my astrogeodetic material in a statistical derivation of the world datum which would go into the merger with the OSU version.

When I saw Kaula at an AGU meeting in 1977, he reminisced about our old discussion of the meaning of a statistical answer about the Figure of the Earth. He would try to explain then: "If there were a million Earths or more, then the most frequent number for this parameter would be such and such," to which I might have answered but there are no million Earths and, quoting Einstein, God doesn't play dice. And even if there were a million Earths, I am interested in only this one Earth on which I live; and this one may be the odd case, way out. The most frequent value may not at all be the one for this particular individual case."

We also had a friendly banter about doing and writing things in a simple or complicated way. Why be simple, if one can make it complicated? I had managed to persuade him to write a detailed Technical Report (No.24) about his statistical gravimetric work for the benefit of others, and when I read the manuscript I wanted him to add another clarifying sentence or paragraph here and there to make it easier for the reader. He would demur with something like this: I worked hard to straighten these things out in my mind, why shouldn't others also work hard to understand. TR-24 is already much too detailed for a technical paper."

I noticed the same attitude in technical journals in general. German and Russian journals may bring all the details of reasoning and of the derivation of equations, to make it clear and easier for the user, while American journals will cut a paper to the minimum, cutting out derivations of equations and extra illustrations and explanations. A friend of ours who had made similar observations from his perch as a university professor in physics, summed it up at one time: An American scientist would rather bite off his own tongue before he permits his technical work to appear in easy, detailed steps. A little obscurity seems to help raise the status." I myself am rather proud of the numerous laudatory comments about the clear style of my technical papers. It may be my European background or rather my having been a teacher, that I try to make things appear to be as simple as I can make it. I remember fondly that my boy in sixth grade lectured his class on the Figure of the Earth and the idea of the geoid, quoting his mother as his scientific reference.

Kaula's final derivation required the solution of, I believe to remember, 199 simultaneous equations in just as many unknowns, and he pointed proudly to the complexity of such a large system and the machine time needed to solve it. In my own work, I had not put everything simultaneously into one big pot, but had taken interim steps dealing first with countries, then continents, and then hemispheres, with time to analyze each step carefully. The final combination of the hemispheres was made with about 300 equations in only three, four, or five basic unknowns (depending on the version), which reduce to three, four, or five equations in as many unknowns. And just as proudly, I pointed out that I could do those with the desk calculator if I had to. Of course, I was finished much earlier. And the Devil put in a finger too: There was some mix-up in the Computation Branch with Kaula's (for those days) huge equation system and he got answers back that looked suspiciously different from mine, which he conceded as being correct. If that had happened to me, it would have been easy to find the culprit by some checks on the desk calculator. Kaula lost weeks or even months, before he found the error in the complex system. Eventually, Kaula's results were used as the Army version to be merged with the Air Force version as ordered by the Bedford arbiters: we took the straight mean between the final AMS and ACIC numbers in formulating the Defense Department's World Geodetic System of 1960 (WGS-60), which was promptly classified.

In the meantime I had also made a classified update of the Tentative World Datum, called "Tentative World Datum II", as an in-house "Progress Report" upon urgent request by ACIC who even supplied the scarce typing power in their eagerness. It was discussed at another Conference for Military Geodesy in August 1959, and then again in October at AMS, where ACIC was represented by Mr. Bela Szabo and Mr. Charles Frey. The only feature of interest today is TWD II's small radius, 6378,132m, small especially in comparison at that time with the OSU result. Yet, such a small radius is confirmed by today's satellite, while the relatively large ACIC-OSU result is obsolete and forgotten.

Among the continuous stream of visitors and requests there were some from the Naval Proving Ground at Dahlgren, from ACIC, from the Johns Hopkins Applied Physics Laboratory, and others. The Vanguard Datum, our first operative geodetic world datum, was still alive, and Holloman Air Force Base wanted us to convert their local White Sands Datum of 1952 to it.

Stanford Research Institute wanted Vanguard coordinates for tracking stations. Other transformation requests involved countries with no geodetic tie between them as yet, but you could not really say "no". If you did, somebody less informed might pull some preposterous numbers out of thin air. (Remember Dr. O'Keefe explaining to me the ways of the world?) So it was still better to use the best available information such as TWD II in conjunction with the global gravimetric estimates and make an educated guess sandwiched between prayers. You could explain to customers that there were not enough data available yet for a reliable answer, but when all was said and accepted, they still wanted a number to use in their project. So, a number had to be produced. Rosemary Phillips worked on missile trajectories with Kaula and I too became somewhat involved again with deviations from free flight due to deflections of the vertical. Mary Slutsky was finishing her report (TR-30) "Isostatically Reduced Topographic Deflections of the Vertical at Selected Stations in Both Eastern and Western Hemispheres," and I intended to announce the availability of this work at the Helsinki Assembly. There were other duties as assigned," but I will tell about those later.

The IUCG Assembly in Helsinki, summer 1960, had announced its approach already in August 1959 with Brig. G. Bomford's reminder to the members of his study group, asking for information about work produced in the three-year period. I had been in correspondence with Bomford all along. De Graaff-Hunter had honored me with an invitation to join also his Special Study Group No.16. In February 1960 it was time to submit abstracts and in April the finished papers went to clearance and mass production. In March 1960 Kaula had left for NASA as expected and Henriksen had taken over without any noticeable break in continuity. In June I left with my family for a ten-week trip of vacation connected with TDY in Helsinki (24 July–6 August) and the International Geographers' Congress in Stockholm (6-13 August).

2. THIS GREAT WORLD

I left on June 15 with MATS (Military Air Transport Service) from Andrews Airfield to Frankfurt, with stops in New Jersey and again in Prestwick, Scotland. There was an amusing incident at the first stop: the passengers were required to leave the plane for the stop and then were called for reboarding according to descending military rank or civilian grade from a list. When they called for Mr. Fischer, I stepped forward toward the plane stairs but was held backs "Wait a minute, Ma'am. It is the GS-12s now; you'll come later." And they called again for Mr. Fischer. With a grin I explained to them that he was me. In those pre-inflation days, GS-12 was quite a good grade and apparently they had not considered that a woman traveler could possibly qualify. MATS flying meant narrow stiff seats with the back turned to the flight direction, not very inducive to sleep and it meant overlong waits for meals which were identical for dinner and breakfast. But I was young and strong and did not mind. I was looking forward happily to go from Frankfurt to Munich where I would meet my husband and son who flew in by commercial airline via Amsterdam.

We had decided to take our American teenager on an extended tour of this great world of ours, to give him a glimpse of what there was outside of these great United States, his native and our adopted homeland. Five years earlier we had gone alone. Our daughter Gay was in college then, and now she was a career girl with AMS, a mathematician in the Computer Services Branch, and currently on a three-month liaison assignment in Boston to ease the AMS acquisition of the new Honeywell-800 computer. Michael had been too little in 1955 and we had to store him with a neighbor. I still remembered the sad but composed look in the little boy's face and his resigned good-by: "Long time no see." I had felt like calling the whole thing off, and we promised to take him along the next time.

In the meantime he had seen a lot on this side of the ocean, been north and south along the eastern seaboard, on a cruise to the Caribbean, and on an extended trip to the West in connection with the Geographers' Convention at

Santa Monica. As a budding geographer in the footsteps of his father, he kept a world map on the wall of his room and lined out there where he had been. He also helped to assemble our travel photo albums and learned to write a travelogue. On our exploration trip to the West in 1958 we had flown with him to Salt Lake City, listened to the Mormon story, witnessed a blast in the copper mine, tried to swim but could only float in the Salt Lake, marveled at the textbook geysers in Yellowstone Park, felt impressed when crossing the continental divide, then enjoyed San Francisco's sights including Chinatown and the Japanese Tea Garden in the Golden Gate Park. We went to Berkeley, the Muir Woods with the huge redwood trees, Stanford University and the Hoover Library Tower. With a rented car we drove down to Santa Monica along the mission trail, aware of the impact of Spanish culture. At the Conference there were the geographer colleagues and friends. Mike and I listened to Dad's presentation and Dr. Paul Siple's talk about his Antarctic expedition, took part in field trips around the vicinity and to the oil fields, heard about the subsidence problem where too much oil had been extracted and water needed to be pumped in to avoid disaster. My Army Map Service identification card did wonders when we visited the Nike station there. They were happy to see a visitor from far away Washington in their lonely outpost and proudly brought up a Nike missile from its underground hideaway for us to gape at.

The way back home was also planned to be filled with unique experiences. We flew to Las Vegas for the night and showed Mike the silly, frenzied expressions on the faces of the gamblers at the "one-armed bandit" machines in the lobby of our hotel. Children were not allowed in gambling locales, but the guard let us watch from behind a cordon, after we explained to him our educational purpose and complete disinterest in gambling ourselves or in the floor show. We must have seemed odd to him. Next morning we drove by bus via the impressive Boulder Dam (Hoover Dam) at Lake Mead through the Nevada and Arizona desert getting a good view of its plants. We spent a few days at the gorgeous Grand Canyon, awed by the 5000 feet abyss and the changing color scheme of the landscape as the weather changed and time went by. The Hopi Indians performed for the tourists. Flying from Flagstaff to Denver, we could see the Great Meteor Crater of Arizona, and contemplated the traffic jam, in the universe where accidents like this could happen, hopefully not again today. Smaller but more real uneasiness was caused by our small plane of Frontier Airlines which kept stopping practically at each street corner between Flagstaff and Denver, delivering milk and newspapers, and in between kept falling into air holes in the hot midday sun to the discomfiture of my stomach. But Mike received a document that he had been commissioned a "Junior Pilot on Frontier Airlines."

Glad to be on firm ground again in Denver, we went on a sturdier bus tour through Estes Park and over the 12,000 ft high alpine road to Grand Lake and back. There even were a few snow patches to be seen in the high mountains, but no glaciers. We saw the Big Thompson River Project designed to bring water from the upper reaches of the Colorado River on the western slopes of the Rocky Mountains through a tunnel to the eastern slope into the Big Thompson River Canyon for electricity and irrigation. It had been quite a trip, seeing the diversity of this great country. Yet, the experiences of the present trip would be so different again. Most would be revisiting for us, to give our son some understanding of his family's cultural background; but some would be new to us too, such as Scandinavia. The first item on the list was to show our young geographer a glacier. Before getting there, we had a sightseeing day in Munich and a couple of days in the Austrial jewel called Salzburg. The high alpine Grossglockner Road and the Pasterze glacier was an exhilarating experience for all of us although the tourist commercialization jarred with our youth memories of divine quietude. A few days in Heiligenblut immersed us in the alpine atmosphere. Then came a day of Innsbruck and a view from the over 2000m high "Hafelekar" before we took off for Italy.

Young Mike sat on the stones of the Forum and the Coliseum, overcome by this intermingling of present day life in Rome and the telling ruins of a long ago past. We walked through the Titus Arc with the marble relief showing captive Jews brought triumphantly to Rome after the destruction of their homeland nearly 2000 years ago. Right next to it, obviously painted by a recent tourist, there were big Hebrew letters on the wall proclaiming, "The People of Israel Lives." Where is Titus now? And what a marvel of vitality is the resurrection of the Jewish homeland and its modern flourishing. The famous Michelangelo statue of the Horned Moses in the San Pietro in Vinculi church evoked one of my favorite childhood memories, the middle-sized replica that stood in my father's study. The horns here and in medieval paintings are explained as a mistranslation of the Hebrew text that Moses' face "radiated" after his experience with God, into Latin "horns" (rays = horns) and from there into vernacular English "horned" but they are also akin to ancient pagan use of horns for gods and kings symbolizing special strength, and to the double-peaked shape of a bishop's mitre. Then we admired the ceiling of the Sistine Chapel in the Vatican, with Michelangelo's painting of God giving the spark of life through his outstretched finger to Adam's outstretched finger, amidst his other famous paintings.

And then there was...but we had only three days in Rome; then two days in beautiful Naples including a day visiting the wonders and catastrophe of ancient Pompeii and the fuming crater of the Vesuvius which had been the villain on one ominous day about 2000 years ago. We took tickets for a combined cab ride up to mid-mountain and funicular from there, but when we arrived at the end of the road, the funicular was broken today, so sorry." Surely, they must have known at the ticket office. What can you do half up the mountain? Return without having seen the crater? Some people did, since the footpath looked narrow and steep. But the three of us would not give up. It was very hot in the midday sun, no tree, no shade. And the pebbles and ashes underfoot made us slide back a step for every two gained. But the view at the top was worth it. The difficult climb made it seem even more forbidding, and seeing the crater at last even more rewarding. The size, the fumes, the smell were impressive.

Then four days in lovely Greece; first Athens, then north by bus through the countryside to Delphi with the Apollo temple and the rock of Pythia where she sat entranced producing obscure yet influential oracles. Here also was the "Navel of the Earth", the midpoint of the known inhabited part of the world at the time of Anaximander of Miletus (6th century B.C.), who had drawn the first map of the Earth, a circular region around the Mediterranean Sea, from the Pillars of Hercules (near Gibraltar) to the Caspian Sea. It is a thumb-like statue in the Delphi museum. For a geodesist, this was an important site as well as sight, and I took a picture of it home and even used it in an article on the history of geodesy many years later.

Going south into the Argolis, we crossed the narrow Canal of Corinth and watched a ship come through, visited the ruins of old Corinth, sat on the stone benches of the ancient theater in Epidaurus, paid homage there to Asclepius, the god of health and medicine, and saw the tomb of Agamemnon in Mycenae whose treasures we had already admired in the museum in Athens.

A two-day trip to Crete took us back several millennia to the fabulous Minoan culture. The plane landed on a meadow in the middle of nowhere. Following the other passengers, commuters rather than tourists, we entered a sad-looking eating locale to ask our way, and found almost unbelievably that it doubled as the airline bureau and that we could take a bus to Heraklion. The contrast of the dismal present to the splendor of the ancient past was overwhelming. The Minoan past was even more impressive than expected. The palace of Knossos with its red and black columns, colored walls and paintings, water system for hot and cold, clean and used water, including bathtubs, gave the impression of sophisticated life with almost modern creature comfort, apparently forgotten in the intervening millennia. A special bonus waited for Mike. When he sat down on the comfortably molded throne of King Minos, he discovered with amazement and pleasure that his feet rested on the ground. So King Minos was not any taller than he, Mike. Small among American boys, Mike would fit well in size into the Minoan royal family; an American six-footer would be oversized and out of place.

The next chapter was a three-week exploration of Israel. Mike's album prefaces this section with two quotes from the Bible: "Arise, walk through the land, to its length and to its breadth, for I will give it to you" (Gen.13:17) and "See the land, what it is and the people who dwell there, whether they are strong or weak, few or many; and what the land is on which they live, whether it is good or bad, and what cities they live in, whether camps or strongholds, and whether it is fat or lean and whether there is wood or not." (Numbers 13:18-20) Obviously, the Bible speaks to everyone on his level, to geodesists and geographers and even to future anthropologists. And here I find another telling note at the end of these quotations: "Veni, vidi, didici" (I came, I saw, I learned), a fitting modification of Caesar's "veni vidi, vici" (I came, I saw, I conquered). And there was a lot to learn, indeed; too much to record it here, as we truly went the length and breadth of it, aware also of the historic dimension which nourishes the present impressive flowering of the modern state. We could not show Mike the old city of Jerusalem, however, which we had seen before the Jordanians took it over, prohibited access, and desecrated its shrines.

A strenuous all-day flight from Tel-Aviv to Helsinki required changing planes in Rome, Zurich, and Copenhagen, where we missed the connection with the last flight to Helsinki. The airline sent us to Stockholm instead where the last Finn Air plane waited for us and finally deposited us about midnight at the already deserted Helsinki airport which hastily closed down for the night after us latecomers. We were exhausted.

Next morning we found ourselves in an entirely different world, like awakening from the spell of a time machine that had shown us glimpses of past worlds on this same planet, intermingling with and shaping our present world. The magnet here was the apparent modern willingness to cooperate internationally, at least on a technical level, as symbolized by the IUGG Assembly of about 2000 participants from 44 member countries. We were placed in a dormitory of the university, and realized in very practical terms the high latitude of our location: we had difficulty falling asleep in bright daylight till 11 o'clock at night, with no shutters on the windows since the dormitory was usually uninhabited in summer. Disregarding the sun's timetable, the town went to sleep around 7 p.m. The streetcar to our dormitory stopped running at 6 p.m., the restaurants stopped serving at 7 p.m., and if you wanted a bite to eat after that, you had to find a nightclub. Otherwise, Finland was a delightful experience: the charming land-scape of graceful birch and spruce forests with the many lakes in between, the col-orful flower and fruit markets, the new striking architecture in the suburbs, the modern churches there with the tall slender crosses built into the landscape out-side the glass walls, the colorful modern apartment houses with flowers on the window sills, the Finnish glass designs with the slender, narrow-spaced lines etched into the glass, the Sibelius concert, and the warm hospitality of the hosts, Professor Heiskanen (Vice President of IUGG and President of the Finnish Arrangements Committee), Professor T. J. Kukkamäkki (Vice President of IAG and Secretary of the Finnish National Committee for Geodesy and Geophysics) and others. They even provided a Finnish companion for Mike of about the same age. Eric sampled the sessions of the other Associations but came also to my pre-sentation, of course.

The Army Map Service was represented by Mr. Mills, Mr. Culley, Mr. Hen-riksen, Mr. Chovitz and myself. Dr. O'Keefe, Dr. Hertz, and Mr. Kaula were there too but they belonged to NASA now. It was a pleasure to meet again with some of the great people of the international geodetic community whom I had first met in Toronto 1957, and to meet several more, e.g., Director Charles Antoun, Surveys of Sudan, who was interested in having AMS help with the tri-angulation of the 12th parallel North in Africa. Henriksen reported on the con-nection between the U.S.A. and Japan by occultations, and on the role of artificial satellites in datum connections between several Pacific islands. I reported on my work of the last three years, specifically the extension of the European Datum into Japan and India, the ice age traces which are still noticeable, and the new world geodetic datum derived from all available data as of now. Mr. Mills was happy about the well received AMS contributions. When I asked him what he foresaw that I should do from now on, he said: "Just carry on the way you have, you are doing fine." He agreed that I should join Mr. Rice's Study Group on combining gravity and astrogeodetic data to span large water bodies, and he stressed as my major duties to take part in these international projects as much as I could and stress AMS cooperation. He himself played a strong role on the inter-national scene.

I was asked to cover all sessions of Section V (geoid) for a report on the Assem-bly in the *Transactions of the AGU*, but there was still a chance to look into other sessions. There was a Symposium on Mean Sea Level where I picked up an intriguing paper by Ilmo Hela (Institute of Marine Research, Helsinki) called: "Causes bringing about the difference between the geoid and the mean sea level surface." It listed twenty-one such causes and made practical suggestions of what to do about them and how to establish that elusive mean sea level surface. Only many years later when I entered the debate between geodesists and oceanographers about this difference, I re-read this excellent paper. Unfortunately, it had never reached publication and, as I was told upon my inquiry, the illustrations to which the text referred were lost in the attempt at publication.

On the weekend, we participated in an excursion to Turku to see the old castle and more of the lovely countryside. We stopped to visit a newly renovated medieval church at Lohja and listened to the wondrous story of the restoration: it had been a simple Protestant church with sober, stern white walls when in the cleaning process mysterious colorful pieces appeared. Closer inspection revealed beautiful paintings underneath the whitewash which now was removed with utmost care. It turned out that the church had been Catholic originally, its walls covered with didactic pictures of Bible stories and graphic tales of what the Devil would do to a sinner. But pictures, whatever their content, were not tolerated by the later Protestant congregation and painted over to complete oblivion. Now the sense for old cultural treasures prevailed and the pictures were restored to life and admiration. In fact, they are enchanting and full of humor in depicting the Devil as a black horned and tailed creature sitting on top of his victim or plaguing him in various ways. Obviously, there are different educational methods, with and without humor. One also gets the feeling that the painter very much enjoyed himself, more interested in his art and joy of life than in the purpose of his employers.

The Association of Geodesy elected officers for the 1960-63 time period. Bomford continued his Special Study Group (SSG) No.10 and I continued as a member. I was a member also of De Graaff-Hunter's SSG No.16 and I joined now Rice's SSG No. 15. Tengström was a member there too and he had specific plans to accomplish the Crete-Egypt connection across the water. Other members were Dr. Atala Wassef of the Survey of Egypt and Dr. George Veis of Greece, who both actively participated in that project.

There was no time for us to take advantage of post-assembly excursions such as the tempting one to Lapland, because we were headed for the geographers' version of an International Congress in Stockholm. We left Helsinki five minutes past 7 a.m. for a 50 minutes flight and arrived in Stockholm at five minutes before 7 a.m., a wondrous stunt to live double time, but explainable by flying west and the different time zone. Our young geographer took out a registration card in his own name for the Congress, which officially entitled him "to attend all Congress sessions and to visit the exhibitions organized by the Congress." After seeing the sights of Stockholm, we joined a geographers' field trip to Uppsala and saw glacial landforms, a rune stone with its ancient writing, and unexcavated grave mounds of prehistoric kings. We saw farm houses with grass and bushes growing on the roof and even a goat grazing thereon. Near Uppsala we saw the country home and garden of Linnaeus, the originator of the plant classification system. In Stockholm (as also in Helsinki before) we took a boat ride through the narrow waterways, the skerries. The visit to the Milles Garden with its seemingly weightless sculptured figures, some in graceful movement, some with a sense of humor, was a unique experience. We resolved to go and see the Milles Memorial Park back home in Falls Church, Va.

But first we had a week in Norway. Two days in Oslo took us to the exhibition of the Viking ships, Nansen's polar exploration ship "Fram", the Kontiki Museum, the outdoor Folk Museum with the wooden farmsteads of old times, and the Vigeland Sculpture Park reminiscent yet different from the Milles Garden. A six-day tour into the countryside (with unbelievably generous and colorful smorgasbords at breakfast times) took us first by bus to another very interesting open air folklore museum at Maihaugen Park near Lillehammer and on the Peer Gynt mountain road to over 3000 feet elevation with beautiful sights; then past more lakes and more high mountains to Olden, from where a "Fjord Pony" and cart ride (with official fancy certificate) and a hike brought us to the Brixdal glacier, part of Europe's biggest ice field, the Jostedalsbreen. Earlier on the way we had stopped at a little medieval church in lovely scenery at Lake Vaga. It had been rebuilt, but a beautifully carved side door showed pagan symbols. It had been preserved from a pagan temple: one can never know what might help. Such ambiguity is known from many other examples in Europe and elsewhere. It is not always so clear why long revered symbols or notions should suddenly lose their validity or why double loyalty should not give double insurance.

In Balestrand on the Sognefjord we changed from bus to steamer and sailed a whole day through the fjord and the skerries to Bergen, wondering at the loneliness of some farms on the narrow shores with the steep mountain slopes right behind them and neighbors few and far between. What a contrast to lively Bergen. A day here, and then back to Oslo with the spectacular mountain railway that reaches over 4000 feet in elevation, goes through numerous tunnels of various considerable lengths, and needs extensive screening against snow danger. It must have been an engineering feat.

One more brief stop of two days in Copenhagen. I had been there as a child and was looking forward; better a short stay than none at all. Already the arrival at the airport was different from anywhere else: every American passenger (I don't know about others) received a cordial welcome letter from the Prime Minister of Denmark and a pin of Danish porcelain. Cordial, informal behavior of people on the street, bus drivers, police, just anyone, was the normal way of life. In the middle of town is the Tivoli, the amusement park for every age. In the bakery, there are cakes and goodies that look and taste as if from Vienna (the highest praise there is). In the shops there are humorous carved figures in all sizes. Mike bought a palm-sized carving of a little rotund man with a precious good-natured smile, as a sort of talisman-companion. I bought a couple of mobiles to hang from the ceiling, and some silver and enamel spoons for mementos. People smile and enjoy sitting in the afternoon sun. No rush—enjoy your day while you can. A young woman pushes/pulls a cart filled with many round-faced little children: a nursery school group? A cluster of a little older children is tied with colorful ribbons to their kindergarten teacher, pulling her along, yet forming an orderly line, hand to hand, when crossing the street. Everyone waits, everyone smiles. Picturesque narrow row houses, a boat ride on the canal, Thorvaldsen museum, the graceful statue of the "Little Mermaid", the Hans Christian Andersen statue, a quick trip to Kronborg Castle at Helsingor of Hamlet fame, from where one can see Sweden across the water, and it was time to leave. And time to say good-by to each other in the afternoon, hopefully just for a short while: I had to go alone to Frankfurt to my MATS return flight. Eric and Mike went back by commercial airline. They would be across the ocean in a few hours and home still today, much quicker than they came on their overnight flights, another very practical lesson for the young geographer, not to forget the time difference for longitudes. I would take a little longer time difference plus the slower MATS time; and an overnight stay at Frankfurt to get that MATS flight. A memory from early childhood about a geodetic wonder suddenly wells up in me as I wave good-by to my two men, who were here just before and are gone now: does their plane know where Washington is? I am for a moment the little girl at the Vienna railroad station, waiting with my parents for the train to move and take us to Ischl for the summer. "Does the train know where Ischl is?' The wonder of geodesy.

3. "OTHER DUTIES AS ASSIGNED"

We came home from this memorable trip by the end of August 1960, refreshed and batteries recharged with *joie de vivre* from seeing the fullness and diversity of this great multi-facetted world. And the whirl of office life was waiting. A formal briefing on the use of gravity in geodesy had to be prepared for the benefit of ACSI (Army Chief of Staff for Intelligence) generals; it had to be written out so that anyone, not yet determined, could present it. Somebody from Staff had to write a speech on the Figure of the Earth for the Commanding Officer, and came up for help. By the end of 1960 we had to rewrite job descriptions for the upgrading of the administrative levels: the Geodetic Division became the Department of Geodesy, the Research and Analysis Branch became the R & A Division, the Geoid Section became the Geoid Branch, and the respective chiefs received higher grades. Everything else seemed unchanged. O. F. Mattingly had replaced Henriksen as Chief of the Astronomy Branch when Henriksen became Chief of the Division. There were several visitors and requests for information about the latest on the Figure of the Earth: from Formosa, from Indonesia, from Brazil, from the *National Geographic Magazine*, from GIMRADA (Ft. Belvoir, Va.) and others. GIMRADA is the acronym for one of these long names of government agencies, impossible and unnecessary to remember. If you are one of those who have to know, it stands for U.S. Army Engineer Geodesy, Intelligence and Mapping Research and Development Agency. The much more palatable name "Gimrada" became so entrenched that someone once inquired about a paper written by a Mr. Jim Rada.

"Other duties as assigned", a catchall phrase in the job descriptions, covered happily for me also language courses at that time, Russian to take and German to give. How did that happen? We had an excellent translating and abstracting service at AMS, but the increasing interest in Russian journals made it advantageous to be able to read at least the titles and authors' names and to scan the subject matter to pick out articles or mere paragraphs for closer translation. The request for the Russian course may have originated at the library, and a few of us were permitted to join a beginners' course. The request for German lessons originated with us. Classic geodetic literature was mostly German. Some people in the Astronomy Branch were involved in lunar problems which interested me too and I had been helping them all along with clarifying German paragraphs in pertinent literature such as Schrutka-Rechtenstamm on the reduction of 150 points on the Moon, and Schmidt on the mountain ranges of the Moon. Also other colleagues asked occasionally for quick help in exactly understanding and thus adequately answering correspondence from German geodesists. In such cases, the translating service would have provided more than needed and taken longer than useful. I offered a geodetic reading course and, with the precedent of the Russian course, it was approved. The first Russian course was given by Mr. Efraim D. Ackerman, two hours per week, from September 1959 to April 1960. I had had a lot of linguistic training in my young years (French and Hebrew by private teachers as a pre-schooler, with German as my native language; then in school eight years

Latin, six years English, four years Greek), but I had never learned a Slavic language. Thus, I had no trouble whatsoever with the Russian grammar (is it not like Latin and Greek?) or syntax (is it not German?), but for the Slavic vocabulary there were in general no associations, and that required memorizing from scratch. Long sentences were intelligible much quicker, I found, by translating them first into German and only then into English. A second "advanced" course was given by Mrs. Alina Truhan from September 1960 to March 1961, and concentrated on reading technical literature. It was very useful, indeed. Even many years later, after my retirement, when asked by the AGU to check and edit their translation service of Russian geodetic articles, I remembered and could revive enough of these excellent courses to accept this task. For my German classes I had asked for a slight modification in place and time. Instead of one hour twice a week in the Ruth Building's training school and wasteful waiting for the shuttle bus, I asked for half hour class times four times a week (greater frequency is better for any language study) and the use of a small conference room in Erskine Hall to save the time of transportation back and forth and minimize the absence from work. I had everything arranged very quickly with the people immediately involved, but this very fact incurred the displeasure of the training school who felt left out rather than assisted, to my naive surprise. German beginners' classes (November 1959 to May 1960) opened with two sections of about 12 people each, and were followed by a short "advanced" course (October to December 1960), specifically designed for reading and translating selected geodetic material. The courses were open to everyone, of course, and other departments participated. I still have the class list of the three courses. I myself felt like a fish back in the pond, having been a teacher all the years of pre-AMS times. In fact, my teacher's career had started at age 14 when I tutored a 10-year old in Latin and mathematics and rescued her from flunking which would have meant the repetition of the year. The grateful parents presented me with a beautiful ivory pendant which I still like to wear. This early entrance into the labor force enabled me proudly to cancel my parents' allowance for pocket money and even to buy that long desired "Lodenmantel" for my hiking trips (a specific raincoat used in the Austrian Alps). My nostalgic teaching activity was vicariously extended to a college textbook on earth sciences, which Professor A. N. Strahler of Columbia University was writing. He had seen my two articles in the *Journal of Geophysical Research* and wanted some help in using their content in the book. "I believe that this will be the first time," he wrote, "that geodesy has been introduced into a basic college text in science at an elementary and non-quantitative level. Heretofore geodesy seems to have escaped attention in the general education." Good that he wrote; so I could advise him of the recent changes due to the satellite value of the flattening. There followed some more exchange of letters for clarifications, and in 1969 another update for the revision of the book. I was glad to help geodesy get some attention in a general college course.

"Other duties as assigned" could possibly be construed to include also safeguarding AMS prestige and achievement credits, and resenting infringements. We did a lot of such resenting in the face of some pirating activities by ACIC, as we saw it. Their liaison officers, resident at AMS, came by quite often, practically looking over our shoulders for everything we did and had and then some more. Sodano complained to Mr. Mills that they bothered us, took up too much of our time, and built up duplicate holdings, which would lead to work duplications and infringement on AMS functions. While we could not refuse specific requests (after all, we were all working for the same government), Sodano was particularly irritated when he found some of his work and ideas used in ACIC reports without mention of the source, and he began to stamp every single page of what we had to transmit, with big letters saying Army Map Service" in the middle of the page, so it could not be cut off or erased. I don't think it did us any good other than letting off steam in a visible protest; but Mr. Mills at least stopped the over-eager liaison people from walking in uninvited. You might smile at such inter-agency forays if you are not involved yourself. But I forgot to smile when it happened to me directly: I was pleased at first when an ACIC researcher gave a speech on the difference between the development and projection methods, supporting my contention that it was necessary in today's long-distance geodesy to discard the development method (my old battle ground). He showed the magnitude of the discrepancy in two numerical examples taken from my works, from the 30th meridian through Africa, and the other from my North American geoid study. But he never mentioned my name nor did he list my well-known published work in his references, although that was clearly the source of his realistic numbers. Instead, he dressed his examples up as hypothetical, with the realistic numbers plucked at random from thin air. Sodano, Henriksen, and Mills decided on a formal complaint with the request that in the publication of that speech my work should be listed in the references. As you might have guessed, a letter came back from ACIC pretending that the paper had nothing to do whatsoever with my work. When it was published about two years later, it did not contain the requested reference.

Acknowledgement of achievements was freely accorded from all other sides: Dr. Max Kneissel, President of the "International Permanent Commission for the Readjustment of the European Triangulation" wrote in spring 1961 for summaries of my early articles pertinent to the European Datum for inclusion into their archive of documentations. The establishment of this archive had been decided at their Symposium in Lisbon in spring 1960 and endorsed by the IAG in Helsinki. As mentioned earlier, Mr. Hough's work was to be included, and I sent brief German summaries of the papers on the deflections in the Mediterranean, the Hough Ellipsoid, and the influence of the distant topography on deflections. The Arctic Institute of North America wrote for my North American geoid chart. Dr. Fubini at the Pentagon, scientific advisor to the President, wanted to be briefed about the world datum. An invitation came from Professor Heiskanen to give a paper and be a panel member of a symposium at OSU in February 1961. And another invitation came from Mr. Browne, Cambridge, England, to attend a symposium on gravity there in July 1961. I felt very much honored but had to decline partly due to travel cuts at AMS, but mostly because my primary interest and energies had been swallowed up in another direction: the mystery of the parallax of the Moon, and the paper I planned to present about it at the AGU meeting in April 1961, and time was getting very short.

5

THE DISTANCE TO THE MOON (1961–1963)

1. THE LUNAR PARALLAX

It had started innocently enough with a simple little assignment at the time of Dr. W. D. Lambert's visit in March 1958, to insert our new geodetic knowledge into his old paper on the lunar parallax and see what happens. The lunar parallax is the angle under which the radius of the Earth at the observatory is seen from the Moon; or viewed from the Earth, it is the difference in the direction to the Moon if seen from the observatory or from the center of the Earth (Fig 1). If we knew this angle, then the observed direction from the observatory to the Moon (another angle) would make it possible to establish all parts of the triangle: observatory—Moon—center of the Earth, by elementary geometry. (All angles known of a triangle give its shape; one side is known as the radius of the Earth at the observatory, and thus gives the scale.) Then the distance of the Moon from the observatory as well as from the center of the Earth would be known. The problem of determining the distance of the Moon is therefore equivalent to the problem of determining the lunar parallax angle. How can we get this angle? Suppose we had two simultaneous observations from two different observatories on the same meridian, then we could compute the same geocentric lunar distance in two ways, from either observation, if we knew the correct parallax. We can tentatively use a theoretical value from lunar orbit theory, derive the two supposedly equal lunar distances and use the discrepancy to correct the theoretical value.

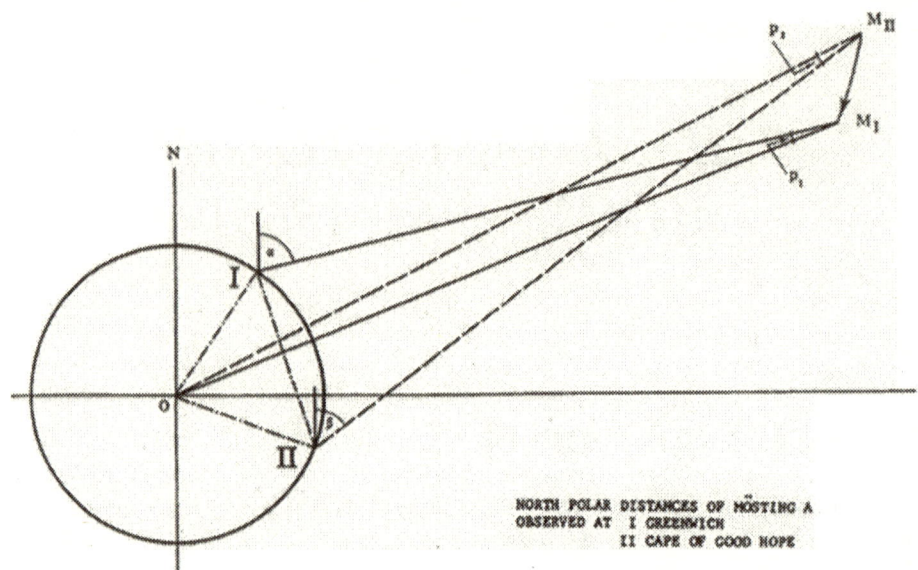

Ch. 5, Fig. 1. North Polar Distances of Mösting A.

This was essentially done in the years 1905 to 1910 in a common enterprise at the observatories at Greenwich (W. Christie) and the Cape of Good Hope (D. Gill), and a mean correction was determined from 104 pairs of observations. Since these observatories are not really on the same meridian, the observations of a pair are not simultaneous, and lunar orbital theory must be used also to correct for the Moon's movement during the hour and a quarter time interval. The computations are reported in a 1911 paper by W. Christie, D. Gill, and A. C. D. Crommelin, which were used in Lambert's 1928 paper and was still widely quoted as the classical parallax determination. The terrestrial connection between the observatories had necessarily been an educated guess up to now, which we could replace by our recent geodetic information about the African 30th meridian and the better earth model. I had done this at the time, using my "Tentative World Datum", but put it aside since the satellite-derived flattening would require a modification of that world datum.

After my return from Helsinki, I picked up this work again to repeat the simple little task with the newest version of the world datum. The fact that this work on the Figure of the Earth would have an impact on lunar studies was inspiring, as it widened the scope of its usefulness. And the connection with the historic work of 1911 gave me a thrilling sense of responsibility to do the best I possibly could. But in trying to understand the background of the various discussions, assertions, and achievements in the wider and complex inquiries concerning the lunar parallax, I raised questions in my mind, sought answers in the literature, which raised more questions and the need for more literature. The little assignment was blooming into a fascinating labyrinth that held me captive, almost obsessed, for more than a couple of years. A certain urgency was given by the increasing interest in the Moon at the time, and a lunar distance based on the best geodetic data was desired. Ben Yaplee at the Naval Research Laboratory bombarded the Moon with radar to achieve an unprecedented hypermodern measurement of the distance from the Naval Research Lab to somewhere on the Moon's surface—he was not sure in the beginning, at which point. Marvin Marchant of our Astronomy Section was establishing a first order control network of lunar landmarks in relation to crater Mösting A as a datum point, comparable to a mapping on the Earth. There were peculiar difficulties on the Moon, such as the absence of a natural zero line for heights as is the sea level on the Earth, and the dependence on the limited number of crater observations in the literature; Marvin had just finished Technical Report No.29, "Horizontal and Vertical Control for Lunar Mapping, Part I: Methods."

I had kept in contact somewhat with this pioneering work as I helped Howie Genatt and Marvin read the pertinent German literature; and I got an appreciation of the uncertainties of locating crater Mösting A in relation to the Moon's center—just as we on the Earth had uncertainties in locating ourselves with respect to the Earth's center. In that literature I happened across a casual mention of the mean lunar parallax value which the great German astronomer P.A. Hansen had used in his lunar tables, a basic reference value for lunar orbit theory—and it was slightly different from what I had gathered from the English literature by implication, not by a clear direct quote. I needed this value to properly evaluate Crommelin's result, and I had inferred it from a comparison with Lambert's paper which gave results in a different form. Had I made a mistake in this inference? Where did Schrutka-Rechtenstamm, the German author, get his value? Where did Lambert get his and which did Crommelin really use without spelling it out? I wrote to Dr. Lambert, but he could not remember it anymore. So I searched the astronomical literature further without much luck. I discovered a peculiar trait among most astronomers: apparently, they find it boring and superfluous to repeat numbers which should be well known to other astronomers of the inner circle (and obviously never thought of me trying to ascertain such a number). They won't quote Hansen's parallax, but they'll say that someone else had raised Hansen's parallax by so much and that they have computed a correction to that raise. Even in a record of observations (and I had requested and received copies of the records of the 1905-1910 observations at Greenwich and Capetown) they won't write down systematically what the observation was and what numerical operation they did with it and what for, but you may find on the edge or even on the back of the main record a little scribbled number by itself, maybe just the tail end of a longer number, and you may recognize it as a correction to some previous correction of some previous observation, if you become attuned enough to the size and shape of numbers you are looking for. I should not complain too much about the tail end of a long number because we often do the same in geodesy when talking about the radius of a reference ellipsoid; for instance, it is 6378 388m for the Hayford ellipsoid, but only 270m for the Hough ellipsoid and 160m for the Fischer ellipsoid. Everyone knows the first four digits 6378, so why repeat them; only the last three are important for identification, right?

The 104 pairs of observations in the Crommelin paper which I recomputed with the new world datum information, referred to the same crater Mösting A, that Marchant had made the datum point for his moon mapping, as it is the best observed and documented lunar feature. Observation of a well defined small crater rather than points on the limb had been the big innovation towards greater accuracy at Crommelin's time, made possible by the annual publication of the crater's ephemeris in the *Berliner Jahrbuch* since 1894. Upon application I could get a six-months loan of the *Jahrbuch* for the years 1905 through 1910 from the Library of Congress. It was not likely that the U. S. Congress would need it during my study. In fact, the U.S. Congress fortunately never needed the books I borrowed from there over the years and I very much appreciated this borrowing privilege, not available outside Government. I am also very grateful to the efficient people in our library who procured what I needed from the "L. C." or another library in town or elsewhere, and even found sources for which I had only scanty references.

Although Dr. O'Keefe had long left the AMS, he still took interest in my parallax work and he advised me to see Dr. G. M. Clemens (U.S. Naval Observatory) and to write to Professor Dirk Brouwer (Yale University) who worked with Ben Yaplee on the parallax problem using the radar measurements. Both responded kindly, but Hansen's value, identified by Crommelin only as the one "which Newcomb had raised by 0.45," and to which Crommelin in turn determined an improvement, was still elusive; and I kept searching big volumes of German as well as English literature for the one little number. When Dr. Clemens suggested that my recomputations of Crommelin's work in terms of the new world datum and in kilometer distances would be of interest in themselves and should be published, I decided to submit an abstract with the title "The Geometric Parallax of the Moon" to the Annual Meeting of the AGU in April 1961, and to mention in my paper the mystery of Hansen's parallax value and the uncertainty it caused: The as yet unexplained discrepancy between the ideas of J. Peters, 1895, and Schrutka-Rechtenstamm, 1955, on the one hand versus S. Newcomb, 1880, on the other about what Hansen considered the correct mean parallax, would lead to an error of about 6km in the mean lunar distance: 384 413km versus 384 41km. Maybe someone in the audience or readers of my subsequent publication would know the answer.

In March, I had already a request from California by a group of scientists working on a system of fundamental constants, for several preprints of my paper announced as No.9 on the AGU program, and also for reprints concerning the newest dimensions of the Earth. That gives you the good feeling that your work must be useful to someone. I remember that I was in high spirits at my presentation, because I had such interesting things to tell (another reliable trick to fight stage fright). And apparently, I did quite well as there were many kind comments afterwards; but no one came forward with an answer to the Hansen Mystery. One person kept shaking my hand vigorously and introduced himself as Dr. Jesse W.M. DuMond, California Institute of Technology. He was the one who had asked for the preprints for himself and his colleagues, but now he was here after all. A few days later I received a letter from him: Dear Dr. Fischer, it was indeed a pleasure to listen to your paper on the distance from Greenwich to the Moon crater and to note you care about the six kilometers! I could not resist the desire to shake your hand and congratulate you on what you modestly referred to as your 'stubbornness'. I like that kind of stubbornness. We need more of it!...I am a little stubborn too..."

2. THE MYSTERIES OF THE PARALLAX

My intensive peregrinations through astronomical literature chasing the Solution of the mysterious value of Hansen's parallax paid off eventually. While the elusive value was nowhere quoted directly in Hansen's major works, I happened across a letter to the editor of the journal *Astronomische Nachrichten* where Hansen elaborated on what he was doing. This was Newcomb's source for the value he attributed to Hansen and also for his claim that Hansen had contradicted himself when using two different numbers in the same article. But here I could put the finger on a definite misunderstanding Newcomb obviously did not know German well enough to understand Hansen's terms. Here I had found the solution of the historical mystery and I felt like Sherlock Holmes. The number quoted by Schrutka-Rechtenstamm was the correct one (no language barrier there, of course) and the one used by Newcomb was incorrect. So Crommelin had only picked up someone else's error without checking the way I did. He trusted the great Newcomb's authority, while I trusted my native German. This unquestioned trust led to Crommelin's incorrect conclusion about the flattening of the Earth, which in turn influenced the subsequent astronomic lunar orbit theories.

Hansen's value was only one of the parallax mysteries; the search for it had grown on me as a personal obsession. But there was yet another much bigger mystery that worried the experts since many decades. The geometric parallax as described above (observation of a well defined point on the Moon from well defined observation points on the Earth) is only half of the parallax story. The other half is the possibility to determine it in a different way, namely, from orbit theory (the dynamical parallax). The two should agree, but mysteriously they did not.

The derivation of the dynamical parallax focuses on the gravitational. The mysterious discrepancy between geometric and dynamic parallax forces shaping the lunar orbit. The gravitational interaction between the Earth and the Moon binds them into a unified system, and their interdependent motion is described by the so-called "two-body problem". Its constituents are the masses of the two bodies (or the Moon's mass expressed in a ratio of the Earth's mass), the mean motion, the semi-major axis of the Keplerian elliptical orbit, and some constants. Since the sun must also have its finger in it, there is actually a three-body problem, resulting in a perturbation factor messing up the beautiful Keplerian ellipse. Its modified semi-major axis is the mean lunar distance. The corresponding mean parallax is the largest angle under which the radius of the Earth is seen from the Moon; it is the small angle in a huge right triangle with the mean lunar distance as the hypotenuse and the radius of the Earth as the opposite leg. The other leg may be interpreted as the distance of the Moon from a surface point on the Earth when seen at the horizon. For graduates of a trigonometry course it is obvious that in computations the parallax angle will mostly (though not always) appear as sin p, that is the ratio of the opposite leg to the hypotenuse. To confuse things a little more or rather to connect them with data obtainable on the Earth by geodesy, the mass of the Earth is replaced by the flattening and equatorial radius of the Earth, and mean gravity. In short, the dynamical parallax depends mostly on these four constituents: the radius and flattening of the Earth, mean gravity, and the ratio of the mass of the Moon to the mass of the Earth. Since ideas about the magnitudes of these constituents have changed over the years, conclusions drawn in the literature have changed too. But the discrepancy with Crommelin's geometric parallax persisted.

It was the reason for Lambert's refinement of Crommelin's paper, and for my assignment to refine Lambert's and then Crommelin's paper. But my attempt to duplicate, e.g., Lambert's number for the solar perturbation of the lunar orbit sent me from one reference to another, where I had in turn to deal with numbers which I could not duplicate without looking up more references; and speculating about their differences and how I might compute what I was looking for by inference from what the various authors had written. It was a maze of little mysteries which all hung together so that one could not skip any of them. If such a mystery unmasks itself as error, it causes extra anxiety; for it is very disquieting to find an error in a printed book by a respected author. It takes extra careful checking and double-checking before one can work up enough courage to say and publish that, for instance, a lengthy derivation of a numerical factor is erroneous in the otherwise exemplary textbook by Jordan-Eggert, or that the great Newcomb misquoted Hansen's writings.

Eventually, I came up with a table of comparable numbers which the various authors had used directly or by implication for their determination of the dynamical parallax, and I augmented it with the best current numbers including my own world datum. Although this table was of interest in itself the most meaningful interpretation for my purpose seemed to be a comparison in terms of lunar distances in kilometers. It turned out that these were all different, but all were definitely smaller than the distance of 384,413 km from Crommelin's geometric determination (even using the correct Hansen value). The newest determination, using the most recent value for the mass of the Moon, gave 384,400.06 km. But now, about half a century after Crommelin, there were two new unprecedented determinations of the lunar distance which could also be called geometric: the occultation method by O'Keefe and the radar measurements by Ben Yaplee. O'Keefe had used observations of four occultations from nine stations in the United States in 1950 and had come up with a lunar distance of 384 408km, which changed to 384,400.9 km after I inserted the new geodetic information from my North American geoid map and my world datum. Yaplee came up at first with 384 402km, using a lunar radius of 1740 km. When I asked him and his assistant Stephen Knowes where this radius came from and why it was different from the conventional value 1738km which O'Keefe had used in the occultations, it turned out that there was no apparent reason. For a fair comparison it would be better to use the same value. After 1738 km was inserted, the lunar dis-

tance resulted in 384 400.2km. The agreement between the three modern determinations, one dynamic and two geometric ones, was remarkable and exciting. It
changed the scenario of the big mystery from dynamic versus geometric parallax
to "everyone against Crommelin." Poor Crommelin, whose paper is nonetheless
an important milestone in the history of the parallax, needed rehabilitation.
Unless one can put the finger on a definite error, the judgment is not really final.
One should find out exactly why his result is consistently larger than the others,
before judging it incorrect. I spent much time in studying the distribution of the
104 observation pairs within the lunar orbit to check for possible bias and unac-
counted influences. There also had been a question of the illumination of the crater and the possibility that it had caused an unintended focusing on a far point
instead of the center of the crater. Dr. Clemens and Dr. Vaucouleurs who had
worked on this problem before, suggested that I check the observations from
their original records. So I worked on a further paper with the tentative title "The
Place of the Moon at the Instants of Christie and Gill's Observations 1905–
1910," where I intended to report all such investigations in the hope that they
would close in on the explanation why Crommelin's lunar distance was so much
longer. But in view of the agreement between the three other determinations of
the present time, this investigation lost its priority value and had to give way
more and more to other, higher priority work. So this mystery remained.

3. COMMUNITY INTEREST IN THE LUNAR DISTANCE

In November 1961, I was invited through Dr. O'Keefe as the program chairman,
to talk to the Washington Philosophical Society at the Cosmos Club the following January, about my parallax adventures. I planned it as a light "after dinner"
talk, and I think everyone, including myself, felt entertained that evening. I enliv-
ened the discussion of the various parallax determinations and of the world
datum with Plato's famous philosophical cave story and with Jules Verne's science fiction about the hundred-year old dream of the African geodetic arc along
the 30th meridian, which my agency, the Army Map Service, had been able to
fulfill. And I dramatized the mystery of the parallax, the frustrations when simple
problems grow into vexing difficulties at every turn, and the wonder of science
being able to find out things up there from down here. The nicest compliment
came from some scientist in this high-level audience: "Here I have learned so
much, in such a short time, in such a delightful way." I was happy, naturally.
There was some publicity in the *Science News*, the *New York Times*, and the

Washington Post. The talk was published in the Society's *Bulletin* (vol.16, no.2, 1963) which had been revived with this volume under the presidency of Dr. François N. Frankiel, after half a century's sound sleep.

In December 1961, again on Dr. O'Keefe's initiative, I was invited by Mr. G. F. Shilling, Rand Corporation, to participate in a five-day Working Symposium on Fundamental Astronomical Constants, in Santa Monica at the Rand Corporation, in February 1962. Like the group of scientists around Dr. DuMond, so this group at Rand and several other groups of space scientists, astronomers, and geophysicists were concerned with the recent refinements of certain basic numerical values such as the equatorial radius of the Earth, the ratio of the mass of the Moon to that of the Earth, the velocity of light, the parallax of the Moon and of the Sun, and many others. The problem was their interrelatedness and the danger that an independent update of one value would create an incompatibility with others. The aim was, therefore, to construct a system with some independent values chosen as "fundamental constants" from which the others could be derived by formula without creating inconsistencies.

Connected with the problems of such a system was also the problem of which recent numerical values should be considered as "best" and recommended to users, with special recognition of two types of users: the space scientists with their immediate needs for "best" or "most recent" numbers, versus the more cautious and long-term oriented astronomers who would prefer waiting a hundred years or so for the proper long-term credentials of such a "best" value. There were also cautionary voices that we should not discard traditional methods in favor of a new apparently superior method of a certain time, because it has happened repeatedly in the history of science that overconfidence in the latest results and methods was not always justified and a systematic error was discovered at a later time. As many different methods as possible (new and traditional ones) should therefore be considered, and that might help to turn the inevitable systematic errors into accidental errors.

It was very stimulating and informative to listen to the informal discussions. Among the participants were such experts as Dirk Brouwer (Yale), Gerald Clemens (Naval Observatory), Paul Herget (Cincinnati), Gerard de Vaucouleurs (Texas), Samuel Herrick (IT), Y. Kozai (SAO), Robert Baker (UCLA), W. Kaula (NASA), Mary Francis (UCLA), G. Shilling, and Al and Donna Wilson (Rand), and several others.

Professor Brouwer was also the editor of the *Astronomical Journal* where I had submitted my parallax paper for publication, which by now included the detailed story of the Hansen parallax, the recomputation of O'Keefe's occultation paper, and the comparison with Yaplee's radar result. Brouwer had written to me earlier that he considered the paper excellent and accepted it for publication but wanted it shortened considerably and the notation changed to conform to his own and the journal's notation. He had brought my manuscript along in the editing stage. It meant some rewriting. Dr. Clemens and Dr. de Vaucouleurs and others showed kind interest in my paper and I profited from discussions and advice. R. C. K. Johns had seen the publicity in the *New York Times* about my moon paper at the Washington Philosophical Society, and he telephoned from California to my office that I should send him that paper and talk to his group at Aerospace in El Segundo after the Rand Corporation conference. He also phoned Dr. Mills for permission and sent a formal request letter. I received permission in my travel orders. R.C.K. kindly met my plane late at night and we had a cup of coffee and talked and talked. I had already gone once through midnight on the plane from Washington due to the time difference, and here I was going through a second midnight in the same night. Crazy world; I was dead-tired but grateful. The talk at El Segundo was very pleasant too. I also was invited to R. C. K. John's home and met his delightful family—young children then.

On the way home I briefly visited my daughter in Palo Alto. She had left AMS a few months earlier to pursue an M.A. degree at Stanford University, and it was good to see her again.

Back home in the office, it was decided upon Dr. O'Keefe's suggestion with Henriksen's concurrence, that the well-known habit of the *Astronomical Journal* to cut submitted articles to a half or less, should not keep us from preserving in other ways the details of my work which had been cut out and also to preserve our familiar notation for the benefit of interested geodesists. This advice led to the preparation of TR-34 and, upon solicitation by Mr. Hough as program chairman, to another talk at the AGU meeting in April 1962 on "The Lunar Distance".

The readership and circulation of the *Astronomical Journal* and of an AMS Technical Report are quite different and neither document would necessarily reach the international geodetic community. So, a year later when I presented this work at the tri-annual IUGG Assembly in Berkeley, it was published in full detail in the *Bulletin Géodésique*, with the most recent values inserted. For an entirely different readership again, I was asked to write a brief article on this subject for *The Military Engineer* (Nov.-Dec. 1964,p.6) under the title "Triangulation to the Moon;" this was unsigned, as usually requested there.

In 1970, my husband and I had the opportunity to visit Cape Town, and we went to see the observatory there. A young man was very willing to show us around. When I mentioned my name and added that years ago I had done some work with their records, he looked at me and then said: "Irene Fischer? I happened by chance last month to re-read your article in the *Astronomical Journal* of 1962. That was very interesting work. So glad to meet you". It makes you feel good to know that even "down under" your old article is still kept and re-read.

6

NEVER A DULL MOMENT (1961–1964)

1. VARIED SIMULTANEOUS BUSINESS

In 1961 we had a chance to take a computer course at the Bendix Corporation downtown, since we had a Bendix G-15 computer at the office. George Dudley was in charge of the computer located in the Americas Branch in the Ruth Building. Its use was easy enough to learn, but the big problem was to get access to the computer when you needed it. There was a schedule, meant to spread the goodies around, but to get on the schedule, stacked with "priorities," was an art.

Work on the long-term projects in which we were involved, was limited time-wise by the steady stream of visitors and customer requests taking priority in a service agency. Service to customers, governmental and non-governmental, national and international, was in itself interesting, as it gave a feel of the relevancy of your work and also provided stimulation and exchange of views through personal contact. Telling only of the major research would give a wrong picture of office activities; so here is a more explicit description of the type and variety of its daily foil in these years

NASA visitors asked for transformation formulas between our Mercury and Vanguard datums. Other agencies wanted copies of my papers which explained why the conventional datums of the northern hemisphere did not fit the southern hemisphere. ACIC wanted more copies of my North American geoid map and all the improvements in the various areas accumulated up to now. Because of its importance and usefulness, they wanted to convert it also to the classified DoD world datum. They also wanted all the background details of our TR-30 on topographic-isostatic reductions. A visitor from the Air Force needed help in changing geoidal heights from one datum to another and suggested, as did also other customers, that I re-issue my North American geoid map on a larger scale, easier on the eyes of the users, with the background detail of the base map repressed and all the new information included. I was going to pursue this suggestion and to figure out a way to beat the manpower obstacle. An Army captain came by on his way to an overseas assignment, a graduate from Ohio State University with a master's degree in geodesy, who, as Mr. Mills told me with a grin, had never been taught there that one could determine the Figure of the Earth by astrogeodetic methods. So Mr. Mills introduced him to "Miss Figure of the Earth" as he sometimes liked to call me, so that I would tell him the news. It was a pleasant briefing and the captain asked for all my reprints. Mr. Hiromiti Suzuki from the Geographical Survey Institute (GSI) in Japan visited AMS after a six-months study at Ohio State University. I had used the GSI journal for the newest deflection values when constructing the Japan geoid chart. I had met the Chief of the Geoid Division of the GSI, Mr. Okuda, in 1956, when he measured gravity in the basement of the Coast and Geodetic Survey here, while Suzuki measured it in Tokyo to establish the difference; he also made measurements in Singapore and Cape Town. Suzuki had participated in the Second Japanese Antarctic Expedition of 1957/8 for gravity measurements, and he told me about the bad weather they had run into.

Col. Kruppa of the Pentagon came to visit and commend the big relief pre-presentation of my North American geoid which featured the impressive ice-age hole around the Hudson's Bay. He had seen it hanging in Col. Conard's office and came up from there just to tell me that he thought it was very interesting. This exhibition piece which had been made in connection with the *Life Magazine* article about AMS activities in 1958, had become an endangered specimen, as someone had wanted to dump it into the ash can. Why? I had appealed to Col. Conard for help and/or shelter and he had kindly hung it on the wall of his own office for safekeeping and as a sure conversation piece. I received it back when the danger seemed to have passed, and it served beautifully in many future briefings and exhibitions. I fondly and gratefully remember Col. Conard's kindness and understanding in saving this masterpiece, as I also fondly remember the first impression he had made on me when arriving as the new assistant to our Commanding Officer, Col. Diercks, in February 1961. After the ceremonial walk-through with the usual entourage to look us over, he had come back alone the next day and sat down by my desk:

'Now tell me in detail what you are doing." It is a pleasure to work for interested superiors.

There was an opportunity to meet Sir Harold Jeffries, the well-known British geophysicist, known to me until then only as a difficult textbook and several difficult articles. He gave a talk on the "Strength of the Earth" at the Washington

Philosophical Society, and there was a dinner party in his honor at Dr. O'Keefe's house.

Dr. R. K. C. Johns was chairman of a new AGU Committee on Scientific Papers to encourage people to write geodetic papers for publication, and I was asked to be a member of that committee. I never quite understood how I could help, since it seemed to me that everyone who had some work done would gladly publish anyway. But I stayed with that committee for many years, serving also at the request of the subsequent chairmen: OSU Professors Ivan I. Mueller and Richard R. Rapp. Its activities were later extended to select Russian papers or abstracts for AGU translation and publication. In a related activity I served repeatedly as a referee for papers submitted for publication. Also, abstracts of foreign articles prepared by our own library needed some overall checking and I was asked to do that every so often. In November 1961, Sodano left for GIMRADA (Ft. Belvoir, Va.) and I took over the Branch. There were not many people then. Rosemary Phillips had moved to Atlanta with her family; I was sorry to see her go. There were only Mary Slutsky and myself for geoid studies, and Thelma Robinson to continue Sodano's work on geodetic tables and on support for map production. But within a few weeks I had persuaded some people in other branches to come and work with me on much more interesting projects than they had there: Mr. Gerharz and Mr. Gardener to work on the parallax, and Mr. Peat to work with Miss Slusky on the geoid in Libya and on my new plans to update my North American geoid chart. Some time later I picked (or was picked by?) one of the new recruits, Mr. John Suroddin, because he had welcome experience in gravity work with oil companies. My plans for a uniform recomputation of the whole North American geoid map tackled the critical manpower problem through a complete change of approach: from graphical to computer methods. We would start from the datum point in the center of the country (Meades Ranch in Kansas) and use the 35th parallel and the 95th meridian as the best determined backbone profiles and spread out into the four quarters from there. A detailed geoid profile along the 35th parallel would be completed by the Coast and Geodetic Survey (Mr. Don A. Rice) before long; and I knew that Mr. Rice planned another detailed geoid profile along the 95th meridian where he had collected closely spaced gravity anomalies in a 300 mile band for the purpose of interpolating between the sparse astrogeodetic deflections. I called Mr. Rice to find out the completion schedule for this work so that I could plan my use of it accordingly. Mr. Rice said that I could have the data of his 35th parallel as soon as it was finished. As to the 95th meridian, he could not tell me a completion date. He had done the part south of the 35th parallel but had to put the northern part aside indefinitely due to other priorities. But he was willing to let me have the material if I wanted to do the northern part myself. What a question! Of course, I wanted that, and I was happy and grateful that he offered to let me have it. It meant for me an opportunity to plan another interesting study to establish the best practical way of gravimetric interpolation, which in itself led to several publications in time. Similarly, other aspects of my new map construction required specialized studies before the whole map could be reliably put together. So there was plenty to do and to think about and to learn. In any case, I would need all newly available deflection values on the whole continent, and I asked Mr. Fuller of the Americas Branch and wrote to Mr. Lilly, the Dominion Geodesist in Ottawa, to keep me supplied, which they did.

Several invitations to speak had come by the end of the year, which always required getting permission to accept the invitation, writing, and getting clearance in good time. I already mentioned the invitations to speak to the Washington Philosophical Society in January 1962, to the Aerospace Corporation in El Segundo, California, after the conference at the Rand Corporation in Santa Monica in February, and to the AGU meeting in April. Also for April was Kaula's invitation to speak in his section within the First International Symposium on the Use of Artificial Satellites for Geodesy, sponsored by COSPAR (Committee on Space Research) and another one to speak to the Fifth National Surveying Teachers' Conference in Dahlonega, Georgia, in August. In October, there was to be the centennial celebration of the International Association of Geodesy in Munich, and Mr. Mills who always stressed my role as an international liaison and good will ambassador, wanted me to go there as our representative. But that was still far away.

More immediate was an in-house seminar that Henriksen tried to organize in December 1961 in order to foster communication and interest in what your colleagues were doing. I volunteered for the first session, prepared a formal talk with slides on the parallax problem, happy to share my findings. There were a few more sessions with other speakers, one with Archie Carlson and one with Vic Lewicke, but the idea unfortunately petered out for lack of enthusiasm. Similar attempts were made repeatedly in later years, but they never amounted to much.

Dr. Tengström had already completed the Egyptian part of his project of an Egypt-Crete tie in spring 1961, supported by the Survey of Egypt. He had established a gravity net around a new astrostation near Alexandria, and also made gravity ties to Cairo, Beirut, Athens, Rome. Now, his Christmas greeting card said that he planned to do the counterpart in Crete in February, and that the loan of a helicopter and a jeep from AMS would be of great help there; he had already asked Mr.Rice, the chairman of our SSG-15. I called Mr. Rice and talked to Mr. Mills who promptly started procedures for the loan; I was glad to convey the good news to Dr. Tengström and to keep track of the progress. My son Michael was interested in this project from a very different angle. Dr.Tengström sent him Egyptian, Russian and Turkish stamps for his collection, and made it a habit to collect a lot of interesting stamps for him for years to come. Michael and my stamp-collecting colleagues in the office profited also from my other international correspondence. To our dismay, some Commanding Officer, I don't remember which one, was also interested in stamps, and for some time my mail came with the stamp cut out. Complaining was of no use, of course.

In April 1962, I accompanied my husband to a geographers' meeting in Miami. I remember sitting on the beach and working on my geometry book. The draft for this book took up all of my free time, weekends and evenings and vacations. Writing this book for intelligent senior high school kids to learn and enjoy geometry, was a pure delight and the book seemed to write itself, just as these memoirs seem to write themselves. I had to leave Miami earlier than my husband, however, because I had to get back in time for my papers at the AGU meeting on "The Lunar Distance" and at the satellite symposium at the Naval Observatory on "Comparison and Combination of Satellite with Other Results" including comments on the notion of the triaxiality of the Earth. Tengström was at these meetings, also George Veis from Athens (who wanted to borrow a gravimeter and a ship for his gravity survey in Greece), A. Caputo (then Assistant to Marussi), C. Morelli (who was doing a gravity survey of the eastern Mediterranean), R. C. K.Johns, H. Eichhorn, NASA people and others. Professor Morelli wanted me to come to the conference on three-dimensional geodesy in Cortina d'Ampezzo, end of May. I had received an invitation in February but had disregarded it in view of all the things I was involved in. Now came another personal invitation from Professor Cassinis, President of the Italian Geodetic Commission, apparently initiated by Professor Morelli. Mr. Mills wanted to send me and talked to the Commanding Officer, then Col. R. C. Miller, who had repeatedly shown interest in my work; but it turned out that the remaining two weeks were rather short even for an extra push to start travel procedures which then normally took about seven weeks. I did receive a picture postcard from Cortina with several signatures.

A phone call from OCE (Office of the Corps of Engineers) wanted me to explain what Tengström and Veis were doing in the Mediterranean and who was financing it. It turned out that OCE was eager to help and would arrange that the Navy provide a ship for George Veis.

My mail was usually varied and interesting. Professor Max Kneissl in Munich, Director of the German Geodetic Research Institute, wrote for permission to translate my Helsinki paper, published in the *Bulletin Géodésique*, into German for distribution to interested German colleagues as an issue of the German Geodetic Commission. It was translated by L. Kolb. Ivan Mueller of Ohio State University was writing a textbook and wanted for it an advance copy of my AGU paper on the lunar distance. Also General Electric Company (Apollo Support Program) wanted that paper, and while at it, wanted all my other papers on the subject. The Survey of India in Dehra Dun asked for a reprint of "The Influence of the Distant Topography on Deflections", as it "would be of immense value to us." Tengström needed classified Greek maps on which he had plotted his Crete gravity survey, but which he was not allowed to take along. So we wrote (over Col. Miller's signature) to the Greek Army for permission. Professor Helmut Wolf, University of Bonn, sent a congratulatory letter on the "excellent results" in my parallax paper. There were also "well done" letters from my "mentors" Dr. O'Keefe and Mr. Hough. A letter from Col. D. Macdonald, Director of Survey, Australian Military Forces, asked for data and information to help them adopt a Figure of the Earth for the Australian adjustment. And there was the correspondence with Sadler in Greenwich and Stoy in Capetown about the records of the old parallax observations there.

All this time I had been only detailed into the branch chief's job without its grade and salary. When I asked Mr. Mills about it after the first 60-day detail had passed, he said it would be better if the change came out of the annual routine survey rather than by individual personnel action. When I asked Mrs. Iola Rust, our job analyzer, she said I would surely get the GS-13 on my individual merits, independent of what would be decided about the Branch. When I asked Henriksen and again Mr. Mills after the second detail (not to exceed 120 days) had run out in May, they apparently pushed action. My argument for Mr. Mills was that I needed the authority of the position to tell the people what I wanted them to do. "I could not spank the kids if I needed to, because they would rightly say: now, who are you." To which Mr. Mills replied: "I give you the authority to spank them. But I am going to see about the position anyway." In June, finally, I was asked to write up my job for Mrs. Rust and she came for interviews three days in a row. In July, her evaluation was completed, but it took till September before I saw results. I happen to have a copy of Mrs. Rust's evaluation sheet, which is interesting because it continued the grading philosophy mentioned earlier. It refers to the Guide for Evaluation of Positions in Basic and Applied Research, and states that "the professional qualifications and scientific contributions of Mrs. F. contribute principally to the grade of the position." There was a point system for weighting four factors: the level of the contributions, the freedom in choosing and planning the branch programs, originality, and especially the personal professional standing in the worldwide geodetic community. The latter is given maximum weight citing "the opinion of the Division Chief." It concludes that "this places evaluation of position at the upper range of GS-13." (Nonetheless, the pay level assigned was the lowest.) A Branch chief's grade in the department structure was not mentioned as an evaluation factor at all; that appeared only in the job description. Thus, the intent to distinguish between a personal research career and a routine administrative career became clear, even though these can be combined. It feels good to acknowledge the support, encouragement, and good will of my superiors at that time.

2. DAHLONEGA, MUNICH, AND THE SVEN HEDIN ATLAS

The invitation by Dr. Milton Schmidt, University of Illinois, to speak to the Fifth National Surveying Teachers Conference at Dahlonega, Georgia, in August 1962, had already come in November 1961, addressed to the Commanding Officer, and Col. D. Conard had accepted for me. The conference was sponsored by the American Society for Engineering Education and the Georgia Institute of Technology, and would take place at the Georgia Tech Surveying Camp at the North Georgia College in Dahlonega. Dr. Schmidt explained that the surveying teachers, that is, the various civil engineering departments in the United States and Canada, wanted to hear something about the connection between the geographically limited concerns of their surveying courses with the global scope of contemporary geodesy. Since that fitted my work, would I talk about something such as the "Recent Determinations of the Earth's Shape"? Of course, I would, and Mr. Mills was all for it. Also for it were my two geographers, husband and son, since we had never been in Dahlonega, Georgia, yet. Also, it would give Michael, 16, a newly licensed driver, ample driving experience as we planned to take the Blue Ridge Parkway all the way down and back. It was a beautiful drive.

The week in Dahlonega was very pleasant. I listened to some-not all-lectures about teaching surveying in civil engineering departments and I watched some-not-all-lectures demonstrations. My own contribution was very well received and elicited many questions and further explanations of the mysteries of geodetic datums, as may be gathered from a lively photo taken at the conclusion of my paper and published along with it in the "Proceedings". One of the participants confided his special satisfaction with my talk because his department had a mysti-fying government contract, mentioning different kinds of geodetic coordinates; they had not been willing to ask questions and thus to admit their ignorance. Now things were clear. Another participant from Canada confided a few days later when we said good-byes "I have to apologize to you." What for? When I first saw the program, I was annoyed and could not see why the United States Gov-ernment would send a woman to talk to us on geodesy. But now I know why and I thank you very much for your interesting talk."

Soon after returning from Dahlonega, including a few days vacation along the way, it was time to go for shots and passport to the Pentagon for the trip to the IAG Centennial Celebration in Munich, in October. I did not relish the idea of going to Germany, but Mr. Mills wanted it and so I went. As an enticement, he allowed me ten days vacation on my travel orders for a subsequent hop over to "nearby" Israel to see the family there. I had to go alone to Germany this time and I felt rather uneasy, seeing a Nazi in every German-speaker. I kept to the English-speaking group for comfort, especially Mr. John S. McCall from my agency, and Mr. and Mrs. Whitten, Mr. Buford Meade (Coast and Geodetic Sur-vey), Mr. Walter Lambert, Professor Heiskanen and others. Several friends of the international set were there and many kindly inquired after my husband and Mike: Brig. Bomford, Levallois, de Graaff Hunter, Vening-Meinesz, Carlo Morelli, A. Marussi, and others. Gradually, I had to admit to myself that the Ger-man-speaking colleagues also were very nice to me. Professor Max Kneissl went out of his way as the host to look after me, introduced me specifically to the state prime minister at the official reception at the "Shack-Galerie" (sent me a photo-graph of that reception later on), placed me next to Martha Näbauer at the opera performance so that the two lady geodesists could become acquainted, and did whatever he could. A beautiful bus trip took us to Hohenpeissenberg to see the oldest mountain weather station of the world, in continuous operation since 1871. On the trip back we stopped for a beautiful organ recital at an interesting pilgrimage church, where little paper "thank you" notes were stuck into the crev-ices, noting divine help in all kinds of predicaments including school examina-tions of most recent dates. There also was a guided visit to the picture gallery, the "Pinakothek", and a beautiful performance of Mozart's *Marriage of Figaro* in the charming rococo Cuvilliés theater on the last day. The centennial celebration had been combined with a Symposium on the New Adjustment of the European Tri-angulation. Brig. Bomford reported there on the adoption of a new ellipsoid (mine) for the current adjustment of South Asia by the pertinent IAG commis-sion. He suggested that also in the new European adjustment the new satellite-derived flattening should be considered and the same ellipsoid used. The Symposium responded with a resolution recommending to the IAG to create a special study group for studying such a change of reference ellipsoid along with a corresponding change in the international gravity formula.

Some people seemed surprised that I joined the technical meetings instead of the ladies' program, but I was amazed how many other people knew my name and my work. I was kindly introduced to apparent strangers and said: "Fischer".—"Irene Fischer? Oh yes, of course; I do know your papers." Dr. Gottlob Kirschmer (Munich) said he had translated several of them into German and planned soon to translate my parallax paper. I also remember meeting friendly Dr. Fritz Leschner (Salzburg, later Aachen) and, of course, Miss Näbauer, with whom I exchanged Christmas greetings for many years afterwards. I met there Dr. Helmut Wolf (Bonn) who sent me encouraging notes off and on through the years; also Dr. Walter Grossmann (Hannover), Dr. Karl Ramsayer (Stuttgart), Dr. Hellmut Bodemittller (Darmstadt), Dr. Erwin Gigas (Frankfurt), Dr. Walter Hofmann (Bonn), Dr. Karl Ledersteger (Vienna, Austria) whom I had met already at previous meetings, Professor Heinz Draheim and Professor H. Lichte (both Karlsruhe), and many others. Their friendliness eased somewhat the eerie feeling of unreality, which colored my stay in Germany about twenty years after the Nazi time. The brief stay in Israel was a welcome antidotal relaxation.

The expanding international Christmas greeting mail included this time a letter by an army colonel from Taiwan. He had been at AMS with group of Chinese surveyors for several months to help with old Chinese maps. Also a group from Thailand had been here for some time. They spoke English sufficiently well to communicate in daily life but not enough to really converse. They came around every so often to ask questions or just be sociable. Most of them were field surveyors back home and theoretical geodesy was not their specialty. This colonel, however, came quite often to my office, obviously interested and trying to learn more about what my Geoid Branch was doing. He read the publications, came back for explanations and asked for reprints. He was rather unhappy when their assignment was up and they had to leave. He kept writing me for many years and they were sad letters. He was teaching at a Surveying College but complained about the drab routine of having to repeat always the same basic courses, of the lack of research possibilities and even lack of references, the lack of technical stimulation and the general isolation. He looked for help to come back and study here, but saw no possibility to do so. He must have been very lonely, had lost his parents and elder sister and brother from sickness, as he wrote, and he had not found a wife either. He honored me by calling me "Elder Sister" and "teacher"; he dreamt of a way to return and work in my branch on the Figure of the Earth, and even of my finding a wife for him then. There was nothing I could do for either dream, and it was hard to say so in the face of such loneliness. It was chilling to realize how much your fate is influenced by the circumstances of your whereabouts and how little choice of change is possible. The correspondence seemed to cheer him up a little. Since a while he has stopped writing. Someone told me he may have retired by then from the College. I wonder where he is and what he is doing.

In November 1962, that is soon after Dahlonega, came an invitation from Professor Peter Wilson, a young South African geodesist teaching at the University of New Brunswick, Canada, whom we had met in Dahlonega. He wanted me to come and talk to his students there "to introduce them to a different branch of work from that normally encountered in Canada." He sent along the catalogue of courses at that University and it showed an impressive college curriculum combined with geodesy; yet it did not offer much on the Figure of the Earth, which apparently more people besides me found fascinating. Pleasant as this Dahlonega fall-out sounded by itself, it turned out that their school year closed already in April, and the idea of traveling to the cold north in winter did not appeal to me then. Maybe I was getting lazy or tired of traveling so much. I did get to see the University of New Brunswick many years later in summer, at the 1974 International Symposium on the Redefinition of North American Geodetic Networks. In the meantime, there was a much more convenient lecture assignment near home, namely for both Mr. Henriksen and myself to lecture at the Naval Oceanographic Office to a class of foreign naval officers and Oceanographic Office employees. Not much preparation was needed for such talks by then, since I already had accumulated a collection of good slides to choose from. Both of us received a nice letter of appreciation through channels.

In February 1963, I received a puzzling letter from Professor Erik Norin, University of Uppsala, asking me for help to retrieve some materials from AMS which belonged to the Sven Hedin Foundation of the Swedish Academy of Science. He had written a request to the AMS Commanding Officer several months earlier and had received no answer. Now, his friend, Dr. Tengström, gave him the advice to write to me for assistance. The whole story was very complicated, went back to 1946, long before my time at AMS, involved some people who had left meanwhile and others who still were here but did not remember clearly what had happened, when and how, and who else may have been involved. It took me a while to unravel the various aspects and to follow up the tenuous threads from one person to another and from one geodetic memorandum or letter to another; but Mr. Mills said I should try to pursue the matter until it was set straight. It emerged that in 1946, the Army Map Service had agreed to compile a Central Asia Atlas from the field maps and notes of the Sino-Swedish Expedition to Central Asia from 1927 to 1935, which had been organized and led by Dr. Sven Hedin, whose scientific team had included Professor Norin and his brother-in-law, Dr. Nils P. Ambolt. In 1946 and in 1947 Professor Norin had spent a few months at AMS to provide guidance for the Atlas work. Some of the field material had been returned in 1948 and some in 1961, but not all; or at least Professor Norin had not received everything. Where were the missing items? Norin was working on a book to accompany that Atlas with all pertinent details of the maps, the survey, the geographical features, and the history of the expedition; so he urgently needed the remaining items. In 1947, he also had lent AMS his own copy of a rare pamphlet of the Geological Survey of China and he wanted that one back since it was nowhere available anymore.

With the delegated authority of Mr. Mills, I managed to have people all over the plant search their files and their memories, until some of the material turned up, requested data would be scheduled for computation in the immediate future, and even that special pamphlet had surfaced and been reluctantly surrendered. The latter took some moral persuasion; the pamphlet was a rarity indeed and had been carefully laminated for protection since it was being used. Eventually, it was agreed to make a couple of photostats, send the original to its owner together with an extra photostat as a work copy and a peace offering. Happily, I wrote to Professor Norin telling him I had found out so far and that three shipments were underway. I also traced Norin's unanswered official request to a drawer in some office, and answered it now over the signature of the Commanding Officer. Needless to say, there came a happy acknowledgment from Sweden for packages received; it came even before the official AMS letter had wound its way through the various offices, back again, out again, and to Sweden. Would you believe it—that highly overdue letter, written and forwarded through channels on March 19, came back for retyping, because its six paragraphs should not have been numbered in a civilian letter and the ribbon, somebody thought, was "soupy" (dirty). It was retyped and sent out again on March 27, came back again from Staff on April 2 (my diary does not say why, and there is no change in the copy), went out again on April 3, was back again on April 22 (my diary says: "for no reason at all"). I had thought that letter would have been in Sweden a long time by then. Why had it been held back in Staff these three weeks again? In the meantime, Norin had already answered a question in a private letter to me, so I took that question out of the official letter now to avoid confusing a civilian addressee with the time sequence of government letters. On April 26, that letter went finally overseas (I thought), with contents already obsolete. Norin's answer, however, mentions that letter as dated May 1.

As I write about this presumably silly incident, not at all isolated but paralleled by similar incidents in offices throughout the land, the U.S. Congress debates President Carter's Civil Service Reform bill to strengthen the firing power of management in order to cut down on waste. In this little incident, the only non-management player was the typist. The waste in time, paper, unnecessary interruption of other work and loss of office prestige, if multiplied by the number of such happenings in all offices, is quite considerable. Who caused the waste? It is not the typist that should be fired to reduce waste. Many years later, things of that sort had not changed. A talented secretary had this little poem attached to her desk:

"The Office.

> The office where I spend my days
> I do my work so many ways;
> I type it over and over, too—
> Can't tell you why—just know I do!"

(Anna B. Counts, July 1970)

Dear Mrs. Counts, thank you again for letting me have this marvelous poem of yours, which I kept and cherished through the years. I wonder where you are and what you are doing; and I hope you don't mind my using your poem here.

Back to the Sven Hedin project: Norin's book was planned for publication in time for the Centenary of Sven Hedin's birth in February 1965, which gave him a printing deadline of beginning 1964. He was worried about the time our computations might take and wanted to arrange a vist to AMS by Dr. Ambolt returning from the IUGG Assembly in Berkeley at the end of August 1963, to discuss the remaining technical questions. Fortunately, he sent a copy of that letter to me so that I could respond by return mail to say that Dr. Ambolt might consider coming to AMS before rather than after the Berkeley trip because the people working on the project would be on vacation at the time of the return trip, and also to confirm that AMS would respect the deadline. An official answer to that letter might well have reached him long after the event.

In a sequel, my husband and I met Professor Norin personally at the occasion of a geodetic symposium at Dr. Tengström's Geodetic Institute in Uppsala in 1965. We spent a very pleasant evening as Professor Norin's guests. His book had not come out yet, alas, and there were a few more questions and puzzles which I managed to clarify for him upon my return. In fact, a file number on a computation sheet copy in Norin's possession was identified as an AMS file number when I showed it at the office, and it led to the discovery of a whole shelf of pertinent material and also to the recollection of microfilms stored away. All this material was now copied for Norin. It was not an easy task since the material (no originals) had deteriorated from age and obsolete copying methods to a hardly legible state. But cooperation at AMS in lifting AMS prestige after all this evidence of bureaucratic neglect was great. Mr. Ray Lorah, familiar with the old microfilming methods as well as modern rescue techniques, took particular pains with each individual page to restore the maximum legibility, and he succeeded to the most grateful satisfaction of our Swedish customer. Three big parcels were sent out in December 1965 and their receipt was acknowledged as a "magnificent Christmas present." The "Memoir of Maps" was published finally at the end of 1967, and its Introduction graciously acknowledges the assistance of AMS. In February 1968, we had the pleasure of Professor Norin's visit, showed him around Washington and had him meet Dr. and Mrs. O'Keefe and Dr. and Mrs. Charles Milton (geologist) at dinner at our house. He was interested then in exploiting some NASA photos of northwest Tibet taken during Mercury and Gemini missions, to improve his maps of that region.

The episode of the Sven Hedin project at AMS shows clearly the difference between individual and organizational involvement in a long-term effort. The organizational facilities and manpower resources were needed for such a huge undertaking, but the personal continuity and involvement of Professor Norin prevented the scattering of the material and loss to oblivion. Although the project created great interest and even enthusiasm with all AMS people while they worked on it, the fact that people left or were moved to other positions prevented a personal creative attachment to "one's own" project that would insist on seeing it all the way through. Today, the tendency to move people around has become even stronger and is being encouraged and rewarded as so-called mobility and flexibility. It prevents the formation of a personal interest and identification with a solid, long-term build up of purpose and strength. What do employees expect e.g. of a department chief, exchanged for another model every year? They have no reason and don't care to even remember his name.

3. PERSONNEL, COMINGS AND GOINGS

Faced with a dire lack of manpower for my branch's obligations when I first took over in 1961, I explored two ways of remedy: to recruit new employees from the outside, or to raid other branches for qualified employees with the bait of more interesting work, pleasant working conditions, and the hope of more pay upon accomplishments. The first way was open only from time to time, as it depended on the agency's recruiting policy and success, on a one-time interview, and sometimes a waiting period of several months for the completion of an introductory course for new employees. Without that course, one generally calculated at least one year's need of special attention on the job before a beginner would be able to make independent contributions. The second way was clearly more effective immediately. The prospects knew already at least some of the geodetic subject matter and were known in turn. I managed soon to lure three people into my branch for their own good—and thus ease my immediate manpower needs.

Mr. Gerharz was a casualty of a RIF (Reduction in Force) action in another branch where a high-grade experimental physicist position could apparently not be justified, While looking for an appropriate job in another agency, he agreed to wait things out in my branch as a geodesist at a considerable cut in grade. Despite the strain of being in the wrong place, he tried to contribute to my lunar parallax work by investigating a possible correlation between the large spread in uncertainties in Crommelin's paper and the high sunspot activity in the years of those observations. Unfortunately, a car accident kept him out for several weeks. In March 1962, he left for the Geological Survey. The second one was George Gardener, the best mathematician of the raided Branch at the GS-9 level. He was bored with his routine jobs and welcomed challenge. This he received plentifully in my Branch, and he blossomed into a very capable, resourceful, and reliable project leader who earned his promotion to GS-11 within a year. His extra energies were spent in evening classes to achieve a Master's degree in mathematics. He stayed with us for about two years and then transferred to the Defense Communication Agency. The third was Dick Peat, a geology major, whom I discovered doing dead-end geodetic routine work in another branch instead of bringing his background into play for the benefit of the Figure of the Earth. He too found his way around quickly, grabbed the opportunity of a weekly training course on office time for three months, responded happily and excellently to my invitation to give a talk on his experience in a geodetic field team in Greenland, and soon suggested and worked on a project of his own: to challenge, from his geological viewpoint, statements in the literature of a simple global correlation between elevations and gravity anomalies. He did not have time to finish it, however, because he had so many irons in the fire at the same time. I was happy to promote him to GS-11 as soon as legally possible. One of the irons was helping the Field Survey Division (under Mr. Jack Lewis as Chief) with their gravity reduction problems, which soon after his promotion netted him a three-weeks T.D.Y. mission to Samoa for a gravity survey. Soon afterwards he succumbed to the lure of the greener pastures on the yonder side of the fence—a promise of another quick promotion in the classified chambers called ESPA (Engineer Special Project Area) in Governmentalese. The promise, predictably, did not materialize for quite a while since the same pay regulations applied everywhere; but the experience was worth the move as it opened career avenues. In time, he as given the opportunity of a government-sponsored M.A. study in geodesy at the Ohio State University, came back to the Geodesy Department as a branch chief, served in Staff for some time and made it to department chief in 1975. For his happy appointment celebration I baked him a "Guglhupf" (a special Viennese birthday coffee cake), pleased with the success of a one-time kid launched from my Geoid Branch.

To get a replacement for Mr. Gerharz, I persuaded Mr. Mills and Mr. Henriksen, to let me pick Mr. John Sureddin, Jr., from a group of recruits that was looked over and picked over by needy supervisors. John (or Sandy) Sureddin had worked for oil exploration companies and was curious to know what one could do with gravity measurements such as he had made for a living. My branch was the logical place to find out. He was a very intelligent, versatile, and knowledgeable person, full of fun, good to have around. In my judgment, he was under evaluated at the usual GS-5 entrance grade and I promoted him as soon as legally possible. He stayed with us about two years, including a seven-months T.D.Y. gravity mission to Jordan. Upon his return he helped with the gravity reduction of his own measurements and started to learn about their interpretation. I still have a delightful letter of his from the T.D.Y. adventure, extolling the desert as "the best place to sleep at night", and noting that "some days make the nights seem two weeks apart". He was definitely an outdoor man; he said he was "going nuts inside a building behind a desk" and had to look for an outdoor job. In March 1964, he was kidnapped for fieldwork by Mr. Lewis (Field Survey Division).

Mr. P. Mallalieu had been assigned to Geoid Branch in February 1963 as a replacement for Dick Peat. During the year and a half of his stay, he helped with the lunar parallax, but most of the time he worked for Thelma Robinson. He was then transferred into the Astronomy Branch and worked on the lunar project with Marchant. Mr. William Eunice, while waiting for his job clearance, was detailed to us in December 1963 as a temporary, three months' replacement for George Gardener. Mr. Phil Wyatt III was hired in December 1963 and Mr. Ray Shirley was "raided" from another division in March 1964. Both of these men developed into permanent assets of the branch.

The three women (including myself) were naturally the core and mainstay of the branch for the simple reason that they had been there the longest. Mary Slutsky (who had service seniority over all of us) worked with me hand in glove on the Figure of the Earth and all its extensions, as already mentioned. Thelma Robinson preferred to carry on and expand Sodano's line of interests on her own. She had an independent position comparable to my position under Sodano's reign. She had help in the same way from the other people off and on when she wanted it. I sometimes called the two my "crown princesses" (under me as the "queen") to underline their senior standing and authority, compared to the men in the branch who, in these early years, came and went. Both earned recognition in early 1964, one a promotion, the other a monetary award, and both presented brief papers at the AGU meeting that spring. The alignment of a female establishment versus a male shifting membership at that time was fortuitous and immaterial, of course. It so happened that the three of us were already entrenched in our careers, while these men were still looking for their niches; and there were practically no female qualified prospects for me to pick. My experience contradicts what can be heard occasionally from certain male supervisors that they prefer to hire men over women because (?) the latter don't stay long. I found over the years that my subordinates as well as colleagues in other units stayed or went according to their individual plans, independent of being a man or a woman. One also can hear a certain type of male supervisor express or try to cover up their uneasiness about supervising a female (other than a lowly typist or secretary). At one time, a girl transferred into my branch because her supervisor admittedly "could not supervise a woman". At another time I have heard comments such as: "This girl (a researcher) is the kind I want; she does what she is told to do without any questions asked" (is that a researcher?) or "I cannot tell this employee what I think, because she is a woman" (Really? So difficult to use civilized language?). Even in later years when anti-feminine comments started to become unfashionable, a girl applying for a government-sponsored long-term training course was told by a male supervisor: "We don't really like to send women, because they might get married or pregnant, and then are less likely than men to come back here to work." Such was neither borne out by experience nor was it the agency's policy but it reflected on that man's attitude. The discrepancy between this man's and the agency's attitude appeared also in another incident when a woman was offered a slot in an outside management course: "It may be rather inconvenient for you to go there, so don't feel you have to take it. In fact, I'd like to give it to a young man, but I have got to offer it to you first because of your rank and grade."

For me, people in the office, whether subordinates, peers, or superiors, are first of all individuals and colleagues, partners in a common undertaking, the "who". Their being men or women is secondary; it sets some rules of daily behavior, the social graces, the "how". Of my many male subordinates I was aware of only one who seemed bothered about serving under a woman, although he appeared most friendly and appreciative to me personally. I noticed it now and then with a sort of sad amusement. Some of the male supervisors mentioned above would probably also squirm in the imagined indignity of working under a woman. Some men still have a lot of growing-up to do; some may never do.

Personnel movements in the outer circuit had relatively little impact on my branch activities. Henriksen, as the division chief, interfered very little, probably because he had his hands full with his own projects and because he knew I could manage very well on my own. To help with the ubiquitous paper work, Miss Joyce Trickett entered the scene, presumably as secretary to Mr. Mattingly, Chief of the Astronomy Branch, but practically to help the division secretary, Mrs. Dorothy Davis, the only secretary up to then. The first thing that Joy (as she wanted to be called) typed for me was one of my lunar papers. After Mattingly left, Henriksen tried the experiment of reigning without filling that vacancy nor that in the Satellite Branch, since he was personally involved in the projects of both Branches and wanted to have more immediate access. He hired Carl Lucac and Fred Fallon (astronomers) and Bill Googe (mathematician). He also tried to stress the scientific character of his division by moving the Datum Branch with its more routine-type jobs out of his division and into the Foreign Control Division in 1963, which some people in that branch resented as a threat to their egos. Vic Lewicke was to assist him in subduing the administrative ogre that kept interrupting and increasingly annoying him. The two over-burdened secretaries were to give typing priority first to the demands of the division chief, then the branch chiefs (since I was the only one by then, I had no quarrel), and only then to others. Henriksen had many things going: there was the occultation method, expanding Dr. O'Keefe's work and leading to a voluminous Technical Report (No.46) in three separate parts; general satellite work to be monitored as chairman of an international study group; and primarily the special AMS tracking venture called SECOR (sequential collation of ranges), which he introduced and had to worry about in its initial difficulties and changing fortunes; a star catalogue project; the lunar control project; and probably some more. Henriksen was very active, very knowledgeable and many sided, an excellent speaker with a great intellectual sense of humor. It was a pleasure to listen to him. He was a colorful personality and, as many such individuals, he had friends and foes. Fortunately, I belonged to the friends. I was told, but never had a chance to witness, that he had a very colorful language when he was angry. There also were some amusing little stories around, of which I am not sure that they were really true. It was claimed that in his anger at the ringing telephone, particularly when the secretary was not there to pick it up, he answered it with a very high voice saying: "This is the secretary. Mr. Henriksen is out sick today" and slammed the receiver down. Or he answered with a deep voice: "Hello, this is General Su" which hopefully left the caller with the impression of a very wrong number, and unlikely to try it again. Or he advised a recipe of how to teach that telephone to be quiet: "Just let it ring; after a while it will understand and stop." But not only the telephone, also Staff got colorful treatment. He reproached them for letting valuable AMS people be hired away to other agencies: "These people produce the golden eggs for AMS, but if you give away the golden goose, there won't be anymore golden eggs."

The frequent security checks on classified documents drove him to drastic precautions. He certainly did not want to find himself in a security violation due to other people's negligence for whom he was held responsible. So he decreed that the classified map flat which contained other than his own classified documents, would only be opened on certain days for a short time. This made things a little difficult for the rest of us even though we did not have to work much with classified material on this side of the building. When I did need something, however, and could not get to it, I happened to hit the right approach to his own good sense of humor with a formal memorandum (I kept it as a souvenir). Therein I respectfully requested the support of the Division Chief in reducing the holdings of the safe by permitting it to be opened for one minute this morning, so that I could remove a certain document, then have enough time during the day to study it and also pass it on to the classified library; this would reduce his responsibility from n documents to n-l. The memo came back immediately with the comment: "An excellent suggestion".

The lunar group around Marvin Marchant and Marian Hardy, with whom I had a little more contact than with other groups due to my own lunar interests, did not enjoy such easy sailings with Henriksen as I did. Whatever the reasons were—and such reasons did not necessarily have to be rational—Marvin and Marian had trouble getting sufficient manpower for their ambitious undertaking of establishing the best first-order control network on the Moon. They had completed Part II of their TR-29: "AMS Selenodetic Control System 1964", which was the largest lunar fundamental system available then, but should be further extended and refined. The group was enthusiastic about their project—just as involved as I was in the Figure of the Earth, which made me particularly sympathetic. Mr. Mills was ambivalent. He wanted to give support to a brilliant project but he did not want to antagonize the division chief by ordering him to do what he did not want to do. So he used me as a tool to find qualified people in other divisions who might be persuaded to transfer. When Henriksen was out for more than a week in February 1964, I substituted as "acting division chief" (as I always did in his absence after Lewicke had transferred to the Department of Computer Services in September 1963). Mr. Mills asked me to seriously recruit for Marvin and start personnel action for transfer and immediate "detail" assignments. When I countered that Henriksen might kill me when he found out upon his return, I received another lesson in government regulations from Mr. Mills: "A legally appointed Acting Division Chief has the responsibility and the power of the Division Chief for the time he is appointed; otherwise government would stop. You are the Acting Division Chief this week, so act as a Division Chief." And I acted, although a little worried. Henriksen did not kill me, he did not even say a word about it when he returned; obviously he knew the regulation too. Among the interviewees was also Ray F. Shirley who was eager to transfer out from where he was but not into the lunar project. So, I approached Mr. Mills with a request for a "commission for mission accomplished under risk of life." "What do you want?" Mr. Mills smiled, and I replied: "I want Ray Shirley to replace John Sureddin in my branch." And that is how Ray joined us.

Around this time, my husband's division (ESID = Engineering and Strategic Intelligence Division) was transferred as a whole to DIA (Defense Intelligence Agency) at Arlington Hall, Virginia. That was the end of our pleasant drives together to and from work. I joined a car pool then that persisted, with some variations in passengers, until and beyond my retirement date. Mr. Joe Murr was the driver, and the passengers for the longest time were Mrs. Specht (our nurse), Mrs. Mary Mango (Personnel), Mrs. Edith Walsh (travel), and in the last years, Miss Laureen Sugden, now Mrs. R. Pond (library). It was a wonderful door-to-door service with built in news exchange and grapevine information; it was the famous "AMS Special".

7

FROM BERKELEY TO GUATEMALA AND UPPSALA (1963–mid 1966)

1. IUGG, Berkeley 1963

Preparation for the UGG Assembly in Berkeley, August 1963, began with requests for contributions to the U.S. triennial report which would be distributed there and was also to be published in the June 1963 issue of the AGU Transactions. The AMS part for Section V (Physical Geodesy) contained a brief description of my parallax work. Mr. Ken Lawyer, who worked at that time in the Publications Office under Mr. A. Trunfio, was very helpful in preparing the two hundred copies of the handout for Berkeley. He introduced new attractive covers, gray with red lettering for conference papers, and gray with black lettering for Technical Reports, to make us proud of the appearance of AMS publications. The Technical Reports were a numbered series, meant as a permanent record of our work.

My husband and I left for California a few days earlier on annual leave, to visit our daughter who had just received her Master's degree from Stanford University and had decided to stay on for a while and work full time. She had not had a chance yet to see the countryside and the people around her, she said, what with having had her eyes buried in the books for study and on her part time job at the University all the time. At least she wanted to take a look at where and among whom she had been these past two years. So she moved from Palo Alto to a place with a swimming pool in Menlo Park, to live it up a little, California 'style. We wanted to see her and her new place and then we wanted to have a brief but good vacation trip together into Yosemite Park. She took us in her car to all the impressive sights, up to Glacier Point along Tioga Road, to the redwood trees in the Mariposa Grove, and to Tenaya Lake. The night before the Assembly started, she deposited us at the Berkeley campus; and she came over again some other times to meet some of the people there.

Berkeley was a pleasant experience, with the many people from all over the world whom we knew already and those we met now. There were Dr. and Mrs. Tengström from Uppsala who planned to hop over to Hawaii after the Assembly to celebrate their 25th wedding anniversary. Hawaii was "nearby" considering that they lived in Sweden. I remembered the occasion with greetings at several returns in later years, specifically in 1978 when it was the 40th. There was Dr. Nils Ambolt from Stockholm as expected, Professor Norin's fellow explorer, who would visit AMS on his way back. And there was Professor H. Lichte from Karlsruhe who had just visited AMS on his way out here. Others from Karlsruhe: Professors Heinz Draheim and E. Kuntz. The Canadians: Mr. and Mrs. J. Lilly, Mr. and Mrs. A. Hamilton and children, Mr. D. Nagy, and Mr. L. Gale. There was the whole family of R.C.K. Johns including dog. And many others, not to mention individually the big names in the Association and the profession. At the meeting, I gave a very brief presentation of my parallax study, referring for details to the handout but drawing attention to five important aspects: (1) the application of my geodetic world datum (presented at Helsinki) in combination with three current estimates of the mass of the Moon; (2) the insertion of this geodetic knowledge into the Christie, Gill, Crommelin parallax paper of half a century ago, which removed the geodetic uncertainty of that paper and showed its discrepancy from modern results in terms of kilometers in the lunar distance; (3) the insertion of geoidal. heights into the ten-year old paper by O'Keefe and Andersen, which updated their result to excellent agreement with other results; (4) the sensitivity of certain derived results to the adoption of several numerical values considered "best" at a given time which at a later time were recognized as in error; and (5) the little known historical detail that Crommelin had used an incorrectly quoted value of Hansen's parallax (through S. Newcomb's mistranslation from German) and thus derived a much too large flattening of the Earth, which influenced later developments.

Brig. Bomford presented the final report of the Commission on the Adjustment of the Triangulation in South Asia, and explained why he had chosen my ellipsoid with the radius 6378 155m for the new "South Asia Datum". He also reported on Special Study Group No.10 (of which I was a member) concerned with the Astrogeodetic Determination of the Geoid.

S.W. Henriksen, chairman of S.S.G.-26 on Geodetic Connections by means of Artificial Satellites, presented the work of that group, which included a report by C.W. Williams on experience with the first geodetic satellite called "ANNA"

(Army, Navy, NASA, Air Force, launched in 1962), which carried flashing lights for photographic tracking against a star background. There was an animated discussion about questions such as the optimal altitude of geodetic satellites and the distribution of tracking stations.

A. Marussi and M. Hotine explained the merits of their so-called "intrinsic geodesy" (discussed also at the Colloquium on Three-dimensional Geodesy in Cottina d'Ampezzo in 1962), where the directly observed physical realities (such as astronomical latitudes and longitudes and gravimetric observations) should be used as the basic references rather than abstract constructions of artificial concepts.

E. Sodano presented a method of determining the azimuth between two non-intervisible stations by means of interposing an intense light on board an airplane which flies sufficiently high to be tracked by these stations.

Don Rice, President of Section IV (Gravimetry), talked about the need to update the international gravity formula for scientific purposes, and the equally important need to forego change in the existing data collections to avoid confusion. The need to revise the gravity formula had been discussed by the International Gravimetric Commission at its Paris meeting in September 1962, and also by the Permanent Commission for the New Adjustment of the European Triangulation in Munich, October 1962, along with the need to update the reference ellipsoid. Space scientists were discussing "best" values for astronomic constants in general (as at the Rand Corporation Symposium in 1962), and the International Astronomical Union (IAU) at a symposium in Paris in May 1963, had decided to do something about that. Was it not necessary that also the IAG in Berkeley took a stand on that problem? It was not in the cards, however, to reach a consensus at the Berkeley Assembly; it was decided to set up a Working Group on Geodetic Constants as liaison to the IAU, with Dr. A. Cook (Great Britain) as chairman and four appointed members: Dr. L. P. Pellinen (U.S.S.R.), Dr. M. Caputo (Italy), Mr. W. Kaula and myself (U.S.A.). The elections of IAG officers for the 1963-1967 period made Brig. G. Bomford the IAG President (following C. Whitten), A. E. Marussi and J. D. Boulanger two Vice Presidents. In the elevator, somebody congratulated me out of the blue for having been elected Secretary of Section V. I did not think he could be serious; for once, such a possibility had not occurred to me as I still considered myself a relative newcomer, and then it would be an odd way to let me know. My reaction must have shown my dismissal of the idea, since the well-wisher insisted: "It is a great honor, you know." Yes, it was, and eventually I was notified officially. Professor Tengström became President of Section V (following W.A. Heiskanen) and M. Bursa moved up to first Secretary for his second term, making room for me as Second Secretary. I continued as a member of Bomford's SSG-10, Rice's No.15, and de Graaff Hunter's No.16. The latter was now chaired by Tengström because of de Graaff Hunter's poor health.

The nominations for the elective offices came from a small Nominating Committee, but there was a certain succession pattern. In general, one could be reelected once for a second term; Second Secretaries moved up to first Secretaries and then to the Presidency, provided they were present and there was a vacancy. Thus, Ledersteger as First Secretary of Section V, who was not present, did not follow Heiskanen as president, and Tengström was elected instead. Bert Asplund, Secretary of Section I, became its president when Marussi moved from section president to Vice President of the Association with a bid to the Association presidency the next term.

Among the many Resolutions, there was one to take steps to establish a Central Bureau for Satellite Geodesy. A couple of years later this was organized at the Smithsonian Astrophysical Observatory in Cambridge, Mass.

From Berkeley we went to Denver to the geographers' national meeting, where my husband gave a paper. (Listening to each other's papers is part of our marriage contract.) There, someone I knew, namely P. T. Hoshide of AMS (Chief of the Asia Pacific Branch of the Map Analysis Division), told the geographers some things I knew too, namely geodesy from the times of Eratosthenes up to AMS activities. He told them about long distance determinations, Sodano's long line azimuth determination, the occultation method, satellite tracking, flare triangulation, the geodetic satellite ANNA, even about the pear-shape of the Earth. I felt quite at home. There were talks about the Vema expeditions across the Mid-Atlantic Ridge by Dr. Bruce Heezen of the Lamont Geological Laboratory, and about research ships in the meandering Gulf Stream, by Dr. Iselin of the Woods Hole Oceanographic Institute. I did not guess then that I would have a chance of getting geodetically involved in these topics many years later. Then I heard Professor Walde Tobler of Michigan University claim that he had developed a computer program with options for thirty different map projections which he could compute and plot in a few minutes at low cost! And any other map projection he could handle in a half hour. What a wizard!

The usual reward for attending a geographers' convention are the field trips under expert guidance. This time we joined a spectacular trip to the Mesa Verde National Park with the unbelievable Navajo cliff dwellings, and from there to the impressive two thousand foot deep Black Canyon of the Gunnison River.

2. NEW CHALLENGES

The official adoption of my world datum, the "Mercury Datum", by NASA and by the Department of Defense had brought innumerable repetitive inquiries from users about practical details for implementation. To save our own time and reduce the boredom of repetitions, Mary Slutsky and I put out a Technical Report (No.51) containing several graphics for quick transformations between the conventionally used datums and the Mercury Datum, along with all pertinent explanations. Originally, we had planned this information to be an enclosure to Mr. Muller's intended booklet on the Mercury Datum, but Mr. Mills wanted it to be a Technical Report by itself. When I wondered whether so few pages would be enough content for a Technical Report, he said: "It is not the number of pages that make a useful Technical Report, but the importance of the information." So far so good; no one would expect any difficulties along the way of such a straightforward assignment, and we completed it with beautiful graphics and in good time. When we passed the finished product on to Henriksen for release and further channeling, the curious thorns between government roses began to show and prick. Henriksen would not sign because so few pages were not worth a Technical Report. I sent an "S.O.S." to Mr. Mills who at first seemed to have forgotten that this was his own assignment. But then he remembered and straightened out this hurdle. Another one appeared in the shape of Mr. Phil Schwimmer, representative of our bosses at DIA-MC (Defense Intelligence Agency, Mapping & Charting), on the afternoon before Mary was scheduled to present the paper to the AGU Annual meeting. DIA clamped a "Confidential" classification on it to prevent the presentation. But why?

The information was unclassified and given out as such for three years by now. An unguarded comment gave rise to the hunch for a reason, namely, that ACIC was in the process of doing something similar, but we had come out first with this useful paper. Finally, a compromise was reached: Mary would show the graphics very briefly but give no handouts. The written report would be classified. In order to salvage its availability, Mary later arranged the report into a general unclassified main part and a classified appendix in a pocket in the back cover and thus removable. Mr. R. Galvin in the Publication Office, was very helpful with this rearrangement. It took years before we got this curious, unjustifiable classification removed.

As a member of the new IAG Working Group on Geodetic Constants, I soon received informative circular letters about previous discussions and resolutions by groups of space scientists and astronomers pushing for formal action. COSPAR

(Committee on Space Research) had an AD-HOC Committee chaired by Professor S. Herrick, and the IAU had a Working Group chaired by Dr. W. Fricke (Heidelberg), This group was charged by IAU resolutions of previous meetings to report proposals to the IAU General Assembly in Hamburg, August 1964; and our group had been charged in Berkeley to make an input about the constants of geodetic interest. The IAU group welcomed our comments. I made an elaborate analysis of the geodetic values connected with my work. Dr. Cook, the chairman, combined the information and suggestions from all his members and passed them on. I was pleased to find that the IAU Working Group had accepted my suggestions for the equatorial radius of the Earth (6378 160 m, the middle value of my Helsinki triple) and for the lunar distance (384 400 km, from my parallax paper presented at Berkeley). In February 1964, I received a request from the Joint Board of Science Education of the Washington Academy of Sciences to give a lecture within their "Frontiers of Science" program to about 150-200 high school students of Greater Washington. They suggested a lecture about measurements, a tentative title: "How far is it from here to there?" and a date of 9 May 1964. Having been a schoolteacher for several years, I responded with pleasure. I began designing a talk for a young audience that would take them from pondering the essential arbitrariness of measuring units to the application of indirect measurements in scientific endeavors such as my own work on the distance of the Moon. I started out with demonstrating that the length of the desk from one corner to the other can be measured with a tape measure in inches or, if you don't happen to have one handy, with the spread of your hand. You get a larger number with inches than with hands for the same length, but that mystery is easy to resolve if you consider that the larger number comes from using a smaller unit. Then I told them the ancient story of the Sand Reckoner by Archimedes who tried to see whether there exists a huge enough number if he measured the universe in tiny units such as poppy seeds and grains of sand. Then I discussed the difference between direct and indirect measurements and pointed out that Eratosthenes derived a measurement of the size of the Earth by measuring not the Earth directly but the length of the sun's shadow on a sundial. How could he do that? Similarly, in my own work of determining the distance of the Moon, I could not span a tape measure from here to the Moon, but I combined other, seemingly unrelated measurements with mathematics. (Published in *The Mathematics Teacher*, February 1965).

This request fitted well into my general interest at the time, as all my free time at home and wherever I was on vacation, went into finishing my geometry textbook for high schools, which was to be published the following year (Allyn &

Bacon, Inc., 1965 and 1967). Mr. Dunstan Hayden had a three-volume contract and he wrote the two algebra volumes, with my daughter as coauthor, while I wrote the *Geometry* to which he supplied the Teachers Manual. It was pure fun for me, but naturally also a lot of-work. It followed new recommendations of the College Entrance Examination Board to raise the level of the high school curriculum, in response to complaints by university professors and research employers that the high school graduates were so ignorant of mathematical knowledge and unprepared for abstract thinking. My directives had been to address the (intelligent) student directly (without relying on a busy and all too often poorly qualified teacher's explanations) and try to arouse his curiosity and sense of achievement and enjoyment, by anticipating questions and possible difficulties and by providing a wealth of elective materials. I translated these directives into a rather innovative type of textbook of about 580 pages, that should entertain while teaching. Conceding the facts of school life, I made it a three-track text: for a minimum, a recommended, and an advanced course. I remember with nostalgic amusement some of the reactions of various editors, a sequence of editors who came and went during the several years of the project. I still have the whole correspondence which at times was rather spirited and colorful. One of them wrote: "It has been an exciting experience to read Mrs. Fischer's material, and see how striking original work in geometry can be! Much of her work, particular that on locating a point in three dimensions and the discussion on longitude and latitude, is really remarkable." But then he went on to say that the level might be too high for the average student (he should have said more fittingly: for the ignorant teacher raised on and accustomed to the old-fashioned dull textbooks). Another editor tried to delete my explanations of why we only admit ruler and compasses as instruments in geometric constructions. He thought it was sufficient if the teacher said so. But that drew the ire of my most authentic and most severe critic, a representative of the student body, namely my son whose memory of his geometry course was quite recent. He wrote to the editor that this question (about the reasons for the classical instrument restriction) was of major importance to the student; it had come up as one of the first questions in his class, and "if it is not answered, many students may feel a disinclination to accept these arbitrary rules of geometry and consequently 'do not give a damn' about the course. (A brush-off) answer may silence the questioning which some teachers find embarrassing, but it dies not really satisfy the students. May I point out that high school students are very critical of their teachers and are very quick to recognize their teacher's failings;…they recognize that the teacher is unqualified to answer.…It is imperative that such important questions be answered here (in the book). I have suggested that my mother expand her reasons…" And Mike won the day with the editor, and a much expanded version rather than the abbreviated one went into the book. Another editor (they did not last very long with the Company, it seemed) tried to omit my little drawing of a half-open nutshell showing a couple of equations slipping out. It accompanied the title "Equations in a Nutshell" of a very brief reminder of how to solve equations. The editor could not get himself to smile at it, and I could not get myself to give up that smile. After a long drawn out battle, I won, using as the last effective punch a comment by my good friend and colleague, Marian Hardy: "The fact that some grown-ups (some teachers and editors) don't understand this kind of humor in teaching is the very reason why a thing like this should be in the book itself to reach the children directly and relieve the tension." As soon as I received word of my victory, I sent each of the four people involved a box of Barrancini chocolates, without any accompanying note other than a drawing of my little nutshell, this time spilling "THANK YOU", "THANK YOU" instead of equations. I fondly remember friendly Mrs. Blanche L. Blunt, the queen of the arts department, who handled my jillions of complicated drawings with so much care and efficiency, responsive to my anxiety that every one of them should be reproduced just so.

When the manuscript was completed, I fulfilled my clearance obligations as a government employee by notifying the Technical Liaison Office (Mr. Earl M. Collison and his assistant, Mr. K. R. Stunkel) and received confirmation that no further clearance beyond this notification was required for a private effort. They added their good wishes for success. Upon publication in early 1965, I was invited to meet a group of math teachers in Richmond, Va., and there was a pleasantly animated question period about ways to teach geometry. The office Bulletin brought a pleased and laudatory announcement about the book and there came a letter of congratulations from the Commanding Officer, Col. W, H. Van Atta.

Compared with the efforts of keeping editorial errors out of my "masterpiece" book, the request by Mr. Collison, Chief of the Technical Liaison Office, for two unsigned articles for *The Military Engineer* was just peanuts. I quickly wrote one on my lunar work: "Triangulation to the Moon", and the other on "The Mercury Datum", which appeared in the consecutive issues No.374 and 375. An invitation from NASA to a satellite meeting was welcome and its attendance interesting but did not take much time or effort. The requests from Staff (C. T. Williford, A. Rutscheidt, Joe Bernard, and others) to help with justifications of projects, letters, explanations, requirements and the like, had become almost routine over the years and were quick and easy to handle. On very short notice, due to a mysterious change in AMS representation at the "DoD Geodetic Objective Symposium at Cameron Station, Va., October 1964. I was asked to fill in and give a paper and serve on a panel there. The paper "From the International Ellipsoid to the Mercury Datum" had to be written in a hurry, be submitted in a hurry in final form, and was telephoned to OCE for quick approval. Its presentation at the Symposium was beautifully received, and so was Marian Hardy's on her and Marvin's lunar work. An appreciative comment in Army correspondence (Col. H. B. Pillsbury) specifically mentioned "the two ladies from AMS who presented their geodetic and moon topics so well" and came down the channels with pleased congratulatory cover letters by General T.J. Hayes and Col. W. H. Van Atta.

A pleasant invitation (also for October 1964) from the Geophysical Institute of the Czechoslovak Academy of Science in Prague to the Symposium on the Determination of the Figure of the Earth, had to be declined, but I did get another chance to go to Prague to such a symposium a few years later. Also pleasant was the attendance at the Ohio State University Symposium on the Extension of the Gravity Anomalies to Unsurveyed Areas, November 1964, convened by Professor Heiskanen and sponsored by IAG. In my new capacity as Secretary of Section V of the IAG I had the privilege of welcoming the Symposium. That was a new role for me, played for the first time; and it was fun. Then Mr. Mills had a two-hour briefing on AMS activities coming up for the "Inter-agency Geophysical coordinating Group" at the Army Research Center in Arlington, Va., on 3 February 1965, and he asked me to give there a half hour presentation on the Figure of the Earth studies in the past ten years. I enjoyed that.

But most of my time and absorbing interest, balancing the interest in my book at home, was concentrated on three topics which began to grow into major projects: the computational scheme of the new North American geoid map, methods of gravimetric interpolation of astrogeodetic deflections, and the inclusion of South America into the global figure of the Earth study. How could I have been so intensely involved in so many things at the same time? One answer is the old saying: Ask a busy person to do something special and he/she will find time to do it. Another answer is that I considered "doing something different" as relaxation; my work on the book was relaxation from office work and housekeeping, and the office work was relaxation from the other two; only housekeeping remained a chore, though pleasant in its results. The third, very important answer is that I was so lucky with my subordinates. Mary Slutsky's special standing I have already mentioned. But now there were two new people, Phil Wyatt and later Ray Shirley, both true godsends. Although both were new-comers to my type of work and needed much of Mary's and my time for explanations, that time was a good investment. Both responded very quickly to the spirit and excited involvement with the various problems. They took over from John Sureddin and continued his computer work on the Bendix, plotted the new data for the new geoid map, and learned to question everything unless double and triple checked in different ways. Since we had trouble getting access to the Bendix, the two men as a team proposed that they would use machine time on evenings and Saturdays, although I pointed out to them that they could not get paid. I did get permission, however, to give them compensatory time off at their convenience. Joe Cafaro was placed into the branch temporarily for three months and he helped all around, partly on my map and partly on Thelma Robinson's projects. Henriksen then reassigned him into the Astronomy Branch to help Milton Stein with the occultation work. Henriksen had wanted earlier to have Thelma help Milton Stein meet his deadlines, but naturally she did not care to leave her own projects and her autonomy. So we put our two heads together to decide on a strategy and formulated an official memorandum which I, as the Branch Chief, sent to the Division Chief. Henriksen gave in. Mary and I worked on new formulas for the computation of long geoid profiles along meridians and parallels which would take the curvature of the Earth into account. These were computed on the Bendix for every degree, forming a 1° x 1° grid like a chessboard across the continent. Phil Wyatt and Ray Shirley took care of that. Since every intersection point would have two geoid height numbers (as computed from the meridian and from the parallel), we would have to devise an adjustment scheme to establish final compromise numbers with due consideration of the accuracy, that is, the strength of data upon which each preliminary number had been based. This plan could be worked on in all its theoretical aspects but the final numerical work had to await the completion of the profile along the 95th meridian. And for that, the gravimetric interpolation of deflections was needed, a major project in itself. With the necessity to economize in sparse man-hours, I set up a method for mass computing gravimetric deflections that would shift the bulk of work from human labor to the electronic computer. It required deriving and checking long involved formulas and putting them into the most economic form for mass production. It further required investigations into getting the best accuracy for the least expenditure in labor and machine time under specific circumstances. both men as well as Mary were involved.

The third project, South America, was at that time not yet a project by itself but part of the perpetual effort to improve the World Datum by utilizing any new data. The story of how it grew into a major, internationally acknowledged achievement in the following years, starts with 1965.

In the meantime, in December 1964, Henriksen had left AMS. A.S. Rutscheidt became Acting Chief, and after several months, was given the position of Chief, where he stayed for about ten years. Joy Ticket had become division secretary after Dorothy Davis had left; and there was also Mrs. Eleanor Ward as both secretary and division librarian. From a memorandum it appears that Mr. Mills had considered a change by filling the vacancy at the GS-13 level instead of Henriksen's GS-14. He may have weighed the difficulties of combining, in one person, a high-caliber technical competence and genuine leadership with administrative skill and diplomacy to grease the wheels for the division projects and ward off the not always well-informed interference from Staff; he may have felt that the administrative part was all he needed at the moment, with himself firmly in charge. He apparently anticipated the development in later years toward such a separation of functions, when the title for that position was toned down to "Scientific Administrator"', but that was not on the books then.

Henriksen's leaving was certainly a blow to the division, although we did not realize it right way. Some people whom he had rubbed the wrong way were glad to see him go, but more were sorry at this turn of events. The fact that this talented man was glad to leave a position he once had wanted very much made me wonder about the type of daily activities in that position, and ponder whether it was worth running after illusionary administrative power in exchange for the satisfaction of genuine productivity. In a way, Henriksen's leaving was the beginning of the end of an era at AMS, which still lasted as long as Mr. Mills, a strong leader of international repute, was around. It was the era of "hand picked characters" as I had been told with a certain pride when I came in new. These "characters" were strong and competent individuals, with creative ideas and the capability and stamina to carry them out, think or themselves and be not afraid to say what they thought. here was truly competent leadership on top, that aroused respect, and encouraged professionalism. One of Henriksen's one-time foes reminisced much later about the change in circumstances: "You know that I had much reason to complain about Henriksen; but he was certainly a man of substance. One could at least discuss technical problems with him. He was knowledgeable and one could always learn something useful from his arguments."

3. THE SOUTH AMERICAN GEOID

The contribution of South America to my geoid studies up to 1965 still consisted essentially only of the long coastal arc that had played such a crucial role in the determination of a new Figure of the Earth, way back in 1956. There was a chart, however, showing the location of South American geodetic activities, but Mr. Fuller said the data were not available. One day, he came to my office with some questions about the geoid in the Northeast of Brazil, and I realized that he had some data there which I did not have. We made a bargain: I would help him in the Northeast, if he let me have not only these but all other data he had of South America. I also urged him to use his connections to the South American agencies to persuade them into letting us have the data shown to exist on that chart, that is, data from Argentina, Uruguay, and Paraguay; but he claimed that that was not so simple since these countries were not connected with the IAGS (Inter American Geodetic Service), our avenue to the other countries. Well, maybe we could interest them into contributing to the fascinating study of the Figure of the Earth? I got Mr. Fuller interested, but that was clearly not enough. Mr. Fuller had proved a friend before, when he put personnel at the San Antonio Field Office at my disposal or constructing my first geoidal map of North America. And he proved a friend again: he talked my plans over with Mr. Mills, and the two of them had a splendid idea: There was going to be an Assembly of the Pan American Institute for Geography and History (PAIGH) in Guatemala City in July 1965 to which both were going as U.S. delegates. If I were preparing a paper about my plans, they would distribute it there and see what happened. Great! From the data, which Mr. Fuller all of a sudden found in his pockets, I constructed a geoidal loop around the northern part of the South American continent. The data were referred to the Provisional South American Datum of 1956, and my geoid computations showed that to be a very poor datum. Not only did the discrepancy between geoid and reference ellipsoid rise to around 300m when carried southward to Chile, but no "Molodenskiy Correction" had been made to counteract the ensuing big distortions. After inserting these corrections, I showed in tables and graphics how these large geoidal heights could be made to disappear by choosing any of several other datums, and why this would avoid accumulating distortions. The underlying message was: I can compute a geodetic datum that fits South America much better than the one you have now; and it would fit even better if you let me have all the data you keep locked up in your files, and maybe even observe a few more where I would need them most. Mr. Mills and Mr. Fuller were enthused. I did not know then that the PAIGH had been trying to determine a well-fitting South American Datum since 1944 and had not been able to agree on what approach to take. And here, Mr. Mills would be able to present to them a practical, common-sense solution to their problem and at the same time reap a big boost for the prestige of the Army Map Service. In fact, I had not even known what exactly the PAIGH was and what the problem was. I just wanted more data of the southern hemisphere for my global preoccupation with the Figure of the Earth. As time went by, Mr. Mills came in more and more often to look over my shoulders, nervous that I would not be finished in time. "Mr. Mills, don't worry. You know that I have never missed a deadline yet. I am not going to miss this one." Mr. Fuller speeded up the adjustment of his data, which was the routine obligation of his division, and also found and offered beautiful base maps for my graphics and had my drawings superimposed on them in color. Mary Slutsky signed as co-author and Phil Wyatt got an appreciative acknowledgement statement for his assistance with the computations. The paper was finished in April and cleared in May. Mrs. Sener of Trunfio's Publications Office made beautiful handouts. Then she asked me to write a synopsis for *The Military Engineer* (unsigned), which was published in the September 1965 issue. Meanwhile we wondered what would happen in Guatemala City.

That meeting was an important one for me, indeed: there was an unusual success for my paper—and I was not even there! Mr. Mills and Mr. Fuller came back from Guatemala City by July 16, and they were beaming. Mr. Mills called to say that I would have a lot of important work to do and he'll tell me more about it. The next day, I received a personal note from the leader of the U.S. Geodetic Delegation, Col. F.O. Diercks, who had been our Commanding Officer several years earlier: "I thought you might like to see this vote of applause for your paper on the Study of the Geoid in South America, adopted at the recent PAIGH meeting in Guatemala. It was certainly the most popular paper at the meeting, in my opinion." Attached was a copy of the official "*Voto de Aplauso*" from the Committee on Geodesy, a copy of Resolution 3 (in Spanish): "Considering (1) the objectives of the Committee, (2) the great interest in the technical and scientific content of the paper 'A Study of the Geoid in South America' by Irene Fischer and Mary Slutsky, and (3) The necessity for a Datum. It is recommended (a) To continue the investigation with results from Chile, Bolivia, Argentina, Paraguay and Brazil. (b) To publish this paper in the *Revista Cartografica*." Also attached was a copy of the Committee's program projection referring to the "Datum" for South America. The latter spelled out specifically (in Spanish): "The Objectives: (a) To choose a datum point in the most convenient way. (b) To recommend a uniform and coherent system for South America. The Phases: (1) To compile existing information in order to continue the investigation initiated by its. Fischer and Miss Slutsky. (2) To amplify this work, and (3) To present the results at the next Reunion" (in 1969). The paper was published also in Spanish as a PAIGH Publication (No.293).

The IUGG Chronicle, a quarterly periodical published by the IUGG General Secretary to record briefly the facts in the mainly administrative life of the Union, had a short report on the two-week PAIGH Assembly by the liaison officer, Dr. M. Maldonado-Koerdell, in its December 1965 issue. The paragraph on the Geodesy Committee said (in French): "The technical debates were largely dominated by the United States and dealt with electro-magnetic and electro-optic distance measurements, and the use of artificial satellites for geodesy by optical (stellar photography) and electronic methods (SECOR and TRANSIT). The presentations on this subject did not contain anything new. But worth mentioning was an interesting study by I. Fischer and M. Slutsky of the shape of the geoid in South America, which showed that the International ellipsoid fits very poorly in this part of the globe."

To understand the impact of my paper, one must consider the history of the PAIGH debates on the subject. As I began to read up on that history, I learned that the PAIGH is one of six "Specialized Organs of the Organization of American States (OAS)" and comprises 22 states of the Western hemisphere for the purpose of coordinating studies in Cartography, Geography, and History, effected by affiliation through their foreign Offices (the State Department for the United States). The Commission of Cartography created a special Committee on Geodesy in 1944, specifically to develop a continental geodetic datum to unify cartographic work and engineering surveys. Triangulations in the individual countries should be extended to their borders on provisional reference datums and joined to an over-all continental system, which should be based on well-distributed astrogeodetic stations in such a manner that distortions in the network would be as small as possible. This plan fit exactly my Guatemala paper, but I did not know that when I wrote it. To me it was just common sense. In 1944 when the plan was formulated, I was teaching mathematics and engineering drawing in Bard College in Annandale-on-Hudson, N.Y., entirely innocent of any geodetic ambition and ignorant of geodetic datums and geoids (what's that?). What happened in all these years? The gyrations in the PAIGH debates between their straightforward recipe in 1944 and their need for a proposal from a relative outsider in 1965 for the same thing are quite interesting and indicative of the proverbial "too many cooks spoil the soup."

The Commission on Cartography and its Committee on Geodesy met every few years in different places for Consultations, and almost each time further clarifications were discussed. After formulating the purpose and method of the South American Datum project in Rio de Janeiro in 1944, the Third Consultation in Caracas, 1946, decided that the common fundamental reference point (the origin of the continental geodetic system) should be in the center of the continent. In 1948, the Consultation in Buenos Aires wanted to stress this requirement and delineated a geographical area between certain meridians and parallels as the datum zone in which the future origin should be located. So far so good, although the choice of the area was made without considering the physical features of the specific region, whose geodetic emptiness and inaccessibility carried the seeds of future frustrations. But then new entanglements were started which led to an inextricable Gordian knot. Rather than following the original practical design of first creating a continental network on preliminary national systems, and then orienting it on the basis of well-distributed astrogeodetic stations, in order to minimize deflections of the vertical and geoidal heights, they decided to start the other way around: they would make gravity observations in the now delineated datum zone to select there a point with negligible deflections, and then start the continental network from there. Argentina, Bolivia, Brazil, and Paraguay who shared the zone were requested to make gravity observations in their part of the zone. This theoretically acceptable alternative was doomed to failure in practice for several reasons: (1) a very dense system of gravity data is needed to compute deflections, and that did not exist. Even under the best circumstances it would take time to create a satisfactory data collection, delaying the project; (2) the "best circumstances" were not given in a case where four independent countries were involved; (3) gravity observations alone are not sufficient. One needs to know where they were made, that is, one needed triangulation and leveling, and these did not exist either; (4) vast areas of the zone were practically inaccessible; (5) there was no other incentive, say on economic grounds, to overcome the inaccessibility at high cost; (6) in the unlikely event that all those problems including time and money could be solved in all four countries, and a point with negligible deflections identified in the delineated zone, one could not bank on it that the derivations from this limited area (which by practical necessity were not absolutely error free) would give the desired result for the whole continent.

The practical difficulties of the changed approach soon became apparent and were deplored at the Consultations in Santiago de Chile, 1950, and Ciudad Trujillo, 1952, without finding a way out of the predicament. In Mexico City, 1955, a review of the whole project pointed out that there were only few triangulation chains and level lines in the Bolivian part of the zone and none in the others, and no significant change to be expected in the foreseeable future. The opponents of the switch to the gravimetric method eloquently described its practical pitfalls, and gained in strength. In Cuba, 1958, and in Buenos Aires, 1961, they stressed the purpose of the whole project, namely to provide a uniform basis for the growing engineering surveys and cartographic works, and they pointed to the urgency of achieving this goal. They criticized the La Canoa Datum of 1956 (my encounter with this gravimetrically derived Venezuelan datum was described earlier, in chapter I.4) as inadequate for the continental purpose, because its location in the north and the inaccuracies of the gravimetric derivation caused large distortions in the rest of the continent. What they wanted and needed was a centrally located origin and geodetic ties across national boundaries, and the sooner the better. They still did not come up, however, with a concrete plan of how to get there, nor did they or could they drop the gravimetric plan. There was too much invested, at least in discussions and resolutions since 1948 and also in money and prestige, to say that this was a dead end route.

The next Consultation was in Guatemala City, 1965, where Mr. Mills presented my paper. It was just what the doctor had ordered. If developments over the years had been rigged, the timing could not have been better. The paper gave a practical straightforward proposal of how to reach their goal in a relatively short time, fulfilling the requirements spelled out in 1944 and bypassing the difficulties of the gravimetric quagmire. What was new in my approach, as compared to the twenty-year long PAIGH discussions, were three features: (1) the three-dimensional way of fitting a reference datum to the geoid, computed preliminarily on an auxiliary datum; (2) demonstrating the effects of different datum choices and establishing the mechanism of changing from one choice to another; and (3) establishing the desired features first, and then using that mechanism in reverse to select the datum which would produce those desired effects. The latter were, of course, small geoidal heights to avoid distortions, a datum point somewhere in the middle of the continent in a well-surveyed area, and best accuracies in areas of dense habitation in need of engineering surveys and good maps. There was no mention of the inaccessible datum zone and its insoluble problems one way or another; it just was not needed. The quick acceptance and promise of cooperation revealed the relief of seeing a practical way to an increasingly urgent goal. A small international Working Group of four people was set up to push for action: Mr. Fuller as chairman, Ing. Rene de Mattos (Brazil), Ing. Pablo Dragan (Argentina), and myself. But then, lightening struck: Mr. Fuller suddenly died. That was 7 September 1965, the morning after Labor Day.

It was hard to believe. True, he had looked thinner than usual, but he claimed that he was his usual summery self, playing tennis and not eating much in the summer heat. But on September 7, less than two months after his happy return from Guatemala, he was dead. There was a memorial service for him the following week. I missed him. I missed his frequent friendly visits, his tall slender figure unexpectedly coming towards my desk and settling down comfortably for a chat about our common project. Before Guatemala, he came to watch the progress of my paper, offering help where possible such as supplying those beautiful background maps for my illustrations. After Guatemala, he would talk about his obligation in getting me more data from his South American counterparts and persuading them to add new observations. He also talked about his perennial plans to improve his Spanish and Portuguese, so necessary for his role. The Ninth Reunion of the Directing Council of the PAIGH, Mexico City, 1966, issued a special Resolution in honor of Ing. Homer C. Fuller, recognizing his important contributions and expressing homage to his memory and condolences to the authorities of the United States and members of his family.

It took a long time before he was replaced.

4. IAG SYMPOSIUM IN UPPSALA, JULY 1965

In May 1965, there came Dr. Tengström's invitation to a meeting of SSG-16 in Uppsala, Sweden. Dr. Tengström was then the President of AG Section V and also Chairman of SSG-16. The topics would deal with gravimetric determinations of deflections and geoid heights, and with relations between geometric and gravimetric methods of determining the size and shape of the Earth. My involvement with the gravimetric deflections along the 95th meridian and my scheme to minimize human labor and let the computer do most of the work, seemed to fit into these topics and Mr. Mills thought I should go. My South American paper had just been finished, so I could concentrate on composing a paper for Uppsala. With permission to extend this conference trip to a vacation in Israel and in the Austrian and Swiss Alps, my husband and I set out for Uppsala, where we had been on a geographers' field trip five years earlier. Mike did not come along this time; he was seeing Europe with a college geography class. Nonetheless he got a heap of stamps from Tengström.

The group met in the beautiful setting of Tengström's Geodetic Institute in Hällby, his pride and joy. It was a small group, quickly becoming friends on a personal basis which transcended international differences. From the United. States there was only Ken Daugherty from ACIC besides us. There was Dr.

Suzanne Coron (International Gravimetric Institute, Paris), Mr. and Mrs. T. Krarup (Copenhagen), Dr. Milos and Mrs. Bozena Pick (Prague), Dr. L. Stange (Potsdam), Brig. A. Glennie (Great Britain), Professor Carlo Morelli (Trieste) and his assistant and former student Dr. Maria Theresa Carrozo (Bari), Professor L. Asplund, President of IAG Section 1 (Stockholm), Mr. L. Pettersen (Stockholm), Mr. L. A. Haller (Uppsala), Mr. H. Henkel (Uppsala) and from the Institute: Mr. Peter Hodacs, Mr. M. O'Shaughnessy, and Mr. D. Stojkov. The newspaper celebrated the meeting with a picture and a long column about the specialists coming to town.

My paper "Gravimetric Interpolation of Deflections of the Vertical by Electronic Computer" had only been produced in limited numbers as a conference handout, because the Hällby Institute intended a publication of the whole Conference. But copies of the limited handout got around and there were requests for more. Several months later, a copy had found its way to the *Bulletin Géodésique*, and they wrote for permission to publish the paper in their next issue because they thought that it was very interesting and would be of interest also to a great number of their readers (*Bull. géod.* No.81,Sept.1966). My trip Report about the Uppsala symposium, where I tried to catch the friendly, yet intensive flavor of the hustle and bustle there, took the form of an informal letter to my Department Chief, which also reflects the non-bureaucratic personal relationship established by Mr. Mills with his subordinates. "Dear Mr. Mills: A very intensive geodetic week here in Uppsala has come to an end; and I ought to report to you about it before I leave for Shangri-la and forget all geodesy. Even more impressive than the geodetic devotion of this conference is the fantastic Tengström—Swedish hospitality and thoughtfulness. It began with flowers for the ladies at arrival, and a festive reception in a historic setting, the 300-year-old Theatrum Anatomicum of the old university, by the Rector Magnificus Torgny Segerstadt, by the Chairman of the Research Council, Martin Fehren, and by Tengström.

The geodetic discussions were held in Tengström's new and beautifully located Geodetic Institute, and on sightseeing busses (to the horizontal pendulum station in the Dannemora mines), and at gatherings till late at night, even interspersed with a rare treat of exquisite piano playing by Mrs. Tengström. This meeting beautifully served its mission of international good will, and this was also brought out at an interview with the press on the last day.

"Intensive work is being done in the Westalp project, where different groups are trying out their specific methods of recovering astrogeodetic deflections from gravity anomalies under the worst but identical conditions of wild topography: Morelli—Coron, Tengström, Brig. Glennie (De Graaff Hunter), and the Russian—Czechoslovakian group represented here by Dr. Pick from Prague. The idea of including geologic considerations into gravity interpolations is gaining ground generally. Daugherty (ACIC) reported on work done at ACIC with Woollard as consultant and with cooperation of the Woollard group in Hawaii. At ACIC there is a Section of about ten people (not including Durbin and Daugherty) working on these aspects; which made me rather jealous, since I have been asking for a couple of years for one person with geological background for the Geoid Branch, to keep track of and use these new developments. I seemed to be the representative of the geometric method (versus dynamical) and enjoyed myself elaborating on this approach. I presented my paper...I had brought enough copies for the meeting, but Tengström would like to have some more copies for distribution. So may I ask you to tell Mary Slutsky to send by airmail 12 copies to...All papers will be published in the Series of the Uppsala Geodetic Institute in due time.

"One evening we spent with Professor Norin, listened to his reminiscences of the Sven Hedin expeditions and looked at his intensive painstaking work of integrating all records of these memorable years and augmenting them with newer information. He was very grateful for AMS help. But I could not help feeling embarrassed for the sloppy way these unique documents had been treated by AMS. On top of it all it seems that one crate apparently sent by AMS years ago must have been lost in our Embassy here—of course, irreplaceable records. One AMS map sheet shows a wrong route for the railroad, without any explanation for the change. I wonder whether I can still find out the reason, maybe from Mr. Alexander, or whether this was just a gross blunder.

"Asplund asked me about his letter to R & A, concerning cooperation discussed with Rutscheidt at Athens. I told him that I had indeed seen his letter, but preferred to let the answer wait for Rutscheidt's return; and that I felt sure that the Athens proposals would be backed up in writing with all pertaining detail. Petterson was also pleased to hear of our continued interest in cooperation, to be discussed with Iverson in Paris next month. Tengström plans to go to the U.S. middle of August, first to Hawaii and then east to be in time for the Paris meeting by 13 September. He and his wife may come through Washington in September.

"In connection with AMS contributions to the World Datum and specifically with the area of Southeast Asia and the West Pacific, it occurred to me that more than a couple of years ago, Henriksen asked the San Antonio field office to start constructing the geoid there. They were supposed to get new data directly from AMSFE as available and Mary Slutsky had to list literature for them. I never heard of it again. Now that time becomes short, would it be possible that Mr.

Lieberman who offered to work with me on this, goes to San Antonio to find out and bring back what they have done all this time, to avoid duplication? This might be a shorter way than asking them for a report, which would be a time-consuming project in itself.

"With best regards to everybody, and good wishes to-you,…"

Mr. Mills liked an informal way of keeping in touch with his people. He often joined us at the "stand-around" coffee break in the cafeteria, cracking friendly jokes but essentially making himself available for quickie information, questions, guidance, and even gripes. Compared to today's formality, inaccessibility, and indecisiveness of insecure managers, you can't beat the economical and psychological effectiveness of a, say, five to ten minute business conversation over a cup of coffee with a boss who knows what he wants and takes responsibility for his decisions. We or he might bring up, for instance, the essentials of a decision to be made; he would listen to alternative possibilities, maybe give us some thoughts on the general policy he wanted pursued, and only then he would ask for a memorandum or draft letter or whatever to be sent to his desk for his final decision or action. If one had an opportunity to touch base with the boss that way, it certainly cut intra-departmental paper work. There was more than enough paper work for him elsewhere. He might cut short his coffee break with us, saying: "I have got to get back to my desk and shuffle a few more papers."

5. FOLLOW-UPS AD KUDOS

One of my first obligations after returning from Uppsala was the follow-up of Professor Norin's questions which I could resolve, as already told above in Chapter VI.3. Another follow-up concerned the extension of my Uppsala paper to include an alternative method adapted from Tengström's computation of the Swedish geoid, using gravimetric curvatures (the changes in deflections along a path) and also the application of both methods to our project of constructing a geoidal profile along the 95th meridian. Ray Shirley and Phil Wyatt worked much on that project with me and signed as co-authors. Both received awards for their achievements, and promotions in due time.

When Bernie Chovitz as program chairman for the AGU Annual meeting solicited papers in December 1965, it was easy to submit several: "Slopes and Curvatures of the Geoid by Electronic Computer," "A Geoid Profile in North America from a Combination of Astrogeodetic and Gravimetric Data," "A Revision of the Geoid Map of North America," and "A Study of the Geoid in South America." Mr. Mills was, of course, very pleased about this prolific show of productivity, he could not resist teasing: "You geodesists think up such long and boring titles. Why can't you say simply: The Cat."

When the AGU meeting rolled around, I persuaded my co-authors, Mary and Phil (Ray was then in school on long-term government training), to present the respective papers, which they did excellently. I felt proud like a mother showing off her brilliant children, or like a hen parading her chickens. The first three papers were published in the *Journal of Geophysical Research* in 1966. Henrietta, the secretary who typed all these papers on short notice and helped process them most efficiently, received an award too. Even people from the Personnel Office got a whiff of the excitement and requested, through Mrs. Rust, an informative talk about the mysteries of our doings. I entertained them and myself for two and a hall hours straight. The official thank-you letter from Mrs. Bernardine M. Wilson, Chief, Position and Pay Management Division, quotes the occasion as a "Lecture on the History of Geodesy and Advanced Theories and Methods for Determining the Size and Shape of the Earth" (hardly conforming to Mr. Mill's suggestion; "The Cat") and claimed that the audience found it "extremely helpful in gaining an insight into a complex but exciting subject." In fact, I had used the same material for a recent invited lecture to advanced high school students at the Fourth Maryland Junior Science and Humanities Symposium of the Maryland Academy of Sciences, March 26, 1966, in Baltimore ("The Shape and Size of the Earth," *The Mathematics Teacher*, 1967). These lectures and my many briefings for visitors were the prototype of frequent future orientation briefings for groups of new employees including secretaries.

The South American study was recognized in official correspondence as an important milestone in DoD goals, and brought Mary and me the 1966 "Army Research and Development Achievement Award", complete with plaque, pin, and a bundle of official letters with cover letters all the way down the official ladder. Col. W. H. Van Atta was the pleased Commanding Officer then and Col. Daniel Hritzko handed out the plaques and pins. For Mary there also was a promotion a little later.

Kudos were coming thick and fast. During the annual job survey ritual in spring 1965, the legal stamp of the career development program for researchers, begun for me by Dr. O'Keefe and Mrs. Rust and continued by Henriksen, was reached by changing my job title from "Supervisory Geodesist" to "Supervisory Research Geodesist," following a new "Evaluation Guide for Research Positions of 6/64". When I wondered what this extra word in a job title should do for me, Rust explained: "Don't laugh. It puts you in a different pay schedule so that your grade can be evaluated on your personal merits, independent of the position structure whether you are also a Branch chief or not, and irrespective of the grade of your supervisor." That meaning, however, seemed to have been forgotten already a few months later.

In December 1965, I received the Nomination for the Federal Woman's Award from the Secretary of the Army. This was puzzling because I could not see the connection between my work and my being a woman. Why should there be a separate category for women? As if in answer to that question, I received the Meritorious Civilian Service Award from the Army (Chief of Engineers, Lt. General W. Cassidy) for the second time in early 1966, this time "with Bronze Leaf Cluster." In the face of these three higher awards, AMS remembered their own in-house award system and gave me an "outstanding performance" rating in 1966, coupled with a quality step increase. Apparently, it did not occur to them (and not to me either, dazed in the downpour at that time) that this would have been the occasion to make good on their proud promise about the title of "research geodesist" and grant me a promotion. After all, people got promotions for a lot less achievement. To get the same grade as one's supervisor was not a legal obstacle, since that had been the case a few years earlier, and would happen again later, when someone tried to set things straight for me. Sometimes, the government forgets its own rules.

Marvin's lunar group, fondly called "the loonies", had better times coming at last. With Mrs. Rust's help, Mr. Mill's approval, and Rutscheidt's concurrence, Marvin became branch chief with the accompanying promotion, which gave him a chance to organize a Selenodetic Branch for enlarged and intensified activities. Both he and Marian received the Meritorious Civilian Service Award medals. Their group had been housed in a very small space, two people to a desk (some working evenings on the computer), climbing over chairs and legs for necessary movement. It was the picture of a dynamic office (reminiscent of C.N. Parkinson's examples of office vitality) where the apparent mess and cluttered appearance indicated the high intensity level of actually well organized absorbing, with no time for external frills for appearances sake. Now they moved to larger quarters with a whole desk per person and real walkways between them. In their exuberance, they used the renovating occasion to have one wall painted gold in homage to their celestial jewel. But most important to that dedicated group was the chance to get manpower help from another division to handle an iteration procedure in a mass production, by routing specific tasks there and back. Marvin and Marian came in on weekends to prepare the material to be sent to the other division, first thing Monday morning. After four working days the completed task would come back for inspection, evaluation, and preparation of the next step by the next Monday morning. Rutscheidt as division chief agreed to act as the administrative liaison to the other division chief and to monitor the arrangement. The "loonies" themselves would thus have more time during the week to attend to the non-routine problems. For a while things worked out beautifully and the "loonies" worked feverishly, happily, and productively. Marian Hardy prepared a paper on this work and future plans for the Second International Symposium on the Use of Artificial Satellites for Geodesy at Athens in April 1965, and at about the same time P. Mallalieu reported on it to the AGU Annual Meeting in Washington, D.C., and there were other products. But later, some sinister clouds gathered and negative forces began to undermine and then destroy a brilliant project and a productive work unit. For the time being, however, the mood in the Division seemed cordial enough for holiday celebrations. There also was Eileen's wedding, changing her name from Segal to Cantwell. Eileen was generally well liked and many of us were invited and helped celebrate.

Kudos in the wider geodetic community went to Professor W.A. Heiskanen at the occasion of his retirement from Ohio State University in 1965. I was invited to come to the festivities there, but I could not make it. I wrote him a letter and received a gracious letter in return. The office gave the medal holders (Mr. Mills was one too) a name shield in red. The medals were big things on a ribbon and there were little substitute pins for wear. I asked Mr. Mills what occasions there were to show off the real thing, other than in a showcase at home. He said: "Well, from now on, each time you come to my office, you can put both of them on."

8

TOWARD THE IUGG ASSEMBLY IN LUCERNE, 1967 (mid 1966–mid 1967)

1. THE FIRST MARINE GEODESY SYMPOSIUM, 1966

Way back in 1958, at the Conference on Contemporary Geodesy at the Smithsonian Astrophysical Observatory in Cambridge, Mass., I had listened to Dr. Maurice Ewing's proposal of establishing a network of bench marks on the ocean floor to serve as basis for transoceanic ties and eventually for the determination of the size and shape of the Earth. A benchmark should be determined by installing three transponders (devices to receive and retransmit signals) in an equilateral triangle on the ocean floor, and stationing a ship above them at a location from where an acoustic signal would take an equal travel time for a round trip to each. Distances between benchmarks would be calculated from the transmission times of sound signals within the water.

Several years later, in December 1964, Mr. George A. Mourad of Battelle Memorial Institute in Columbus, Ohio, visited AMS among other agencies in search of financial sponsors for a feasibility study to put such transponders on the ocean floor for locating ships and to determine the distance between ships on the ocean surface by HIRAN (a high precision short-range electronic navigation system). It was not difficult to get me interested in this whole concept, of course, since its impact on the Figure of the Earth studies was rather obvious. After all, the oceans represent about 70% of the Earth's surface and separate the land-based geodetic networks. But, alas, I did not have any funds at my disposal. I talked to Rutscheidt and Mills about it, but they could not or did not do anything for George either. I did keep some contact with him from then on. At least I could respond to his request to write a brief opinion on the usefulness of his plan which he could use in search of wider support; Mills and Rutscheidt had no objections to that. At GIMRADA, Dr. Armando Mancini and Col. Max Jonah also showed some interest. I had met Col. Jonah several weeks earlier, in December 1964, when Mr. Nathan Fishl of GIMRADA called me about some information about my work. Col. Jonah, who some ten years later was to be my department chief for one year at the Defense Mapping Agency, was then Assistant to Col. Rall, the successor to Col. Van Atta who had just become the Commanding Officer at AMS (1964-67). In April 1965, GIMRADA planned to set up a Geodetic Research Institute with resident and non-resident members, as a vehicle for better and easier cooperation between agencies. A couple of years later, they finalized these plans and officially Col. Fisch made Rutscheidt and me non-resident members; we went to some meetings and briefings, but nothing more developed in the way of closer cooperation.

In November 1965, George Mourad called again. He had made quite some headway in getting support for his plans and there would be a "First Marine Geodesy Symposium" at the Battelle Memorial Institute in Columbus, Ohio, September 1966, co-sponsored by ESSA (Environmental Science Services Administration, U. S. Coast and Geodetic Survey), and with Professor Heiskanen as co-chairman. Would I come? Of course I would, "if they let me." Rutscheidt and I went. There were about 400 participants, representing interdisciplinary interests of international scope. The increasing interest in the oceans or scientific purposes and for exploitation of economic resources required the extension of geodetic services from land to ocean, to establish—just as on land—where what is and how to get there. The concept is the same, but the engineering feat to carry it out in the ocean medium was a tremendous new challenge. The mutual advantage of pooling efforts and knowledge between physical oceanography and geodesy was obvious, but the global scale of the undertaking involved also international cooperation including international law in defining international boundaries at sea, navigation and exploitation rights, and delineation of leased areas. One of the practical purposes of the Symposium was directed, therefore, towards involving international organizations such as the IUGG. As Secretary of IAG Section V, I was asked to participate in the formulation of the Symposium's Resolution which recommended to the IAG and to the International Association of Physical Oceanography to set up joint meetings during the upcoming IUGG Assembly in Switzerland, and also to establish commissions for promoting and facilitating international cooperation. I was happy to be part of this grand venture and to have the opportunity to thank the organizers of the Symposium, in the name of the audience, for the great experience of formally launching "marine geodesy".

2. THE ORNERY INFANCY OF SATELLITE GEODESY

The grandiose concept of using artificial satellites as intermediate objects in geodetic surveying techniques to span global distances as never before possible, masked the great difficulties in perfecting that technique to geodetic accuracy standards. The learning period extended over years and was fraught with frustrations and also tensions over inter-agency competition. Just as children, viewed as sugar and spice in their Sunday best, can be a pain in the neck at home with their frustrations and sibling rivalries, so the beginning of satellite geodesy looks now like a glorious immediate breakthrough while a close-up view of the working experience at that time reminds one rather of a crazy, mixed-up kid with frustrations and tensions all around. Several agencies had technically different satellite methods in the running for the government dollar. AMS championed the so-called SECOR project as already mentioned, which Henriksen had introduced. It was in competition primarily with the Navy's "Doppler" technique and with the photographic methods of the Smithsonian Institution and of the Coast & Geodetic Survey. For my purpose of Figure of the Earth studies, they were all welcome as new tools to get distances and directions, or coordinates of one point on the Earth's surface with respect to another; but they should be consistent and give the same answer to the same question, which they did not always do. So, which one, if any, was correct?

After Henriksen's leaving, the SECOR project was orphaned for quite a while. Charles Batchlor and Marvel Warden valiantly carried the ball as best they could. Rutscheidt arranged for several visits by experts from the Cubic Corporation which had supplied the equipment, and he himself became involved too. George Dudley was transferred into the Satellite Branch from another Division to be the branch chief. There was not much love lost between the theoretical group and the group in the field making the actual satellite observations. Each blamed the other for the failure to make heads or tails out of the heaps of data periodically invading us. At the same time, our OCE bosses (Office of the Chief of Engineers) were breathing down the neck of the researchers for good numerical results which they wanted to pass on to the super bosses at DIA to make a good showing in the inter-agency competition. Some people left because of this pressure to appear as sugar and spice when they thought they should get outside expert help instead.

Thelma Robinson offered to help analyze the problems for many, many months almost full time. My group supplied coordinates and geoidal heights for their tracking stations and made many comparisons of conventionally derived results with those from satellite calculations of every available brand, in order to test for errors and bias. While we were eager to get and use whatever was or seemed to be correct, we kept our fingers crossed for SECOR out of loyalty to AMS. Years later, some hindsight evaluations insisted that SECOR was basically an excellent measuring equipment, unjustly blamed for losing the competition to the Doppler technique. SECOR was a geometric, ranging system, unaffected by errors in the not yet very good orbit determinations with which the dynamic Doppler system had to contend. The annoying frustrations in the learning period would have been eased by better communications between our field and computer groups, since this entirely new and erstwhile insufficiently tested tool apparently needed a longer time to settle down to reliable measurements than the analyzers suspected who used all fluctuating data from the start—and were upset. There was no leader above both groups, knowledgeable in both the field and analysis work. But the decisive reason for losing the competition to the Navy was an economic one that had nothing to do with the quality of the tool: the Doppler technique could use the already existing Navy Navigation Satellites without charge, while satellites for SECOR had to be put up and paid for by the Corps of Engineers.

In order to capitalize on the new and still competing satellite methods, even though they were still in the experimentation stage, DIA decided to update the classified world datum (WGS-1960) as WCGS-1966 and repeat that process every two years. Originally, it had been the Army's obligation to provide the Defense Department with a geodetic world datum, which we did with the first operative, so-called Vanguard Datum in 1956, later refined to the "Tentative World Datum" and the Mercury Datum. In the Bedford Conference in 1958, the Air Force managed to get into the act, and the WGS-60 was decreed to be a mean between their and our proposals. Naturally, we had felt that our product was better—as future developments confirmed; but that had been the beginning of "political geodesy". Now, the Navy too wanted a part of the action, and DIA set up a WGS-Committee consisting of several people of varying relationship to the actual work, but representing all three forces. The chairmanship, including the task of the final integration of the work into a report, was to be in rotation every two years, with the Air Force (ACIC) to start. They've kept it ever since.

The difference between getting work done in a single outfit or by padded committee in a "Joint Service Plan" was interesting to watch. The argument that there was so much more work involved now with all these satellite results, was, of course, superficial window dressing. At the time of O'Keefe and Chovitz, there also was much work involved, done by other divisions, by the Inter American Geodetic Survey, by outside contracts, and by international cooperation; there were no rivalries involved, no duplications, and thus no purpose or interest in padding. The other argument that we heard was more likely: the military should not again be put in a position to choose between an AMS and an ACIC product as in 1958 (and possibly recognize the AMS product as superior), but be presented with only one product. As it was difficult to justify the exclusion of AMS from its traditional obligation, which it had met so well in the past, the next best thing was to submerge its contribution in committee work and take over gradually. Here followed the need for show business to impress the authorities and committee audience with added presentations of details, and there needed to be committee minutes. These had to be circulated as classified documents, amended, approved, retyped, and locked away. here were periodic, time consuming shows with many actors: the principal researchers in the three forces, accompanied by their assistants; then there were the "official members of the Committee", not necessarily identical with the researchers, but maybe their supervisors who did not want to be left out or who watched over political strings. That included also someone (or two) from an even higher management layer, in our case, the CCE. And then there was a representative of the top layer, the DIA. This relatively recent additional management layer assumed near life and death powers over the geodetic activities in the Department of Defense (DoD). To our chagrin, it was made up predominantly of former ACIC people. The distrust went back to our experiences with ACIC visits and liaison, which repeatedly seemed to result in AMS ideas and work being duplicated there without giving credit and references, and it was nurtured by our jealousy of their generally higher grades and greater resources in manpower and money for experimentations. And now we grudgingly had to admire their political talents to get more of their own people into the new almighty top management layer with power to direct, hinder, and also classify our work even without plausible reason. DIA must have been impressed by their own power since they usually appeared in threes (the "troika" as we called them fondly and irreverently), diluting personal responsibility into group responsibility. Looking back, one may see the little acorn from which grew the mighty oak of collective decision making, that in recent years acquired such absurd forms where some high ranking managers are unable or unwilling to make a decision or give a reason for their own actions without hiding behind someone else.

The "troika" at that time probably thought that they were really streamlining rather than complicating things. At a purely personal level, I considered them friends and pleasant to talk to. I remember a DIA plus OCE visit at AMS in December 1965, where I had a chance to appeal to them for help against an apparent attempt by ACIC to withhold from me requested gravity data in the West Pacific, for several months by then. DIA had designated them as the DoD gravity depository, parallel to AMS being responsible for astrogeodetic deflections. Mr. Mills insisted that we fill every request, even though we had reason to believe that ACIC's voluminous requests were not for project but to build up a duplicate depository. The "troikas" promised to look into the matter with a view to changing the procedure if it did not function economically. OCE (Ernie Galleghos) asked me to write a memorandum on it. After another four months' wait I did receive a tabulation of the data but without the requested plot. A month later I learned by chance that the plots had been sent to our Gravity Division instead of to me, and there they had been filed away routinely since there was no message why they had been sent. And the charts were incomplete to boot. It took another two months before the correct charts came. Interestingly, the accompanying note called it a response to a one-month old request, instead of about nine months. By comparison, the Naval Hydrographic Office had filled a similar request in a fraction of that time by way of old-fashioned friendly inter-agency helpfulness.

3. GEOIDAL PIECES IN A PICTURE PUZZLE

Ever since I had insisted on the three-dimensional approach to determining the size and shape of the Earth by using geoidal heights as the basic input (the way I had done successfully in my 1957 Toronto paper), I felt obliged to enlarge the worldwide collection of geoid pieces wherever I could get hold of pertinent information to construct one regional chart after another.

The aim was to fit these pieces together as in a picture puzzle, and forming larger pieces by incorporating adjacent pieces into the same datum through mathematical transformations. In this way, we had extended the European Datum beyond Europe into Asia to Japan, through Turkey and Iran to India and Southeast Asia, and along the 30th meridian into Africa. If one could extend this process further to cover the whole Earth, we would have it made! We could then derive the best fitting ellipsoid as the definitive Earth model.

As it stood in reality, even this unprecedentedly large chunk of the Earth's land masses, representing three continents in a coherent geodetic system, had large pieces missing inside, and its extension through the same technique was stopped at the ocean shores. We had another big piece representing the Western Hemisphere, even though parts of Canada were not covered and the coverage of South America was still at the stage of my 1965 Guatemala report. But there was now a third piece: Australia.

Mr. McCall, in OCE at the time, had periodic business with the Director of Survey, Australian Army, Col. F. D. Buckland, and the Director of the Division of National Mapping, Mr. Bruce P. Lambert, and his assistant Mr. Anthony G. Bomford (son of Brig. Guy Bomford), all in Canberra. He brought back geodetic data as well as requests for comments on their national datum plans. Eventually, they let us have their collection of astrogeodetic deflections along with their permission to use them for constructing a geoid chart of Australia as part of our Figure of the Earth studies. This task was particularly exciting, since no chart for Australia existed yet and we were eager to see what it might look like. It was also interesting from a technical aspect, because we designed a procedure to fit the particular distribution pattern of the Australian data, while applying the ideas from our new construction of the North American geoid chart.

But how to connect these three pieces across the oceans? As in the earlier studies, I would use again the technique of matching these geoid pieces with their counterparts on a gravimetrically derived geoid chart, because the latter, although not as accurate, was supposed to be earth-centered. But now there were also other means: the satellites. Some results were available from so-called geometric satellite techniques (at the Smithsonian Astrophysical Observatory in Cambridge, Mass.) comparable in a way to terrestrial triangulation; and some from so-called dynamical satellite techniques (the Navy's Doppler procedure) comparable in a way to terrestrial gravimetric information. Results from these various techniques were still limited and not always consistent, but it was worth the try. I planned to use each source separately, study the discrepancies and their significance, and come up with a weighted compromise solution.

In the geometric satellite techniques at the Smithsonian and at the Coast and Geodetic Survey, the moving satellite was photographed against the star background from two stations simultaneously, one geodetically known in position and the other to be determined. Each photo established a line of sight between the camera and the satellite with respect to the known star system. Each pair of simultaneous directions from the two stations to the same satellite (see Ch. 8), position established a plane containing the triangle of the two stations and the satellite. As the satellite moved, these triangle planes differed but all intersected in the line between the two fixed stations. Thus, the direction from the known to the unknown station with respect to the star system would be computed. By repeating this procedure, a net of space triangulation was formed, spanning the oceans. As in terrestrial triangulation, the length of the lines in the net was computed from the very precise measurements of one or more base lines.

DIRECTIONS BY SATELLITE

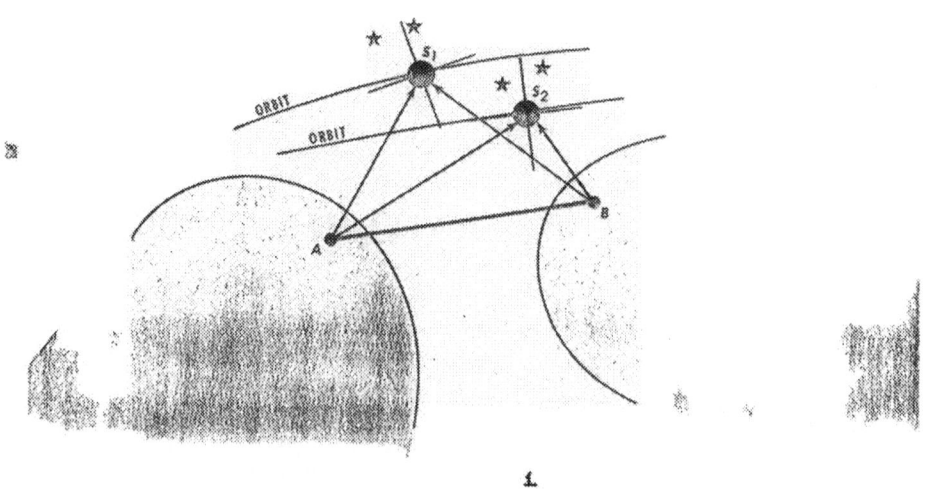

The SECOR system was also a geometric system, but it established distances (not angles) by electronic (not optical) means. Simultaneously measured distances from three known ground stations to the same satellite fixed its position in relation to the stations. If an unknown (fourth) station observed three satellites, each fixed in relation to the original three stations, then its position can be derived in relation to the three satellites and thereby also in relation to the original stations. Repeating the procedure, a space trilateration net was formed, spanning long distances.

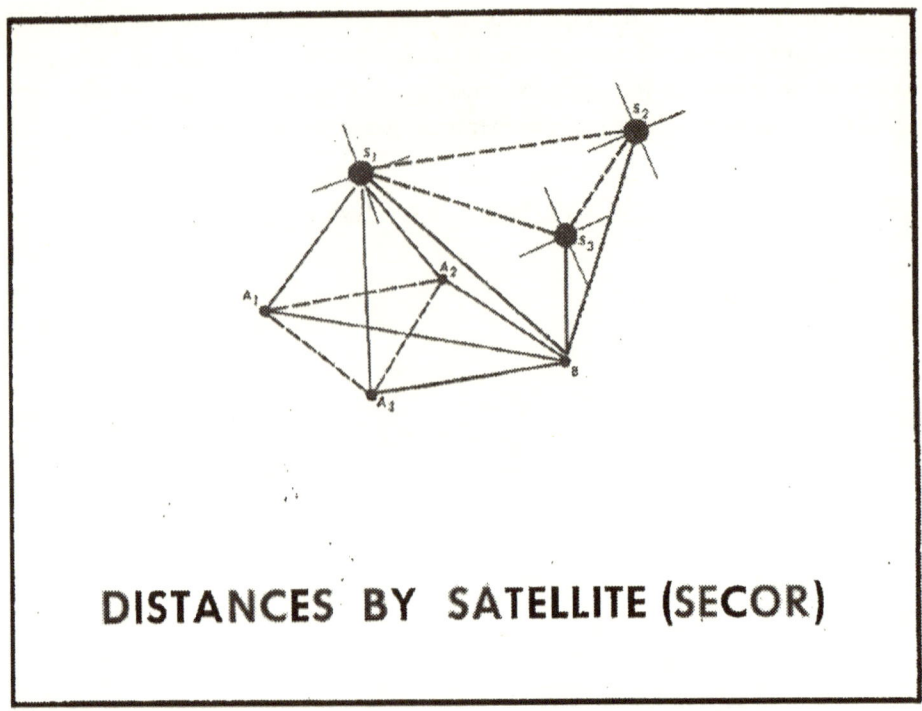

DISTANCES BY SATELLITE (SECOR)

The dynamic systems (of the Navy and also at the Smithsonian) worked differently. The orbit of the satellite is controlled by the gravity field of the earth, and the position of the station which tracks the motion of the satellite can be computed in relation to that orbit and thus in relation to the center of the earth. Thus, stations in different continents are automatically (at least in theory) on the same earth-centered system. Geoid charts from these systems are mathematical extrapolations from the analysis of perturbations at the height of the orbit down to the surface of the earth; they give generalized trends, not details.

My plan was to "ask" each technique that had data available at the time, how it alone would combine my three pieces to a coherent world datum. It turned out that the various satellite solutions were not too far apart, but the gravimetric solutions were different.

Filling the blanks in the South American geoid piece would have to wait for the time being. The success of my South American paper in Guatemala 1965 had opened vistas of getting a load of new data, specifically desirable as data from the sparsely covered Southern hemisphere. Mr. Fuller's passing was a great setback, quite apart from the personal loss. After a while, in January 1966, Mr. Mills found a replacement for our "Americas Division" in Mr. David Byars who turned out to be a very capable and helpful friend. He spoke Spanish fluently and was well acquainted with South American geodesy as well as South American life. He had served with the IAGS and had a charming South American wife, whom I had the pleasure of meeting later on in their beautiful home. It took till August 1966, however, before the PAIGH acknowledged him officially as Mr. Fuller's successor in the chairmanship of the South American Working Group. Col. Diercks kindly sent me a carbon copy of the letter he had received to that effect from General Juan Jose Nano, the Chairman of the Commission on Cartography. I phoned Mr. Byars to congratulate him and I took the letter to Mr. Mills along with PAIGH documents I had just received, which said that $15,000 had been allocated to complete the Argentina-Chile tie at 36° South latitude, referring to my South American paper. We were all very happy that things would go forward at last. It still took some time before Mr. Byars could establish the needed contacts in his new official capacity. In the meantime, his Division worked on an elaborate adjustment of southern Brazil and agreed to gear it to my plans and use one of the datums in my paper as an initial reference; and I started to study the geoid in Uruguay, and also in Central America.

In response to several requests for geodetic tables to aid the widespread use of my "Mercury Datum", the only DoD authorized unclassified geodetic world datum, Mr. Mills suggested to issue a volume of such tables within the well known and much used series of "Department of the Army Technical Manuals", which had identically arranged tables for all reference ellipsoids in practical use. The issuance of such a volume was now of no great expense, thanks to our electronic computer printouts. The Government Printing Office published it with a 1 March 1967 date. The official title referred to the "Fischer 1960 Ellipsoid". When the Mercury Datum had been accepted by the geodetic community in 1960, I had made a strong effort to change the identification in the literature as "the world datum of Irene-Fischer" to the more anonymous "Mercury Datum", but now I felt that the title had its advantages in view of the continuing hassle between AMS and ACIC. A faint memory of a childhood scene also haunted me. I must have shown early interest and capability in mathematics as a child, because I remember my father teasing me that some day I would discover a star and it would be called "Irenaeus". He would have been happy with the "Fischer Ellipsoid" instead; he would have been happy with my entire geodetic career, for that matter. He had passed away before I even entered the AMS service, but I always felt that he was there just the same. His early influence and his insistence on excellence as a moral obligation to do one's best in using God-given capabilities have been a driving force throughout.

The Mercury Datum of 1960 filled a general need for a system unencumbered by classified strings and locks. Laudatory comments by NASA and others were contained in letters to our Commanding Officer who was pleased to pass them on to me with cover letters of commendation of his own. There came a request to republish my world chart on that datum in the French series of popular booklets "Que sais-je?" (H. Dupuy and H. M. Dufour, *La Géodésie*, 1969). The system, however, was now seven years past its derivation and the new satellite results, although clearly not yet definitive, needed to be considered. That is why I planned to fit all the geoid pieces together by means of transoceanic satellite connections of one type only and then of another and then by gravimetric connections only, study the foreseeable discrepancies, and then derive an updated Mercury Datum from a weighted compromise. When my customers in the Navy, Air Force, NASA, and elsewhere asked me about plans for an update, I could reasonably predict an unclassified publication about a modified Mercury Datum after presentation at the IUGG Assembly in Lucerne, September 1967.

All of these many, more or less simultaneous projects, including the perfection of new methods and the priority work on the classified world datum, required a dedicated, hard-working team of people such as my group certainly was. My own excitement and involvement must have been contagious. I used to give much time to new employees, trying to give them a glowing over-all picture of our diverse objectives, then breaking these down into specific activities from which they were invited to choose if possible; or if at that stage they did not show a preference, I would watch their progress and capabilities, and use them where I felt they would do best. This was, of course, a self-serving scheme, because I knew that you get most out of an employee if he likes his work and is capable of doing it. I also tried some psychology that always worked for myself: "Since you agreed to spend eight hours daily of your precious lifetime in this place, you might as well make it worth while. If you invest some interest and effort, you will have fun and a sense of achievement. The more you put in, the more you will get out." Gratefully remembering O'Keefe's and Chovitz's ready availability in my own beginnings, I stressed the same towards my employees. Each one was encouraged to bring his technical or personal problems for informal discussion if he wanted to, at any time at all, even if I was involved in writing or whatever. Sitting down in my visitor's chair next to my desk would give me notice; and if he waited a few seconds to let me finish a sentence or so, he would be assured of my undivided attention.

Our small number ("elite" in quality, as described in Chovitz' time) was enlarged in September and October 1966 when Peter Hintze and Sandra Todd joined respectively. It had been boosted by many temporary helpers in the previous years who either were summer help (college students) or stayed for a relatively short time. Some were hired for the cloistered super-secure unknowns, then called DESPA in Governmentalese (Department of Engineer Special Projects Area), and had to spend the waiting time for their clearances somewhere. I was eager to take them in, even for the usual three months only, because we had so much to do and, with proper arrangement, we could get at least some of the simpler work out of the way. Mr. Mills seemed worried about this steady stream of beginners for DESPA and he wanted me to fill out an application for DESPA clearance with its detailed personal history forms, because "they don't know what they are doing in geodesy over there." I was not very happy about the prospect of working there, as were several other colleagues who did not want to transfer. I suggested that I would not be cleared as I was "foreign born", but Mr. Mills was sure that could be waived. Then I said I would not like to leave my work that was thriving and blooming for something unknown that I would not be allowed to talk about. But Mr. Mills said I could go over there as a consultant now and then, but not full time if I did not want to. So I filled out that long form to please him and then pushed that possibility to the back of my mind. It did not seem to be as important then as it became several years later, when almost everyone was involved there to some degree or another.

In October 1966, Mr. Mills went to the hospital to have "some repair work done". It was not an emergency, but apparently it was as good a time as any to get this out of the way, whatever it was. I saw him clear out his desk and heard him say he might be back in ten days or so, if everything went well. Something did not go well—he never came back. After a few weeks we heard he was going home soon; then we heard that he had a relapse and must stay on in the hospital. Then we heard that things were not going at all as they should, and it gave us a catch in the throat. I used to write him cheerful little letters to the hospital every week or so, to remind him of all the interesting places in the world he might want to visit next year. I remember visiting him in the hospital; Mrs. Mills was there. I had dressed up in bright colors and a big smile, trying to suppress the ache with funny stories from the office, and he smilingly played ball, asking: "and what is Bomford up to?" He was referring to a Bomford letter that Mrs. Mills had sent me to answer for Mr. Mills. So I told him, as if he were coming back to the office, say, next week, that Bomford had written this and that, and that I was going to take care of it meanwhile, and would that be all right?—"Yes, that would be fine." That was the last time I saw him.

He died on 11 December 1966. On the 14th we went to the funeral. The world did not seem the same anymore. Mr. Culley, the Assistant Chief, valiantly tried to carry on, but the odds were stacked against him. He served for almost three years in Mr. Mills place, but was not given the position. When he first started to take hold of the reins, he accepted assistance from the first friendly face that walked in that morning with an offer to help: Mr. Byars. But soon the vultures came. It was sad to hear of a palace revolt of self-appointed crown pretenders who sent a demand to the Executive Office that this interim assistant role must be strictly rotated among more people. The fact that the authorities could not make up their mind about filling Mr. Mills' vacancy for such a long time, did certainly not strengthen Mr. Culley's hand, but created a vicious circle of speculation, intimidation, insecurity and intrigue. Eventually, Mr. Culley gave up and resigned. It was my impression that he was a victim of intrigue. In the end, there were only losers. As far as I and my group were concerned, however, we had a trusted friend in Mr. Culley who helped and backed us where he could; and we needed his help several times in this new world. For a while, however, things seemed to move along as usual, at least at the surface, and we managed to stay out of reach of the turbulence.

4. IAG SYMPOSIUM IN VIENNA, 1967

Dr. Tengström, chairman of the International SSG-16, organized a small symposium on the Figure of the Earth in Vienna, Austria, March 1967, together with Mr. Karl Ledersteger, Österreichische Kommission für Internationale Erdmessung. Mr. Culley decided that he and I should go. Dr. O'Keefe was planning to go also, to continue a discussion with Mr. Ledersteger that had already started at the Prague Symposium in 1964. Both would elaborate on their opposing views about an appropriate Earth model for geophysicists who wanted to study and explain the effects of geophysical forces. Dr. O'Keefe had asked me whether I would prepare a global geoid chart for the Vienna discussion and refer it to the "hydrostatic flattening" which he had championed after analyzing the satellite results a few years earlier (see above, chapter II.4). I quickly wrote (and got cleared) a brief paper "Deviations of the Geoid from an Equilibrium Figure" (*Österreichische Zeitschrift für Vermessungswesen*, Sonderheft 25, 1967). In the last minute, however, Dr. O'Keefe had to cancel his trip. So I added some comments at the meeting in order to connect my paper with the debate on which ellipsoid to use for which purpose.

The one-week trip to Vienna, fortunately with my husband at my side, was like a Rip van Winkle experience. This was the town where both of us had grown up in a circle of relatives and friends, studied, married, and taught school. Was this the same town? It seemed and felt different. There seemed to be two towns on the same spot: the lively town of the Symposium, a beautiful town strangely suggesting that we may have visited here before; and another personal town that was crying with silence, a ghost town. We went to see the places where we used to live, and certain other houses of importance in an earlier life: some looked familiar but also strange with different people about, others did not exist any more, but a plaque told where they had been. A silent world.

At the reception dinner in the Rathauskeller, I was greeted with: "Gnädige Frau, ich habe ein Hühnchen mit Ihnen zu rupfen!'" I had not heard that phrase in decades and its sudden nostalgic emergence stopped me in my tracks. It is a Viennese phrase, picturesque in its imagery. Literally it says: Dear Lady, I have a little hen to pluck with you; and it is the equivalent to the English phrase: I have to pick a bone with you. But there was the additional charm of the grotesque combination of this rustic image with the formal, respectful and courtly appellation: "Gnädige Frau." Dr. Ledersteger, our host, was complaining that at all previous meetings I had pretended not to know German and thus forced him to use his broken English with me, while he had heard recently that I was Viennese. Why did I do that to him? And I had to promise that from now on I would always use German with him—and I did. I then remembered that at a meeting some time back, Professor Carlo Morelli had asked me point-blank where I had received my basic education. I had always evaded previous allusions from various people, but decided to answer Morelli. It was over two decades after the war and I felt secure as an American geodesist. I knew the information would make the international rounds sooner or later.

In the little time that was left free of meetings, my husband and I showed Mr. Culley some of the sights, among them the rebuilt opera house and the Belvedere castle and the dancing Lippizaner horses in a training lesson. In the evenings we went with him to performances at the "Theater an der Wien", the "Josefstadt Theater," a cabaret at the "Kaffee Landmann" and saw "Faust" at the "Burgtheater". The little ornate sandwiches and chocolate arrangements in the windows of little stores were artistic masterpieces as they always had been as I remembered, just as remarkable as those at the official receptions or at the famous and crowded "Dehmel" coffee house. And there were the famous Viennese cakes—but then, I can make those too, with my mother's and grandmother's recipes. We stayed in the "Hotel de France" on the "Schottenring" and marveled at the short distances to everywhere in town, which we remembered as very long distances when we used to live here. Why, in a ten-minute walk you were at the center of town, the "Stephansplatz"! The hotel was near the big "Votivplatz," where as a teenager on the way to school I used to run across, always in a hurry to transfer from one streetcar to another. Today, all the various streetcar lines there connected more safely underground. There were even some little shops underground. The irrepressible Viennese humor and irreverence called this place the "Jonasreindl;" Jonas was the name of the mayor and "Reindl" is a Viennese expression with the typically Austrian diminutive ending, for a round, low pan, a casserole. There was a more elaborate underground shopping area with little restaurants under the "Opernkreuzung," that we had read about but not yet seen. We had lunch there with Mr. Culley, and he said I had been talking English to the waiter and German to him. It was all very unreal to me and confusing.

The Symposium was held at the Viennese Institute of Technology, where I had studied descriptive geometry and projective geometry a million years ago. I still remembered the dismal classrooms and grand but dark staircases, associated forever with tough examinations. It had been an interesting training in pursuing thought structures to their involved consequences. It had also been a sometimes-painful training in perseverance, self-discipline, and perfectionism. I remember the requirement of submitting a number of drawing sheets, each to contain six very difficult theoretical problems of descriptive geometry; these had to be solved, constructed, inked, and lettered flawlessly, with a due date of six weeks. If you then handed it in, proud of having solved every one of the six problems, the instructor might find that the weight of the inked lines was not even enough and he might tear up the whole sheet of six weeks' toil, and you had to start all over again.

And now I came back here as an American guest and participant in an international synposium. The dark staircase and hallways were still there, but Mr. Ledersteger's auditorium was unreal to me in its modern spacious splendor. The symposium focused on topics that would come up in Lucerne such as the question of a new IAG reference ellipsoid, which the Berkeley Assembly had evaded and the Lucerne Assembly should not evade any longer. Dr. Heinz Draheim of Karlsruhe asked me whether I would write an article for the *Allgemeine Vermessungsnachrichten,"* explaining in plain language what the big fuss about a new IAG ellipsoid was all about. There were some more instructions from President Tengström for the Lucerne program, and good coordination with Bursa, the other Secretary. With that and also some talks with a number of participants from all over the world, I returned home to my office, ready to compile my Secretary's Report for Lucerne.

The requested article was supposed to be published in good time before the Lucerne meeting and therefore should be at the journal in May. That did not give me much time, considering the usual two-months clearance requirements. What to do? I appealed to the Technical Liaison Office for advance clearance: "This request is a great compliment to the U.S. Army Map Service, and I accepted. I felt that a refusal because of the two months clearance requirements would make the United States look silly in this particular case." The clearance channels reacted with understanding. These were the days when a Technical Liaison Officer, then Mr. Kephart, dared to make a decision in an obviously clear case without calling a committee or sending the request to the State Department. Impossible today.

With my evenings and weekends thrown in as usual, the paper went out on time, entitled "Do we need a new IAG ellipsoid?" (AVN, August 1967). It was fun to write and it turned out to be a big success. I tried to place the question "to keep or not to keep" the current ellipsoid into the historical perspective of comparable milestones in the long story of geodesy, and to show the changing significance with the changing purpose of the times. For amusement, I included an excerpt describing the haphazard way an international ellipsoid had been chosen the last time around, in Madrid in 1924. Fun also was to tell that geodesy apparently was in the fruit and vegetable business as we read in the literature of comparing the shape of the Earth with that of an orange or grapefruit, pear, and potato. These phrases have become quotes.

With this material assembled, it was easy to respond to two requests for in-house talks: one again for the Position and Pay Management Division in March 1967, and the other within a lecture series for the people in the Computer Services Department, May 1967.

5. MORE INTERNATIONAL INVOLVEMENT AND MORE KUDOS

My duties as IAG Secretary Of Section V were delineated by the Section President, Dr.Tengström, who had already started making plans while on a trip to Honolulu in November 1966, which had also brought him and Mrs. Cajsa Tengström to Washington. We were happy to have them for dinner and reminisce about the pleasant days in Uppsala the year before. Mr. Culley was there too and the Rutscheidts. Mr. Mills was then in the hospital. Dr. and Mrs. O'Keefe, good friends of the Tengströms, were expected to be with us too, but could not make it that evening. Mike, always remembered in Tengström's letters, was in college. I made even an unusual concession to my honored president: remembering the Swedish two to three hour eating sessions with beer orgies in between (an amazing observation for us innocents), I served beer, never done else at our house, but very much encouraged by Rutscheidt. I remember Dr. Tengström explaining to us in Uppsala, that one needed the beer to be able to keep on eating another course and yet another course, and that my inability to eat so much was all my fault because I did not drink.

President Tengström divided the reporting duties between his two Secretaries, Milan Bursa in Prague and me. My assignment said "Investigations concerning the astrogeodetic determination of the geoid on a common datum, combined with gravimetric and satellite interpolations" (a far cry from Mr. Mills's "the Cat"), and that meant summarizing the 1963-1967 activities in all countries on that topic. The presidential addresses, secretary reports, and study group reports are published in the *Travaux* of the IAG, in Paris. For a start, I wrote for information to most of the leading geodesists all over the world who might be in the process of compiling their own national reports, and I rechecked and abstracted the pertinent publications of the last four years. Information and more addresses to write to snowballed, and together with the frequent correspondence with Tengström and Bomford and several others, my IUGG letter file started to bulge. During that correspondence, Tengström and Bomford asked me to chair a new study group to be established in Lucerne. Bomford wanted to close out his SSG-10 which had concentrated on Europe, and see it replaced by one of global scope. After some hesitation, I agreed under the condition that they would let me draft them as members, as an insurance that "nothing could go wrong."

Among the many letters in response to my inquiries of who wanted what included in the Secretary's Report, there was also one from Professor B. M. Jones, Durban, South Africa, who had just completed work on the 30th parallel South latitude. This was welcome news for my world geoid, since geoidal information in Africa was very sparse. Professor Jones kindly acceded to my request for more details of that work for inclusion also into my "picture puzzle" paper.

Mrs. Louise Voelker, then a geodesist at the Francis E. Warren Air Force Base in Wyoming, came by to discuss geoidal heights in a specific area in California that she had worked on. She offered to send us this extra information for inclusion into our new North American geoid map. I enjoyed her occasional visits thereafter and kept in loose touch with her through the years.

Our revised North American geoid chart, in three large map sheets for the convenience of the users, was completed by now and was sent to "Carto" (Cartographic Department, Graphics Section) for finishing. I went to see Charlotte Kubota in the Publications Office under Trunfio, about making a Technical Report from these sheets together with an explaining text, without reducing the useful size of the map sheets. Charlotte was very helpful in designing a re-usable and re-closeable plastic envelope for the package. She also suggested printing the maps in colors and offered to take care of overseeing the production.

The new Australian geoid chart was also in the finishing stage, and Tony Bomford as well as Col. Buckland who visited AMS in July and September 1967 respectively, received advance copies. The paper "A Preliminary Geoid Chart of Australia" with a much-reduced chart was published by the *Australian Surveyor* in December 1967. Mary Slutsky signed as coauthor, and Phil Watt's contribution as programmer was specifically mentioned. Mary Slutsky, Ray Shirley, and Phil Wyatt signed as co-authors with me on the picture puzzle paper for Lucerne.

The voluminous U.S. National Report, 1963-1967, was published in the AGU *Transactions*, June 1967; it contained a brief description of our South American and North American work. It also contained Marvin Marchant's description of his selenodetic control work "AMS-64" and of his subsequent work "Control Integration for Lunar Mapping" where he had incorporated the later NASA and ACIC systems, and a detailed and comprehensive error analysis. Marvin was then at the height of a happy and successful beehive activity. He was nominated for a prestigious civic prize for promising young engineers or scientists, and was celebrated as one among ten finalists in a downtown banquet. We were all very proud of him. It was another "first" for AMS. Marian Hardy's share in the accomplishments was rewarded in a different fashion: she was made Chief of the Astronomy Branch in April 1967.

Mr. Harry Liberman, who had returned from duty in AMSFE (AMS Far East), was made Administrative Officer of the Division at that time. Two new members had joined the Geoid Branch: Denis Popevis in December 1966, and Jane Paul in January 1967. All helped on the various projects, including the classified DoD world datum. Especially, Peter Hintze worked on that and used to accompany me to the three-services WGS-meetings. He was also good company and we regretted his leaving in October 1967 for greener hunting grounds. Sandra Todd, Peter Hintze, and Denis Popevis were given the opportunity to attend a two-weeks geodetic summer school at Ohio State University in June 1967. This was a hectic year indeed. And it was spiced with excitements of a very different type in between. As seen later from the paperwork, I must have been nominated in fall 1966 (that is when Mr. Mills was still around) for the decoration for Exceptional Civilian Service of the Dept of the Army, that is the highest award it can give to its civilians. It was granted in February 1967, and I was immediately nominated by the Secretary of the Army, then Mr. Stanley R. Resor, for the "'Distinguished Civilian Service Award" of the Dept of Defense, again the highest award that it can give to its civilians; and that was granted in May by the Secretary of Defense, then Mr. Robert S. McNamara. It was all very confusing to me, because the informal notifications of the awards, which I did not know even existed, did not come in that logical order. Eventually things fell into place with the two quite ceremonial presentations at the Pentagon in May and June respectively, with lots of pictures. I missed Mr. Mills who undoubtedly must have had a finger in all this. But someone, probably Mr. Culley, had thought of inviting Dr. O'Keefe, and I felt happy and honored that he had come. By coincidence, there also was a childhood friend, Dr. Henry Kalmus, who had come, not for me (since that he would not have known) but for his colleague, B. M. Horton from the Harry Diamond Laboratories of the Army. For the Army Map Service, it was again a happy "first." In the AMS picture booklet about its activities, and in newspaper publicity, it was stressed that I was the first AMS employee and the third woman to receive this honor since the inception of the "DCSA" in 1955. There was a corsage from Mr. Culley and a bouquet of flowers from my Department, another speech by Col. Maberry, and a party for the whole Agency with huge decorated cakes and lots of pictures. A proud husband and son were there too. My daughter was too far away in California and it never occurred to us to have her come.

The people in my team were not forgotten either. During the past year I had written up justifications for monetary awards for several of them and also several promotions. Miss Rosemary McLeod was our job analyzer then and she was very impressed with the high-caliber activities, and made no difficulties with the promotions. On the contrary, it was she who wondered why I had not been getting any promotions myself for over four years by now, despite all these unusual activities and international recognitions, while men all around me were getting promotions, "You know, these men downstairs have no idea what you are doing. And if they had, they would never admit that a woman could do that. You should hear them talk about women. But according to Civil Service standards your work is worth a GS-15. They are using your products and they are proud of the prestige these bring to the Agency. Aren't they? I just laughed that "they" would never let me jump two grades at once. But Miss McLeod insisted that a jump in such a case was not only legal within the special pay scale of a "research geodesist", but that the point and spirit of this pay scale had been evaded in my case, except for the title. She wrote up a "Justification for Additional Upper-Level Grade Positions" for me, which was the first major requirement in a difficult promotion campaign during austerity times. And she evaluated my job according to the "Civil Service Guide Lines of 6/64 for Research Positions" and recommended a GS-15 to make things right; at the same time she asked for an interim GS-14, in case the processing for the higher grade would become involved and delayed.

One day, an administrator walked up to me and volunteered sweetly: "Don't think that you'll ever get that GS-14. There is a man who wants that slot and he warned that he would leave if he did not get it. He must get it, I think, not you." I wondered why this man had told me so much about himself. At least I knew now what he thought. Miss McLeod said: "See what I mean?" An all but forgotten memory flashed through my mind: several years earlier, someone "downstairs" had changed my article about my first North American geoid chart by inserting the names of three men and not mentioning mine at all, so that it read as if my map had been done by one of these men with contributions from the others. Maybe, Miss McLeod had something there?

Our bosses at OCE did approve my GS-14 within record time, to be effective in early July 1967, and they also found a place for the male applicant in another office. In the meantime, Miss McLeod had obtained a GS-15 rating for me from Civil Service. I remember that it had been just before my trip to Vienna when the lengthy application forms for all these purposes had to be filled out and "exhibits of accomplishments" had to be assembled; everything in triplicate, of course. I just did not have the time, what with WGS-committee meetings and reports, award write-ups for several of my people, visitors and requests, passport, Pentagon trip for shots, papers for the Vienna conference and whatnots. But the people in the Position and Pay Management Division closed ranks behind me and did all this for me. I did not even have a secretary at my disposal then. The Commanding Officer, Col. Van Atta, approved the request for my promotion in March 1967, but the Department of Geodesy tied it into a package trying to get a general upgrading for the Department. OCE did not approve that package, and my own case was buried there. Someone told me that it could not be handled separately, ignoring the intent of the "researcher" title that it could and should.

Unfortunately, Miss McLeod left AMS, before she could disentangle my case from the department package. Her successor, Mr. Victor Gensch, tried again and again to overcome the resistance of a key person, identified by then, citing Civil Service standards and the ever growing mountain of evidence of my recognized contributions, but it took five years for the GS-15, and the obstacles were rather ugly at times. Eventually, fair-minded men rallied to my support and prevailed. I was told that when the request finally did leave AMS for OCE, the OCE approval came back to AMS in an unusually short time again. So it was definitely not OCE where an acknowledged merit promotion had been blocked, as someone had tried to tell me.

6. IUGG IN LUCERNE, 1967

Due to the great number of participants at the IUGG Assembly in Switzerland, September 1967, the various Associations were located in different towns: Zürich, Lucerne, Bern, and St. Gallen. The geodesists were in Lucerne. Lucerne is a charming town on the Vierwaldstättersee. The locale for the Association was the new modern "Kantonschule" on the banks of the lake, and it was remarkable that the school activities went on while our Association was there, without any mutual disturbance. During the time of our stay, there was an exhibition of intriguing large plastic sculptures in the entrance hall and on the grassy area in front of the building. The artist, Willi Gutmann, was interested in creating several series of similar geometric forms as cutouts and turning them on the same axis through increasing angles, thus producing three-dimensional, slightly moving figures. The play of light on these differently turned and differently arranged geometric pieces gave the sculptures an increased illusion of movement and an intensified focus on spatial arrangements. A mid-sized prism with a removable cut-out design by this artist had been chosen by our Swiss hosts as a symbol for the Assembly, and miniature replicas were given to each of us as mementos. The same design in green, blue, and white colors was used on a flag and supposed to represent earth, ocean, and atmosphere in the four quadrants of the world (not entirely clear which is which). In a response to the Swiss gesture, the Assembly adopted this design as a permanent symbol for the Union. It appears on the covers of the IUGG Chronicle.

The program of the Sections had been prepared by their presidents in advance by correspondence, but it needed to be verified, adapted, and announced from day to day with the assistance of the Section secretaries. Also, the sessions needed to be covered for record by reporters. I had been a reporter for Section V in previous meetings, and I was asked again. Charles Whitten, member of the Executive Council and head of the American delegation, this time asked to be given detailed reports each evening. So there was plenty to do even before you considered your own mission. Tengström's opening presidential address paid homage to the memory of our great leaders, Vening Meinesz and De Graaff Hunter, who had passed away. It was noted with regret that W. A. Heiskanen (Finland), W. D. Lambert (U.S.A.), and M.S. Molodenskiy (U.S.S.R.) were unable to attend and it was agreed to send a telegram to each.

Then followed the quadrennial reports of the two secretaries (Bursa and myself) and, according to Tengström's plan, I continued with my paper on the "Picture Puzzle" leading to the "Modified Mercury Datum". Then followed the reports of the various Special Study Groups. As planned beforehand by correspondence, Brig. Bomford closed his study group of 16 years, and it was combined with that of Rice's to form a new study group under my chairmanship. A number of individual papers followed, including Levallois' presentation of his gravimetric geoid which he had let me use already for my Modified Mercury Datum. But R. H. Rapp (U.S.A.) had new refined results based on an enlarged collection of gravity data which had not been available to me yet, and which I now planned to incorporate into my work, first thing when I got back home.

The topic of the new IAG ellipsoid turned unexpectedly explosive, and statements were limited to five minutes. Several people opposed any change at all, pointing to the voluminous holdings based on the International Ellipsoid, the enormous cost of changing these, and the danger of the predictable confusions and inconsistencies for the users. Brig. Bomford, with the authority of the IAG president, felt very strongly that we had an obligation to adopt the same ellipsoid that we had helped the International Astronomical Union (IAU) to adopt in 1964. Some others felt that times were moving so fast that results looking good in 1963 and 1964 were obsolete in 1967. My article in the AVN, "Do we need a new IAG ellipsoid?" had been circulating and I was asked to read to the meeting the account of what had happened in Madrid in 1924 in the way of a haphazard vote on a new ellipsoid. The general mood seemed to be a rebellion against the farce of voting on a predecided issue. Eventually, a committee was appointed to draft a compromise resolution and report back for approval. I was asked to be chair, with members Cook, Gaposhkin, Levallois, Marussi, Tengström, and Veis; but Moritz, Mueller, Hristov, and others helped too. The compromise consisted in acknowledging on the one hand that the "best values" at a specific time may rapidly change with new information, may also differ with various authors, and should best be reported in the technical journals; while on the other hand, the Association should represent only major trends and not the oscillations of ongoing research, to avoid confusion and inconsistencies for the users. The significance of a statement by the Association at this time was a recognition that the international Ellipsoid of 1924 did not correspond to today's scientific accuracy requirements and that for these the system adopted by the Astronomers was closer. Moreover, the Association was anxious to avoid an appearance of dictating a new system, and supported the continued use of the International Ellipsoid in the housekeeping chores for the accumulated holdings, but recommended informative instructions for the users.

I personally had no real quarrel with these decisions, since it was the earth radius taken from my Helsinki paper in 1960 that the IAU had picked up and that was being maintained. I would have preferred, of course, to have my more recent, smaller value from the new Modified Mercury Datum adopted, but so did George Veis want his value adopted which was 5m smaller than mine, and others would come up with variations too. In the end, Brig. Bomford's viewpoint made good sense. The IUGG Resolution No.1 defined the new "Geodetic Reference System 1967." Asplund's suggestion of the readily identifying name "Lucerne Ellipsoid" for the corresponding ellipsoid was overruled in favor of the drab and clumsy "Reference Ellipsoid 1967"; its exact numerical flattening together with pertinent formulas was to be computed and published later.

Academician Vladimir Hrstov (Bulgaria) was unhappy about a general preference for the flattening value 1/298.25 to 1/298.3. He had argued (and so had) that two decimals in the reciprocal value of the flattening pretended more accuracy than we had, while a rounding to one decimal would acknowledge that fact. (The official value, published later, had even nine decimals for mathematical consistency!) His main chagrin, however, was the frustration over having for years laboriously computed a whole series of geodetic tables based on that value (rounded to one decimal) and seeing them become obsolete at one stroke. But the sad or happy fact, depending on how you look at it, is that with today's computers you can have such tables computed and printed out in an hour, whereas it used to take years before. Professor Hristov was one of those people who had turned for me by a magic transformation from books and journals into a living person. When I first started out at AMS, I used his publications and long formulas in working for Chovitz. He now turned out to be an elderly, very dignified and pleasant gentleman who was a little lost at the meetings due to his lack of English. He belonged into the era when French had been the world language. He also spoke and wrote German very well. The publications I had used years ago had all been in German. My husband and I now tried to be helpful by translating the English discussions into German for him. When he retired several years later, a Bulgarian committee celebrating his 70th birthday asked me for a contribution to a "Festschrift" in his honor, which I was happy to do with a paper on "The role of Africa in the history of geodetic concepts (Bulgarian Academy of Sciences, 1972).

Another of my missions in Lucerne concerned the interests of the PAIGH South American Working Group. Mr. Byars, as chairman, had been trying for several frustrating months of correspondence to get data from Argentina and to set up a meeting of the Working Group in Rio de Janeiro in November, but the response had been slow. Even an official request for data from the Secretary General of PAIGH to the Director of the Argentinean IGM (Instituto Geografico Militar) was still waiting for an answer. Both the Argentinean member, Ing. Pablo Dragan, and the Brazilian member, Ing. Rene de Mattos, were expected to be in Lucerne, and Byars, who was not there, had asked me to make contact with each one, and try to confirm the date of the meeting. Ing. De Mattos and Mr. Lysandro V. Rodriquez were looking for me too; and we had a pleasant talk about our common interests, and about their troubles with getting sufficient support for their work from the officials and the various organizations involved. It sounded vaguely familiar. They were very cooperative in agreeing to Mr. Byars' date for the meeting. They would invite some of the troublesome officials, and could I not, maybe, work some magic on them to get their support? They thought I could. They also wanted some detailed information about my new method of computing gravimetric deflections by electronic computer, and I offered to bring some reprints to the meeting in Brazil as a contribution. I also met Ing. D. Ferrari, the head of the Brazilian delegation.

So far so good; but now to find Ing. Dragan. I had sent word to him but did not get an answer. Since I did not know him, I needed someone to point him out to me and eventually I did find him. It turned out that he did not speak English, which made any attempt at conversation rather limited, since my Spanish was practically just guesswork. But his handshake was very friendly, and eventually there was some communication about the date of the meeting in Brazil and he agreed. There were a few other Argentineans who spoke English, e.g. Prof. E. E. Baglietto and Prof. A. A. Cerrato from the University of Buenos Aires. Another one asked to talk to me at some length about details in my papers on gravimetric interpolation of deflections, a topic he was involved in. In the course of a pleasant and lengthy conversation, I learned of an explanation why we did not have any success in getting Argentinean data, even though there was no outright refusal. The Argentineans took pride in doing the job themselves, he said. They needed help theoretically as well as computationally, but they did not want to let the data go out so that we or anyone else would do their national work. The officials would not say "no", but the data would just not go out. Here I could see a way out of an impasse. I felt sure enough of the backing from my office to propose tentatively a training period at AMS with the advantage of using our computer facilities. We would be glad to help with instruction and discussions without taking credit for their own work. And I promised to pursue this matter at home.

In another mission requested by Rutscheidt, I contacted the chairmen of special study groups on astronomical and satellite topics, which he wanted to join, and picked up copies of papers that he would be interested in. He also wanted me to discuss with Asplund a specific job on worldwide holdings of deflection values, which had been lying around since Henriksen's time, apparently as an AMS commitment to IAG, Section I, Asplund's Section. It turned out now that requirements had changed, so that we could drop that commitment. I also saw to it that Rutscheidt's abstract on a future paper about SECOR results received wide circulation.

The efforts of the first Marine Geodesy Symposium in 1966 to get international attention and a forum at the IAG bore fruit in Lucerne. A report on Marine Geodesy was given within Section I, and a new Special Study Group No. I-25 under George Mourad's chairmanship was established. Also, another future Symposium on Marine Geodesy was approved for IAG sponsorship.

These had been a very busy two weeks, with talks to colleagues in the little free time left. Claude Gilchrist from DIA stayed at the same hotel with us and there was a little time for a chat at the breakfast table. He was organizing, back home in Washington, a series of interagency meetings on satellite geodesy and he invited me to come. I gladly accepted the invitation, provided I would get permission from my office (an almost rhetorical qualification—or so I thought). Surprisingly, I did not get that permission beyond the first meeting, despite Gilchrist's repeated invitation. Why not? It was the first of many later instances where the "natural" Do as much as you can" was replaced by "don't do this, don't learn that, don't follow things up". Such curious attempts to stifle growth made you wary and wonder. In this particular case, I was told explicitly not to involve myself or my branch in satellite geodesy, and not to attend these informative meetings.

My first duty as the chairman of a new Special Study Group was, of course, to get members. I confirmed the acceptance of membership by Bomford, Rice, and Tengström, and proceeded to invite Bursa, A. R. Robbins (Oxford), H. Wolf (Bonn), and several others, and felt good about their ready acceptance.

On the flight home, the customs officer in New York looked at my official travel orders, which were conspicuously displayed in each of my opened bags. "You still work for that old DoD?" he asked. "I used to work or them too, some time ago. But I can tell you, lady, I am making a lot more money here."

9

ON THE GO FOR AMS
(mid 1967–1968)

1. REFULELING

Before tackling the stack of accumulated mail on my office desk upon returning from Lucerne, I made contact as usual with each person in my branch and others interested in order to get back into the mainstream of things as quickly as possible. In the required Travel Report I told, among other things, of the conversation with the Argentinean geodesist explaining our failure to get Argentinean data and discussing a way out by inviting Argentineans to process their own work at AMS; and I suggested action in that direction. Mr. Byars said there was a good possibility to do just that and he eventually succeeded in bringing two Argentineans to AMS for a stay of several months. Mr. Byars also visited Mr. Hough, our first division chief, now retired in Woodstock, Va., for some thoughts or advice on international cooperation, since Mr. Hough had so much experience and success in the comparable undertaking of the European adjustment. Mr. Hough was very interested in our work. He sent me a cordial note about my activities and these comments: "…It seems a pity that after the more than twenty years of PAIGH we still cannot get the Argentine control with connections to Chile and Brazil triangulation. I believe that the military nature of AMS is what holds them back. It seems to me that if you could introduce the PAIGH into this problem, as we did the IAG in connection with the European data, and have them stress the solely scientific value of the first-order control, that they could make it palatable to the Argentineans…." And this advice worked. Much later when the two Argentineans came to do their work at AMS, they insisted on formally handing over their data, not to AMS, but to me personally for the scientific purposes of the Figure of the Earth and the South American Datum within the PAIGH framework. Obviously, they saw my role in the same light as not only Mr. Hough, but also Dr. O'Keefe, Mr. Mills, and Mr. Culley had described my job to me and its usefulness for my agency: to represent the scientific interests of AMS on the international scene and act as a leader and good-will ambassador. From the same viewpoint I was encouraged in general to present papers at meetings, probably with the dual purpose of stressing AMS continuous presence in the scientific limelight and also keeping me constantly on my toes. Thus, after the trips and papers for Vienna and Lucerne, I found myself refueling for Rio de Janeiro (November 1967), Edmonton, Alberta, Canada (January 1968), the Annual AGU meeting (April 1968), and West Point, N.Y. (June 1968). An invitation to speak at the Annual Convention of the Canadian Institute of Surveying had come already in May 1967 in a long cordial letter by Professor Walter L. Bigg, the program chairman, who kindly remembered the success of my presentation at Dahlonega, Ga., several years back. As usual, I had sent the letter through channels to the higher echelons asking what I should answer. The routing slip came back saying: Suggest she go," and I submitted an abstract for clearance. My participation at the 1968 Army Science Conference at West Point, N.Y., had been suggested by Mr. Albert Nowicki, AMS Technical Advisor, in September 1967, and several weeks later came the acceptance by the Conference Advisory Group for one of the proposed topics. At the AGU meeting I intended to report on the Modified Mercury Datum and the Australian geoid study, and would send abstracts through clearance channels to Chovitz, the program chairman (then at the Coast and Geodetic Survey), when the time came.

First priority, however, went to the implications of the Lucerne activities. After placing the proposal about breaking the Argentinean impasse into Dave Byars' capable hands, I wrote several personal letters to geodetic leaders all over the world to invite them to membership in my new International Special Study Group No. V-29 on astrogeodetic deflections, and was pleased about their ready, cordial acceptance. I tried to make it a truly international group and very soon there were about twenty members, from 13 countries. Next in urgency was the computation and formulation of an "Addendum" to the "Modified Mercury Datum 1967", presented in Lucerne within the "Picture Puzzle", to include the new data in R. H. Rapp's Lucerne paper. Although these did not make much difference in the final results, I wanted to have a truly up-to-date product for my customers at the Air Force, Navy, NASA, and elsewhere. The amended version would be called "Modified Mercury Datum 1968 (MMD 68)" to avoid confusion. In the meantime, Major Mike Mitchell of Patrick Air Force Base, Florida, asked us to compute certain accuracy estimates and transformation formulas for the Eastern Test Range. It was a pleasure to work for Major Mitchell on any of his requests, particularly since he knew how to spice them with flattery. When asked why he did not take his business to ACIC with the Air Force, instead of to us in the Army, he countered "Don't you people know that there are different Air Forces, with not much contact between them? It would take me forever to get what I need from them, but I get always good and prompt service from you." Whether true or not, the flattery worked.

Also George Weston from the Naval Hydrographic Office had us compute specific transformation formulas for his needs. Both he and Major Mitchell intended to adopt MMD 68 as an update of the 1960 Mercury Datum that they were using now on DoD orders, and both felt that using improved coordinates was within their authorization and obligation to provide good service. Our contact with NASA at that time were Jerry Rosenberg and the NASA contractors at Geonautics Inc. (Mr. T. Gunther and Mr. Claveleaux), who also waited to use MMD 68. They published the NASA directory of satellite tracking station coordinates on the various datums, including the Mercury Datum, and wanted the improved coordinates for the next edition.

Urgent requests for the new geoid charts of North America came in from TRW Systems, California, from NASA and others. The charts had been finished beautifully by "Carto," but the Technical Report (No.62) with the explaining text and the useful plastic envelope for the whole package was overdue by four months. We had to send out the maps alone and wondered what had happened to the previous efficiency of the Publications Office. There were other disquieting signs of changes: it took three months to assign the required job number for starting any productions; pre-agreed formats were trivially changed and thus caused time-consuming retyping. In such silly cases, my appeal to the boss, then Mr. Trunfio, always brought relief. Also, there were then Mrs. Ruth Weisinger and Mr. Joe Love in that office, who helped to straighten things out, but their authority in taking action was limited. Mrs. Weisinger had helped me efficiently and pleasantly to get the Lucerne papers out on time; but now she was trying to switch to another office more suitable for her talents. Much more disturbing were a number of exasperating experiences of some colleagues. Marian Hardy, for example, had had a paper at a conference in Athens, Greece, but when she saw the finished handout, the title as well as some crucial paragraphs of the text had been altered without her knowledge; a so-called "editing" had changed the meaning disastrously. There also was resistance to publishing a Technical Report on results of the Star Catalogue Project, which should record our work for the benefit of users. When more and more colleagues had upsetting experiences, Mr. Rutscheidt and Mr. Culley planned a formal complaint through channels. Marian was asked to collect case histories in support of the complaint and formulate a solution. She made the point that the acknowledged purpose of the TR-series was the quick dissemination of newly derived results, as well as a record of our scientific works. Either the Publications Office would honor our due dates as well as the integrity of our technical writing, or we should be given permission to handle our own publications which we had done more economically some years ago. At the occasion of a DIA seminar around this time, the support of Mr. McCall, OCE, and Mr. C. Frey, DIA, for these objectives was informally secured, under the condition that our department would officially invite these higher echelons to step in. But after Marian had spent much time and energy on that task, Rutscheidt decided against a formal complaint. An informal complaint did not much good, however, and case histories of poor service kept piling up. In order to avoid such interference with my work, I decided not to write any TRs any more and to insist on "no editing" of my conference papers. In fact, my articles in the technical journals had much more publicity than AMS TRs. To make it even more grotesque, the Publications Office managed to place a stamp on the cover of the TRs saying that these could be given out only with the explicit permission of that Office. Since it had no jurisdiction over reprints of my journal articles, I really did not need those TRs, which were rendered practically useless by that restriction. I cancelled a several months old request for producing a TR about my gravimetric interpolation work, and made a conference paper out of it for the PAIGH meeting in Rio de Janeiro in November 1967, the paper that Dr. da Mattos had requested. It was published subsequently in the *Revista Cartografica* in Buenos Aires.

Mr. Bill Lane, in the front office of our Department, was in charge of collecting short news items for the DIA in-house information bulletin, the so-called *DIA Notices*. Upon his request, I contributed an item on the IUGG Assembly in Lucerne, the new Earth model replacing the International Ellipsoid of 1924, and the "Modified Mercury Datum 1968". These *DIA Notices* changed later to the *DIA Digest* and, after the dissolution of DIA, to *Items of Interest*. People were asked to contribute newsworthy items of activities now and then. Rutscheidt also asked me for a brief write-up on Army achievements relating to the classified WGS of 1966 and the new unclassified MMD 68.

Personnel changes in fall 1967 included the change of command from Col. W. H. Van Atta to Col. D. B. Conard. In the Geoid Branch there were some changes too (besides the yearly coming and going of college students for summer help). Restless Peter Hintze was wondering about his life's career and felt he needed a longer vacation to think about it and then to try something else. Mary Slutsky had wondered already a few years earlier about the craziness of our hurried, competitive rat race, while she would have preferred a quiet working place without excitements. At one time, she applied to another agency for a routine job, and asked me for a letter of recommendation. I could see her viewpoint and wrote a good letter, but I was glad when she decided to stay after all. But now the pressure of the work days had become much worse. Besides the quick pace of our activities, there were now several talented people in the team who thrived on the quick pace as much as I did, and there also was the increasing impact of computation by electronic computer versus her accustomed hand computing with desk calculators. I am afraid that I added much to the pressures on her in my desire to protect her special status of authority and seniority, by demanding that she be quicker and better than the relative newcomers. Her physical condition apparently was not too good either and she decided to apply for a position in the more routine oriented Datum Branch under Jim Walker. I had a present with a cordial good-bye note for my close assistant of over ten years and I received a nice thank-you note from her. Unfortunately, she could not evade programming jobs and a certain degree of pressures there. I was told of her comments there that in the Geoid Branch at least one got ready recognition for one's efforts.

Mary's switch to the Datum Branch in December 1967 overlapped with a switch in the opposite direction: Foster Walker came from there to us at the same time. I knew Foster well from my German classes where he was quite conspicuous as a very gifted student. He was now a high-grade mathematician and very welcome indeed. A couple of weeks later, Pedro Antonio Santiago came as a replacement for Peter Hintze. Pedro had applied for a job with the Navy at the same time but it had not come through yet. His heart was set at the blue yonder, for a long-range research mission on a ship, both because he loved the ocean and because it was a way to get rich: staying on the ship for months with nowhere to spend his money. He stayed with us about half a year before his dream job came through. In the meantime Louis Jones joined us.

2. WORK GROUP SESSION IN RIO DE JANEIRO, NOVEBER 1967

The first reunion of the Working Group on the South American Datum convened in Rio de Janeiro, November 23-25, 1967, the Thanksgiving weekend. My husband came along, of course. Byars and I left directly from work on Tuesday to get the overnight flight from New York to Rio and my husband would join us in New York, coming from Baltimore where he taught college classes. He had resigned from Government service a couple of years earlier in favor of a college teaching position which, as a dyed-in-the-wool teacher, he could not resist accepting. Probably due to the Thanksgiving holiday, flights were crowded and delayed, and we had trouble compounded with the anxiety of getting to New York and changing to the other airline in time for the only flight, but in the end we made it or, maybe, the Rio plane waited. In the morning we were met at the airport by our Brazilian hosts and there was a grand welcome luncheon for us. General Lima, head of the Instituto Brasileiro de Geografia e Estatistica (IBGE) later gave a dinner party in a beautiful garden setting. Byars opened the working sessions with an excellent speech in Spanish, recalling the mandate to the Working Group, outlining the topics for discussion, stressing the scientific character of the task, and asking for formulation of recommendations for specific actions. He concluded with a quote from Mr. Hough's address to the IAG in 1951 concerning a comparable task, "The development of the European Datum and the subsequent European adjustment of the fundamental geodetic network was possible only through the enthusiastic effort and good will of the geodesists from many free nations working together in harmony and in complete confidence in the good intentions of their colleagues." The task before us had been formulated in Guatemala in 1965 as a mandate consisting of two objectives and three guidelines. The two objectives were: (1) to choose a datum or datum point in the most convenient form, and (2) to recommend a coherent system. The three guidelines were: (1) to compile existing information to continue the study by Mrs. Fischer and Miss Slutsky, (2) to amplify it, and (3) to present results in 1969. Besides the four members of the Working Group, Byars, de Mattos, Dragan, and myself, there were a few observers present: Ing. D. Ferrari, Director of the Geodesy Department at the IBGE where the meeting took place, and Mr. Norman Fassett and Mr. Felix Rabito of IAGS. At first, everything was translated into English for me. But after a while, I seemed to get the hang of things even before translation, considering that I knew the topic and relied on Latin and French to help me guess. Engulfed by Spanish and Portuguese sounds which made increasingly more sense to me, I also watched another remarkable transformation take place: the small group of strangers locked into intensive work for three full days, with lunch bites brought in to minimize time out for a lunch break, intent on completing a historic work coordination, turned into good friends.

In the end, we signed a document of nine Recommendations and two Resolutions. It was recommended that the PAIGH consider the time elapsed without progress or hope of progress since the plan for a datum point was established at the PAIGH Meeting in 1952, and therefore proceed on the basis of existing facts to determine a datum. It was a little masterpiece of diplomatic language, permitting us to switch from the impassable gravimetric approach of finding a suitable datum point, which was on the books, to the practical approach of deriving an overall datum from the existing triangulation, without officially changing the books. It was needed for the legal acceptance of my new and different approach to the whole project. The other recommendations dealt with the specific geodetic ties across borders, additional fieldwork including especially more astrogeodetic stations, and the exchange of data. The latter (Recommendation No.4) was of particular importance with respect to the release of Argentinean data. It tasked the Chairman to establish contact with the authorities of PAIGH and the states to set up effective means for such an exchange. Since Byars intended to go to Argentina for just that purpose, this Recommendation No.4 was meant to give him utmost support. Of the two Resolutions, one was an expression of honoring the memory of Homer Fuller, the originally appointed chairman, and the other gave thanks to the hosts.

Byars had brought along the new adjustment of the geodetic control in southern Brazil (Rio Grande do Sul) which AMS had just finished by binational agreement with Brazil. While the transmittal of the finished product at this session was coincidental, its obvious interest for the Working Group prompted permission to show it to the Group. So Byars discussed the details of this work and the role of the electronic computer which made a simultaneous adjustment of such a huge area possible. The interest shown by the Argentinean member was unmistakable.

During the three days of sessions, my husband saw the sights of Rio, but I did not have a chance other than seeing the beautiful setting of the Sugar Loaf mountain in the bay from the distance. My new friends took pity on me and arranged for a tour into the surroundings of Rio on Sunday and a two-day stay in Brasilia on the way home. A car and driver for all of Sunday was placed at our service by the generosity of the IBGE, and we drove along the shore to Copacabana and further, and into the mountains to the view from the Corcovado and the tropical wonders of the Tijuca Forest. On the way, we also saw the other side of beautiful Rio the unbelievably miserable favellas next to affluent housing.

After a short flight to Brasilia Monday morning, we were met at the airport by Mr. Pericles, surveyor, and Col. Lelio Graja, who kindly showed us this remarkable city. Brasilia was then only about seven years old and unfinished, but the starkly modern buildings were impressive and exhilarating. We liked it all very much: the artificial lake with the Palacio da Alvorada and the Supreme Court Building, the grand layout of the Government Center, the Cathedral sunk into the ground and its unique, widely visible superstructure, the wide avenues, the modern apartment buildings, and the bold architectural features and sculptures all over. People either liked it or disliked it; one could not be indifferent. We were happy and grateful to have had a chance to see it. But even this attempt to build a new clean city in virgin country did not succeed in preventing slums. At the outskirts there already was a lively, crowded shanty town. An attempt was made to replace it by an area of very small clean houses for low rent, but the pace of these constructions was far too slow. When we left for home and waited for our plane, people approached us with a request to take some letters along for their relatives or friends in the United States and mail them there. We saw other passengers being approached the same way. Why? They did not trust their own mail service.

Byars had gone from Rio to Buenos Aires to see what he could do to get the data released from the Argentinean Instituto Geografico Militar (IGM). This visit had been suggested in the official request for the data by the PAIGH General Secretary to the Director of IGM in June, which had been answered only in November saying that compiling the data would cost extra money, but Mr. Byars was welcome to visit. The Working Group's Recommendation No.4 should add some weight to Byars' mission; and even more weight and urgency might be added, so it had been suggested in Rio, if Byars would continue on to Chile and report directly to General Tomas Opaso Santander, President of the Commission on Geodesy. After all, the Recommendations were addressed to this president for action, since we were a Working Group under this Commission. Moreover, in the particular case of the Argentina-Chile tie, Recommendation No.4 obliged Byars to deal directly with the authorities of the countries involved. Since this was a new angle and not contained in Byars' travel orders, I would ask upon my return to AMS that his orders be amended for the detour via Chile, and that the authorization be sent immediately to Buenos Aires.

At the IGM, Byars pointed to the short time left until the 1969 PAIGH Assembly when the Working Group should report results, and mentioned the possibility of speeding things up with the large electronic computer in Washington, where one or maybe two Argentinean technicians could participate in the procedures. Eventually, a "Memorandum of Understanding" was formulated saying that within about six months two Argentineans would hand carry duplicates of the Argentinean data to Washington, be trained there in utilizing the electronic computer and participate in all phases of processing their data. The turning point in the discussions had come with that offer of training two people at AMS in the witchery of a large computer and programming skills for future Argentinean needs. Byars then proceeded to Chile to brief General Santander who concurred with the recommendations of the Working Group and so notified each of the member states. General Juan Jose Nano, President of the PAIGH Commission on Cartography, to which the Commission on Geodesy belonged, asked Byars to prepare a report on the Work Group activities for the PAIGH Executive Council Meeting in August 1968. A report went also to Dr. Arch Gerlach, the Chairman of the U.S. National Section in the PAIGH. As I was returning home earlier than Byars, he had asked me not only to take care of his travel orders, but to brief Mr. Culley and Col. Conard and bring up the point of financial support for an Argentinean trainee (augmented to two trainees in the final agreement at Buenos Aires), that is, travel costs and per diem, in exchange for their data. I sent a note and the whole draft of the Rio minutes to Col. Conard and briefed Mr. Culley in detail. Mr. Rutscheidt was curiously not interested, but Maria Rutscheidt, his wife, was and so I briefed her. She was the geodesy representative for Latin America in the library and quite knowledgeable. Bill Lane wanted a brief account of the Rio meeting for the *DIA Notices.*

During the following months, the correspondence between the members of the Working Group dealt with the technical details of practical work. I asked de Mattos and Dragan to give first priority to new astrogeodetic observations to be made on their respective portion of a route between the Brazilian and Argentinean national datum points, so that I could construct a reliable geoidal profile between them; and I received a positive response from each. Byars arranged for support with additional astronomical observers, equipment, and funds through IAGS. General Santander was kept informed. It was a marvelous experience of international cooperation in getting a specific piece of work done in time for the due date in 1969.

3. MANAGERIAL POLITICKING?

The paper "From Pythagoras to the Modified Mercury Datum 1968" for my next assignment, Edmonton, Canada, at the end of January 1969, had been cleared and was in the good hands of Joe Love of the Publications Office for producing handouts. Joe Love was given to philosophical musings, mostly about the hopelessness of this world and Government Service in particular, since he did not see any chance for advancement nor for interesting work. Naturally, that got me up in arms, preaching optimism and "never say die." Obviously, we did not find a solution to the world's ills.

There were long-distance calls about details of MD 68 from Mike Mitchell and from his contractor, Pan American World Airways, at Patrick Air Force Base; also calls and requests for specific MMD 68 coordinates from the Navy's Pacific

Missile Range, Point Mugu, California. All these kept us jumping, but we were happy to have something useful to supply to the various units of the Defense Department and also to NASA. Then something odd happened: Bernie Chovitz (Coast and Geodetic Survey) called to tell me that DIA and OCE people had visited his office and seen there the program which he put together for the next AGU meeting. They had seen my abstract on the Modified Mercury Datum and acted surprised, because, so they said, they were negotiating with NASA to develop a modified Mercury Datum for them. I could not believe it. The DIA people knew what I had, they had cleared my papers, they had been in Lucerne listening to it. Well, Chovitz said, he just wanted to let me know that something must be going on; these people were quite upset.

This was on Thursday, January 4. On Monday morning, the DIA troika and someone from OCE appeared at AMS to discuss my MMD 68 with Rutscheidt. I was not permitted to attend this discussion, although (or because?) I had all the facts at my fingertips. This in itself made the whole thing very suspicious. It turned out that indeed they were negotiating with NASA to develop an unclassified world datum as a combined effort of NASA and "GIMRADA", leaving AMS out, and ignoring the fact that I already had such a product and that customers considered it good and useful. Apparently, the presentation of the finished product at the AGU was embarrassing under the circumstances, although it had already been distributed internationally and to interested parties nationally. They must have seen the absurdity, since a face-saving compromise was reached: DIA would "sponsor' (say "bless"?) my MMD 68 if the WGS-committee sponsored it and if NASA wanted it. This sounded rather odd, since NASA had been asking me for such an update, and since the WGS-committee's task was the three-service derivation of the classified world datum under the chairmanship of ACIC, and should not have any jurisdiction over AMS' long-standing Figure of the Earth studies. But you don't argue with City Hall. It now looked as if, besides the curious deal with NASA, there was another motive: annoyance that it had not occurred to AMS to ask DIA's explicit blessing for doing good work in an authorized project although that blessing was implicitly given by clearing my papers. Rutscheidt and I called Lou Decker, ACIC, St. Louis, by phone for his blessing and he readily agreed. Lou had been in Lucerne, had listened to my paper, had a copy, and probably wondered why we needed his, or rather the Committee's, blessing. We called Jerry Rosenberg at NASA and he said that the negotiated project with DoD was something quite different: to find a way of checking the influence of observations immediately while the operation was still in location, and this did not affect MMD 68.

We thought we had met DIA's conditions, but there was more to come. The following Monday, DIA asked that the name of my work (which, they said, sounded too similar to that of their own plans for this year) should be changed from the by now well-known Modified Mercury Datum 1968" to "a modification of the Mercury Datum (Fischer 1968)" to identify it as an individual effort in contrast to their own multi-agency plan. This affected the already mass-produced handouts for Canada, the AGU abstract, and the *Bulletin Géodésique* where my Lucerne paper plus the "Amendment" was being printed; notwithstanding that all these papers had been cleared already through channels. How childish can you get? We were not sure whether to laugh or get angry or both, but we complied and thought that somebody's idea of authority had gone awry. They even asked me for a carbon of my letter to the *Bulletin Géodésique* to see whether I really had written it. All the affected papers had to be cleared a second time, but that was done in a jiffy, the Canadian paper even by phone. Col. Conard, our Commanding Officer, took a courageous stand against such high-handed treatment of AMS. He first asked me through Rutscheidt to summarize all the pertinent technical facts for his information. The main facts were (1) that the repeated requests of our customers and even the curious DIA deal in itself proved that an unclassified Modified Mercury datum was needed at this time, (2) that my work was a natural phase in the AMS Figure of the Earth studies, continuously authorized since more than a dozen years (before DIA even existed), and (3) another such product this year by another agency could not possibly be significantly better, because AMS held all the pertinent data, and these were included in my MMD68; a duplication would thus constitute a clear misuse of government funds. Col. Conard's assessment of the situation was spelled out in a note to me "You just beat them to the punch with a timely finished product." He then asked me to draft a letter to DIA for his signature protesting the DIA plan for improving the Mercury Datum of AMS through another agency, and officially submitting my work as the finished product. It occurred to me that the timeliness of my products in the past had repeatedly been made a point of appreciation and awards, which fact made DIA's reaction even more strange. At the conference with Col. Conard, Mr. Culley wanted to cheer me up and wrote on the blackboard [a quote from Mark Twain]:

> "A quitter never wins.
> A winner never quits.
> It is not the size of the dog that wins the fights
> It is the size of the fight that is in the dog."

On the same day, Lou Decker, ACIC, called back saying that he had phoned DIA about our agreement, but had been directed to call a formal meeting of the WGS-committee on the subject; since the Navy member was out of town, this could be done only in February.

Next, DIA wanted references to show that the original Mercury Datum had really been AMS work. Now this was easy to do, since there was enough in the literature about it but it made you wonder. Did the geodesists at DIA not know the geodetic literature, or did they just take up our time while they were casting about for ideas of what to do next in their mischievous, self-created predicament? In the correspondence between DIA and Col. Conard, however, Army Col. Lou Knipling, then at DIA, explicitly praised the quality of my product as "excellent and of high professional caliber." So it was made clear that they did not mean to attack my work as such or me personally, but there were other considerations. At least, this statement assuaged somewhat my personal feeling of hurt and betrayal, and made us all see more clearly what it was: a power struggle in managerial ascendancy.

At this stage, I went off to Edmonton, Canada. My husband came along, of course. Al Nowicki, our Technical Advisor, went too. It was January 1968 and it was cold, colder than usual, and coldest around that time. The weather report from Edmonton made you shiver in advance, as temperatures below zero Fahrenheit were quoted. I packed almost all the warm clothing I owned, to put them on in layers if needed. In Edmonton, we had beautiful sunshine with snow on the streets. At noon, the thermometer said 25 degrees Celsius below freezing, that is 13 degrees Fahrenheit below zero, but I could not believe it. It did not feel that cold as there was no wind. Women did not bother to wear boots or kerchiefs and I did not either. However, each time one used the key to open the door of the hotel room, sparks were flying and one felt an electric shock. I had not noticed such an effect of the cold before, but after that I noticed it back home too in winter when one glides over the front seat in a cold car to open the other door from the inside. When we visited the University and saw the peaceful students in the game rooms or library, our guide commented: "We have no student demonstrations here. Know why? Very simple: it is much too cold outside to find any number of students willing to join a demonstration."

At the Convention of about 800 participants, there were only two U.S. papers besides mine. Nowicki, however, was invited to talk to the graduate students and faculty of the Geography Department on lunar mapping. A lady-geodesist was news to the Edmontoners. I was interviewed by the local newspaper, the local TV-station and the Canadian Federal station. My reluctance to agree to these interviews was overcome by the plea of the publicity chairman to please help him publicize the Convention which took place outside Ottawa for the first time. "And you know, you are somewhat of a newsworthy oddity in a man's world. The larger community wants to see what you look like, maybe a creature with two heads or something like that." So I responded in the spirit of fun. In the interviews, I tried to give the Edmonton ladies a lift, talking about the rather obvious need of a good education for girls in today's world to equip them to earn their living. What about marriage and motherhood? Why, of course, but consider that the child-raising years are just a fraction of a woman's life span, and that today's financial security, if any, may be gone tomorrow.

Back in my office, Bill Lane wanted a note about the Convention for the *DIA Digest*; and there was a lengthy write-up in our little office newspaper, *The AMS Reference Point*, published by the Technical Liaison Office, by Mr. R. G. I. (Bob) Waite as editor. Soon there came a nice thank-you letter from Edmonton, and then requests by participants for more copies of the paper to pass on to colleagues. Some geography departments were interested in the MMD 68 world geoid chart and its correlation to the postglacial uplift. A copy must have found its way even to Spain: several months later, our library found a Spanish translation in the *Buletin de inforacion* Num.6, Servicio Geografico de Ejercito, Madrid 1969, by Capitan de Artillerla, Geodesta, D. Josh Antonio Puerta Navrro. *The Military Engineer* published a version of the Edmonton paper in the May 1969 issue.

The unpleasant incident involving about MMD 68 seemed closed. Major Mitchell's periodic publication about the Eastern Test Range, and also the latest *DIA Digest,* contained technical information about the MMD 68. A beautiful letter by Mr. W. P. Varson, (Chief of Manned Flight Planning and Analysis Division) acknowledged the usefulness of our Mercury Datum, commended our cooperation and direct assistance, and requested to be kept informed about all further refinements. This sounded like a renewed request for MMD 68 details. A delighted cover note from Col. Conard to me said: "well deserved and we'll win out in the end." And so we thought. Ernie Galleghos (OCE) wanted a copy of the NASA letter for reference. As in the case of TR-51 about the 1960 Mercury Datum, so also now an in-house publication was needed (Geodetic Memorandum No.1624) with all the practical technical details of MMD 68 for the benefit of the users, including ourselves. And we settled down to a quiet period of getting things done.

We also were in the process of gathering all geodetic data in Africa with a view to integrating them into a consistent continental reference framework similar to our study for South America: the MMD 68 was the natural medium for that task. Sandra Todd and Jane Paul worked intensively on a global mathematical expression for the MMD 68, and I surprised them with a lengthy congratulatory write-up in the *AMS Reference Point*, which also carried their pictures (May 1968). Their excellent work led to a promotion for one and a monetary award for the other.

An invitation came from George Mourad through DIA to join a new Marine Geodesy Committee of the Marine Technology Society, and Rutscheidt said I should join. George Mourad was working on the problems of a marine geodesic network and presented two papers at the AGU meeting in spring 1968: one on the need for precise positioning of a ship by means of satellites, and another on positioning markers on the ocean floor in relation to the ship and to each other.

This working tranquility, however, was just the eye of the DIA hurricane. Its second part suddenly erupted with upsetting telephone calls from our customers. On March 5, 1968, Mr. McCarthy from Point Mugu, Pacific Missile Range, called long-distance wanting to know what the snafu about MMD 68 was all about. He needed this improved system, yet DIA had prohibited its use. Three days later George Weston called to say that he had almost published their application of MMD 68 when DIA prohibited its use. Five days later. Major Mitchell called long-distance to find out what we knew about the situation. He was going to write to their representative at DIA that he needed the MMD 68 and needed it now. By contrast, a letter from NASA requested specific information in using MMD 68; NASA was outside of DIA's reach.

What should one think of DIA's prohibition on using technical information that DoD's operative units had waited for and needed? The Navy and Air Force units had spent money to install the improvements that they needed; and DIA asked them now to spend more money to undo these improvements and re-institute an eight-year old system. Whose money? The taxpayers' of course. It is a curious notion of the general public when vilifying federal employees, that these employees live and loaf at the expense of the taxpayers; but it is conveniently overlooked that these employees are also taxpayers and just as upset when they see government waste and abuse of power.

It was not the end of the story yet. Even if the Navy and Air Force were ordered to switch back to the 1960 Mercury Datum as the only DoD authorized unclassified datum (actually a flattering compliment that my old 1960 Datum was still good enough to be continued), other customers such as NASA, and scientific investigators, such as ourselves, still required an up-to-date system such as MMD 68 and the explanatory Geodetic Memorandum No.1624. Mindful of

DIA's demonstrated irritation at being left out, and anxious to let then know that we were good children obeying their request for a name change, we sent them an information copy through channels with an official cover letter by Col. Conard. Soon requests for more copies of this very useful document came in from various customers. But the courtesy copy to DIA backfired. (Some days, whatever you do is not the right thing.) It backfired simultaneously from an entirely different corner, with an entirely different motive, and the effects combined to produce more waste—but in the end produced much more publicity and acceptance of my product, which was certainly not intended by either of the assailants. DIA complained this time, believe it or not, that the word "Modification" in the imposed title should not have been capitalized, and then they realized with annoyance that the by now widely used abbreviation MMD 68 fit case, and they demanded a change in all copies. The issue had become ridiculous and irrational. Col. Conard was furious, but as a military man, he was trained to obey. So he ordered that the offending capiltal letter M be replaced by a lower case letter in all places, and that the offending MMD 68 be replaced by the ridiculous AMOMD—F 68 (A modification of the Mercury Datum—Fischer 1968). We were appalled by the pettiness displayed by DIA, and at the same time amused by the (to me) obvious futility, like trying to push a genie back into her bottle: the original name and the abbreviation MMD 68 were so much more natural to say and already in the literature! who of the national or international readers will care or even know of the slightly different form in a restricted Government pamphlet? There was no way to restrict the spoken word or its use in future references.

The other reverberations came from our Publications Office, feeling threatened in their empire building by the fact that this pamphlet was not published as a Technical Report through them, but as a Geodetic Memorandum (a numbered series of very small Government circulation, under our department's jurisdiction). Slow as they were lately, they had not gotten around yet to write new regulations claiming jurisdiction over these also. Capitalizing on our tensions with DIA, they saw a chance to pick up some crumbs in this fight for their own fight with us and they managed to have staff pester us with questioning authorization, costs, purpose, whatnots of the publication. Mr. Culley suggested, rather than make those silly and time-consuming DIA changes ourselves, to let the Publications Office re-issue the whole thing as a Technical Report (No.67). This automatically greatly enlarged the primary circulation, which probably was not DIA's intent, but certainly my gain.

And now there was a last flourish; DIA ordered us to physically destroy useful government property, that is, the remaining copies of Geodetic Memorandum

No.1624, and they even advertised their own government waste and foolishness by recalling also the copies mailed out, in exchange for the identical TR-67s. These were produced in an unusual hurry while I was away on TDY in West Point and a week's vacation. Needless to mention that neither in the AGU handout nor in the West Point paper, both already cleared and gone, was the offending MMD 68 term altered: these papers and the *Bulletin Géodésique* could easily be quoted, and slides and illustrations re-used in the future. My friends at NASA, in the new edition of their Directory of Satellite Tracking Stations, quoted my "Modified Mercury Datum of 1968" by its full name and gave ample references to my various papers including Geodetic Memorandum No.1624. Of course, DIA had no duplicate product to offer the market: they had only gained ridicule. They even increased that, when as a last, last flourish half a year later, they clamped a restriction "For Official Use Only" on TR-67, which underscored the usefulness of my work, annoyed the users who now had to lock it up after each use, but did not prevent its wider distribution upon frequent requests.

4. WEST POINT AND OTHER MISSIONS

West Point, where the 1968 Army Science Conference took place in June, lies on an especially scenic spot on the Hudson River. My husband and I were glad to have a chance to see it. The Conference had about 700 participants, but I knew very, very few among them. The topics of the many papers were strongly war-related. While they were interesting and educational, both in their own right and also indicative of what other Army research units were doing, they made me feel very much out of place with my global, peaceful interests in the Figure of the Earth. My own paper "Geoid Determination" (published in the Proceedings) must have sounded just as strange to most of the audience, although the organizers had chosen its topic. Yet, in due time, there were letters of commendation of its "outstanding quality" from the Army's Chief scientist, Dr. Marvin E. Lasser, from OCE Col. L. R. Yourtee, Jr., and a cover letter from Col. Conard. Among the participants, the only familiar and kindred soul was Professor Ralph M. Berry, Michigan University, whose paper dealt with experiences in geodetic leveling around and across the Great Lakes. The prize paper whose author was expectantly awaited and arrived late and in a big hurry, dealt with the durability of truck tires and their treads.

There were very few wives attending these highly technical sessions as courtesy for their lecturing husbands and none in the general opening session. I was the only female participant and preferred to stay in the background. The keynote speaker started "Ladies and Gentlemen, or rather Gentlemen—since there are no ladies present." 699 persons turned around to me and some called out: "Yes, there is; one." And people around me in the back rows said: "Lift your arm, get up, show yourself." So I lifted my arm and got up, and they clapped their hands. Then the speaker said, "I apologize, and I'll start all over again. Lady and Gentlemen." By that time it was fun. And in my mind there was a quick flashback to my university years in Vienna. My friend and I were the only two girls in a large class on Projective Geometry at the Vienna Institute of Technology. One day, she was sick and absent. The Professor started his lecture, sayings "Gentlemen, and the one lady there in the last row." *Déjà vue.* In summer 1968 a letter from General Robert R. Ploger about reorganizing our part of the woods was circulated for comments and brought much unrest. General Ploger wanted to introduce a supposedly more serviceable mapping concept, and in the process seemed to restrict AMS satellite plans in favor of those of other agencies. We had been through a similar experience of danger to the Department of Geodesy two years earlier when an OCE letter by Mr. Archie Wilson sounded extremely anti-AMS to us in favor of our rivals, ACIC and GIMRADA. In self-defense, then and now, we wrote comments to these reorganization attempts and tried to tear the arguments apart with more or less success. This time, a larger reorganization was brewing, with even—perish the thought—a name change from the world-renowned AMS to an obscure and cumbersome U.S. Army Topographic Command, shortened to TOPOCOM. That really hurt! Several months later, an envelope in my mail told an interesting story: It had been addressed correctly to the Topographic Command, but the post office apparently had not been told about this momentous name change and crossed it out with the notation "address unknown." Then this was crossed out again and corrected with "Army Map Service". Only thus could it be delivered. This definitely made me feel good.

TOPOCOM was organized into centers and directorates, with the military holding the directors' positions. It transmogrified the accustomed picture of our civilian Army Map Service under only one colonel as Commanding Officer. It also created an extra layer of management. People said that one needed a home for the military coming back from overseas duty and AMS was as good a place as any for these many colonels. My own work and immediate surroundings did not seem to be affected, so I did not pay much attention to the new maze of command channels. I was asked, however, to write a draft for the response to General Ploger, list our achievements for the AMS Annual History Summary, write a memo in defense of the AMS SECOR program for the use of OCE, and take part in a briefing for Dr. Martin Swetnik and Mr. Jerry Rosenberg of NASA, among other things. This briefing was arranged by DIA, OCE, and AMS in order to impress them with AMS capabilities and keep them from shopping elsewhere. My many reprints came in quite handy. And so they did also at a briefing for the Interagency Geophysics Discussion Group, arranged by the Army Research Office (ARO) in October 1968, which at that time had funds available for worthwhile projects. You would think that in such times the ranks would be closed in a common effort to put AMS' best foot forward. But the jitters must have affected some administrator's judgment, leading to erratic and irrational behavior. For instance, the colleague who arranged our appearance at the ARO briefing gave me the message that I did not need to prepare anything or bring any materials along, "just bring yourself." That sounded not only silly, but also a little suspicious. Naturally, I did give some thought to possible topics that might come up and I prepared a little paper or handout called "The Role of Geodesy Among the Earth Sciences." It briefly mentioned the goals and methods of geodesy and its interaction with astronomy, glaciology, oceanography, seismology, and geophysics. I would then elaborate on any of these aspects they were interested in, and adapt it to a shorter or longer time at my disposal. I also took a collection of viewgraphs along. My hunch had been correct: while this colleague gave a well-prepared and detailed presentation with a mountain of viewgraphs on AMS satellite efforts to the ARO, whose funds could not possibly change DIA's inter-agency policies, my viewgraphs and possibly worthwhile geophysical spin-offs aroused much interest in the few minutes left to me, and it was suggested, therefore, to continue discussions at a later visit to my office. But when a few weeks later Dr. Zadnik, Dr. de Percina, and Col. Parks came to see me, this colleague appeared suddenly, in the middle of his week's vacation, and managed to sabotage these discussions again with a repeat of his previous lecture. Why? The visitors wondered too and told me so.

An important way to strengthen the performance of the division or any office is to avoid underutilization of employees. I knew of such people in the division and offered jobs to some of them, but had a hard time getting permission for a transfer. Administrative wavering between giving permission and withdrawing it again after work had been laid out for these employees in my branch, back and forth for weeks despite promises that did not seem to mean anything, did not exactly indicate thoughtful planning. These people were sent from one short-term job to another without a chance to stay with one and become involved. It took a long while before Joan Nickless and Willie Nelsen joined my Geoid Branch, where they found plenty of long-range projects with ample opportunity to learn, produce, and advance, commensurate with their abilities. They replaced

Dennis Popevis who transferred to the Department of Computer Services, and Pedro Santiago who followed the ocean's lure. I did not succeed to send Jane Paul to an informative IAG Symposium in Trieste, although neither Rutscheidt nor I could make use of the invitation. Permission was on and off again till it was too late. Nor was it easy to get permission to send some of my people to an obviously useful one-week in-house programming course. Permission was doled out eventually as if it were a favor.

Increasing administrative incapacity to make rational decisions became costly not only in the use of manpower, but also directly in Government expenditures, as for instance in the case of replacing a worn-out computer. Our old Bendix computer had broken down more and more frequently and it took days to get it repaired, only to break down again a short time later. It was under a costly servicing and repair contract, which was automatically renewed unless stopped. The cost of a new, modern, desk-size computer was less than the yearly contract costs and we were asked to do a little research to establish the facts and make a selection among the new machines on the market. Sandy Todd volunteered to do that. But it took much longer than a year's administrative wavering between supporting the obviously needed change and delaying the prepared paper work, before the approved purchase order actually went out. Even after that, the indecision to declare the decrepit machine as surplus and have it removed, kept the automatic maintenance bills coming.

Fiscal Year planning was a little more formal this year (June 1968) and we were asked to submit a listing of all projects, brief yet detailed, complete with names of people involved, current status and plans until June 1969. My branch had an impressive list of twenty projects for such a small group, including Thelma Robinson's two independent studies of LORAN (Long-range radio navigation system) and UTM problems. The other projects were grouped under the major topics of analyzing satellite results produced by different techniques, the South American Datum, the unification of African data, the utilization of gravity data in Australia, and studies of mathematical procedures for automation. Irritation with administrative capriciousness and general unrest accompanying government reorganizations must have been mounting: my last item on that formal list of projects reads: 'Keep Smiling. People involved: everyone." People throughout the plant started to greet each other with: "It is going to be better, because it can't get any worse." "That's what you think. More likely, it will get a lot worse before it can get any better."

In the Geoid Branch, we tried to maintain an island of a pleasant work atmosphere, with plenty to keep us busy. The break-up of the major topics into several different tasks with an abundance of ideas to pursue, made it possible to let people choose their area of involvement. They liked to work in groups of two or three, with one emerging as the leader; but groups overlapped and leaders for one task might be assistants in another, as everyone was involved in at least two, mostly more projects at the same time. This provided variety and flexibility. There also was my obligation to maintain international contacts for the exchange of ideas and information, mostly by mail and visitors and meetings. Two fascinating missions came up over the horizon in the form of two invitations, well in advance, which AMS fortunately wanted me to accept: one from the Geophysical Institute of the Czechoslovak Academy of Sciences to a symposium on physical geodesy in Prague, in fall 1969 and the other by the Conference Committee for the Fourth South African National Survey Conference in Durban, in July 1970, requesting for a paper on some aspect of geodesy, the specific topic being left to my choice. The world was so much wider and more interesting than the commotion at my feet. Keep Smiling!—I knew a wise old lady who refused to give in to a feeling of drudgery in housework. She said: "Before it gets to me, I look away from that dust on the floor. I look up at the sky; it is so blue, and grand, and clean. It makes me feel good."

10

FOCUS ON THE SOUTHERN HEMISPHERE (1968–1970)

1. AUSTRALIA

The excitement in acquiring large blocks of geodetic data in the southern hemisphere was due to the fact that most of the geodetic data so far had naturally been collected in the middle latitudes of the northern hemisphere, primarily in Europe and later in North America. Thus, conclusions about the Earth as a whole by extrapolation and speculation were heavily weighted by the measurements in these areas and needed confirmation or modification from evidence in other areas. Watching the blank spaces in the southern hemisphere fill up in these last few years was a great historical privilege. The two long geodetic arcs, completed in 1954, which reached into Chile and South Africa respectively, had already caused a geodetic upheaval. They had shown that the International Ellipsoid, chosen in 1924, was much too big, that the geodetic datums chosen in Europe and North America did not fit in their southern counterparts, and that the customary neglect of geoidal heights would accumulate to serious distortions in geodetic control over long distances.

Now Australia had come in. Isolated small networks had been connected by new geodetic work, and unified on a coherent national datum. The choice of this system had been guided by Tony Bomford's study of strategically located astro-geodetic deflections of the vertical, so that these little geodetic gremlins would be small in the new system and not cause any significant distortions. And I was the lucky person with permission to translate this work into the first geoid chart of the Australian continent and incorporate it into the global Figure of the Earth studies.

A very densely surveyed east-west arc at 32-degree south latitude had served as a backbone profile for my geoid construction. It served also as a check and scale for the so-called BC-4 satellite triangulation network of the Coast and Geodetic

Survey. (Their basic scale line was across the United States, and another was in Europe between Norway and Sicily.) There were a few tracking stations also for other satellite programs.

A puzzling phenomenon appeared: geoid information derived from satellite observations seemed to indicate an unusually steep slope across the Australian continent from southwest to northeast, on the order of about a hundred meters, which is extreme for regional geoid charts. Why should a whole continent sit on a slope and be tilted so steeply? Was something wrong with these new satellite derivations? Geoid charts derived from terrestrial gravimetry showed a slope in the same direction, but much less steep. My own astrogeodetic geoid chart was based on data strictly within the continent and—because its reference datum had been so designed—the geoid heights were less than twenty meters, mostly less than ten meters. Analyzing these different answers concerning the same territory brought out again the need to keep aware of differences in reference systems, input data, and purpose. For engineering and mapping purposes, one needs a geodetic control that shows all the details and allows computing distances and directions between points of interest. To make a map between, say, Brisbane and Melbourne, one needs to know the surveyed distances and directions in that area, but one does not need to know their distance from the center of the Earth. A geodetic datum that fits the continent as closely as possible to avoid distortions will be the answer: the astrogeodetic geoid chart with small geoidal heights. For scientific studies of the Earth as a whole, by contrast, even the entire continent of Australia is only a relatively small part whose details are not as important as is its location with respect to the Earth's center and to the other parts of the Earth. The steep slope established by satellite geodesy with a maximal geoidal height north of New Guinea gives the geophysicists something to think about. The terrestrial gravity charts are something in between. Theoretically, they should give the same answers as the satellite charts, since both are based on samples of the Earth's gravity field; but they don't because the samples are significantly different. As a consequence, satellite charts give the long-wavelength trends with no local detail, while the gravimetric charts give much more local or regional detail where the data coverage is good, but are weak globally.

The Australian case reminds of the Japan-Manchuria example discussed earlier: the deep Japan Trench causes a large deflection of the vertical in Tokyo, which was unknown or ignored when the Japanese Datum was established under the conventional assumption of zero deflection at Tokyo. This in turn made the Japanese geoid appear as if on a steep slope. The slope disappeared when the

Manchurian Datum was extended to Japan and showed the Japanese geoid to be quite smooth.

These different appearances follow from the choice of a mathematical reference surface, comparable to the choice of a place from where to take a photograph. For an interesting building you may choose a place from where it can be seen as a whole within its surroundings, even though the details of its intricate portal won't stand out. Or you may wish to get the details of that portal without distortions by standing in front of it for a close-up view.

2. THE SOUTH AMERICAN DATUM OF 1969

To understand the concept and significance of the South American Datum project, take a look at a map of South America that shows the distribution of mountain ranges and plains. There is a mountain range along the west coast (where the now famous geodetic arc had been surveyed as an extension from North America through Central America) with a spur through mountainous Colombia into Venezuela. The east is significantly lower and smoother, characterized by the basins of the Amazon and La Plata rivers. Geodetic surveys (Fig. 1) are easier to make in smooth country than in wild mountains and they are needed more in densely inhabited areas. Brazil and Argentina had good national survey nets in their most populated areas, starting at centrally located stations (named Chua and Campo Inchauspe respectively). There were practically no surveys in the upstream river basins. Surveys in the mountain states were not too precise, and neither was the HIRAN connection in the northeast, from Venezuela to the eastern corner of Brazil, as I found out (although the name HIRAN boasts "high precision"). The origin of the so-called Provisional South American Datum of 1956 was at La Canoa in Venezuela, and was affected by the vicinity of the Puerto Rican trench as well as by incomplete computation (see end of chapters II.4 and VII.2). Its extension southward created increasing distortions due to the neglect of the "Molodenskiy Correction'. Forcing this defective datum from the northern end over the whole continent introduced not only an artificial tilt (similar to the effect of the Japanese Trench on the Japanese Datum), but also propagated the uncertainties in the mountains of the west and in the HIRAN stretch of the northeast into the rest of the continent, since these were the only two routes to reach the south. The absurdity of this proposition was not lost on the South American geodesists who never endorsed the La Canoa Datum but always insisted on a centrally located datum point. Yet the mapping agreement that most South American states had with AMS through IAGS extended this faulty datum further and further south due to the routine procedures of cartographers who did not understand or had forgotten or did not care about Mr. Hough's warning in 1956. Argentina, and also Paraguay and Uruguay, did not have a mapping agreement with IAGS and that was why it was so much more difficult, practically hopeless, to get data from there.

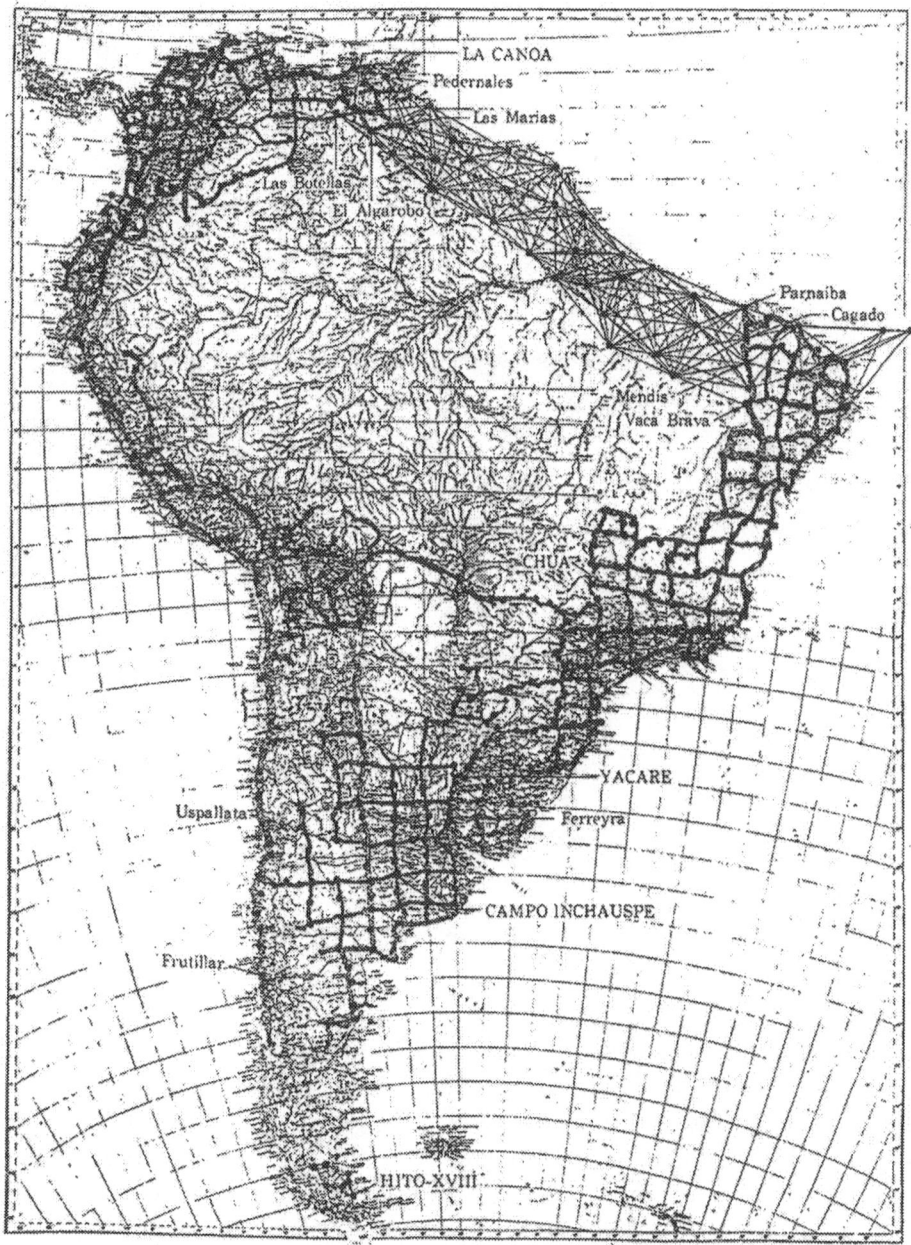

Figure 1 - TRIANGULATION, TRILATERATION, AND ASTRONOMIC
STATIONS USED IN HI STUDY

Ch. 10, Fig. 1: Triangulation, Trilateration and Astronomic Stations

What happened now is shown in Figure 2: The upper part represents the geoid along the coastal arc referred to this "Provisional South American Datum of 1956 (PSAD 56) with a systematically increasing discrepancy reaching almost 300 m in Chile; the lower part represents my goal of a new datum that would fit the geoid.

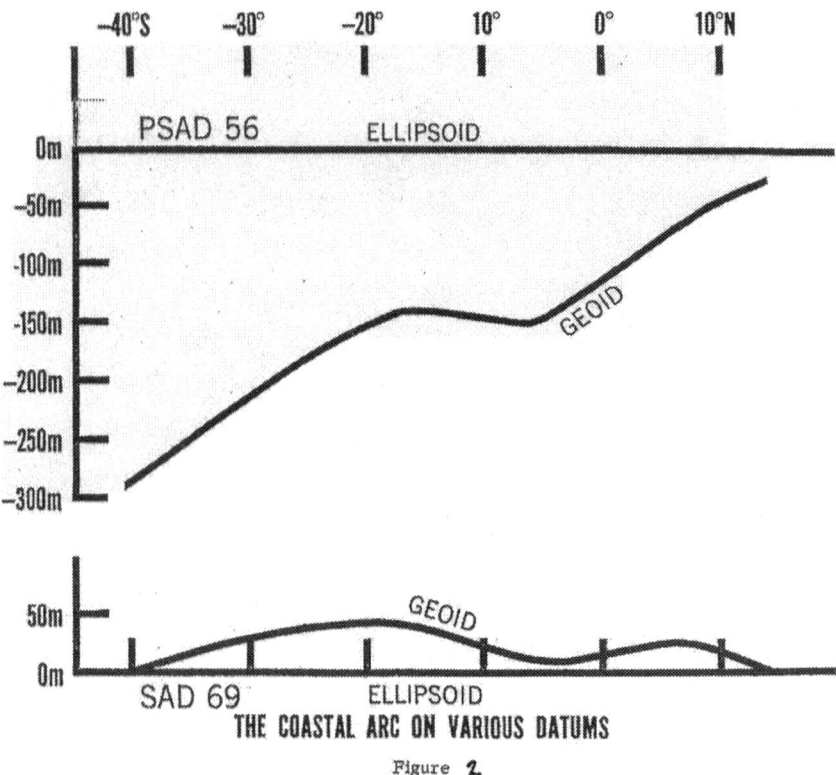

THE COASTAL ARC ON VARIOUS DATUMS

Figure 2.

At the time of my Guatemala paper in 1965, I had only the big northern loop around the Amazon basin and the one line hanging open-ended southward into central Chile. It was sufficient data, however, to demonstrate first the distortions accumulated on the long way from Venezuela to Chile, and in store for Argentina if the PSAD 56 were allowed to extend there too; and then to show how to avoid these errors and how to derive a better-fitting datum. The latter would be even better if the abundant Argentinean data were released to form a second southern loop around the La Plata basin by connecting Brazil (with contributions from Paraguay and Uruguay) through Argentina with Chile, and secure that open-ended Chilean arc. The strategic importance of the Argentinean data was clear. As already told, the logic of the proposal, the need for the product, and the wooing of Argentina to come out and play ball, finally won the day with enthusiasm and marvelous cooperation.

My plan was to start from scratch, that means from the observations as if the PSAD 56 had never been done. Instead of La Canoa in the north, I would choose a more reasonable origin somewhere in the middle of the continent in a smooth area where geodetic surveys were already abundant and reliable (in densely populated areas), so that the unavoidable uncertainties in mountainous areas would be pushed to the outer fringes of the system and not contaminate the good parts starting from the origin. Candidates were, of course, Chua and Campo Inchauspe, the origins of the existing national nets of Brazil and Argentina. Either one was acceptable, so that a choice might have been viewed as a political decision which I wanted to avoid, of course. There was a way out: a very dense and reliable geoid profile between them, strengthened by an extra hundred astro stations upon my request, made it possible to handle these two key points in unison. I would assign zero values to their geoidal heights in the new system and make the whole area between and around them a region of geodetic strength with near zero geoidal heights.

Mathematically, this was equivalent to specifying two out of three necessary conditions for positioning a reference ellipsoid (in this case, the IAG ellipsoid recently blessed in Lucerne 1967). The third condition could be used to assure that geoidal heights in the mountains should not exceed, say, 50 meters. This could be established by a few trial computations. Assigning a tentative geoidal height to a third point (one in the heights of Bolivia, located on the arc from Chua) and quickly computing what that would do for areas in the far corners of the continent, gave me the basis for a reasonable decision. You might try to visualize what happened here by thinking of a plane (instead of the ellipsoid) determined by these three key points, holding the two eastern points with the line between them fixed, and rotating the plane a little about that fixed line to see which position you want to choose. From the geodetic viewpoint there was not necessarily an optimum position but a certain range of acceptable choices.

This procedure was new in the sense that the traditional way of choosing a new reference datum was to formulate the required three mathematical conditions not at the vertices of a big triangle, but all three at one chosen origin in terms of its geoidal height and its meridional and prime vertical deflections. Moreover, in classical geodesy, those three terms were usually and simply assigned zero values. No wonder that such conventional assignments for one single station brought a lot of surprises in the rest of the country when the mathematical-geodetic consequences of that choice were worked out gradually. So I started from the other ends state what you wish to see in the continent and formulate the mathematical conditions to bring it about.

By August 1968, the two Argentinean geodesists, Ing. A. H. G. Christensen and Ing. A. V. Elias, arrived with the promised data (hurrah!) and handed them over to me ceremoniously, for incorporation into our work. Mr. Frank Reynolds of the Americas Division had the job of adjusting all the triangulation data of the two principal loops (with the common arc from Chua into Bolivia forming a figure eight) into a consistent basic framework on a preliminary reference datum which I had chosen already at the time the adjustment of the large area of the Rio Grande do Sul was made. Frank came over frequently for consultation and exchange of ideas with my group and shared our fascination with this project. He did a painstaking and marvelous job. Success breeds attempts at perfectionism. His adjustment of the southern loop, starting from Chua in both directions through reliably surveyed areas first and mountains last (that is, south to Argentina, then west to the mountainous border with Chile, and also going west from Chua into Bolivia and then through the mountains to the same border point), gave a surprisingly small discrepancy at the meeting point considering the long routes. Yet, this discrepancy together with some uncertainties along the way sparked further analyses and speculations and wishful thinking about getting additional enlightening information, which eventually led to Byars' second trip to Argentina in early 1969. Responding to Byars' unusual diplomatic skill, technical competence, and pleasant persuasiveness, the IGM in Buenos Aires opened their files to his search and he succeeded in clearing up some question marks and brought home much useful information that had been denied to AMS up to then. His efficiency was recognized in time by a special commendation from General E. T. Podufaly, our Commanding Officer at a later time.

The adjustment of the northern loop was similarly planned to go from Chua in two routes, one toward the west and the other toward northeast through smooth territories first and the mountains last, and meet in Venezuela. But this loop did not close at first try: there seemed to be a gap of about 80 meters, which despite the great length of the loop was unacceptable. What had happened here? The puzzle brought out my detective instincts and heightened an excited sensitivity to all possible clues. The new geodetic tool, satellite geodesy, was put to work although it was not perfected yet: the different systems still gave different answers to the same question. But within a range of values, we could still squeeze out some information that was not available before. Three satellite techniques

(SECOR, BC-4, and Doppler) gave significantly larger distances to the HIRAN stretch, and roughly closed the gap between the endpoints of Frank's two adjusted routes outside of the HIRAN, and thus seemed to place the error into the HIRAN stretch. This impression and the overall correctness of Frank's continental adjustment were confirmed by a number of satellite derived transcontinental lines which braced the two big loops. We also initiated a study of the HIRAN routine procedures to hunt down the villain, and in due time showed that they tended to produce too short distances (with Sandra Todd as co-author: "A refined procedure for computing geodetic distances from HIRAN observations," *Surveying and Mapping*, March 1972).

My request for the densification of deflections of the vertical along a specific route between Chua and Campo Inchauspe was met with Byars' help through IAGS who happened to have several astro work teams available. Also our own Field Survey Division had a few teams available and Rutscheidt asked me to pick out twenty desirable deflection stations in the mountains on the South American west coast. I was only too happy to do that. A visit by Mr. O. C. Darling from IAGS in Panama, however, threw some doubts on the feasibility of this plan in time to be of use to us. Mr. Darling had been involved for years in geodetic work there and had made investigations on the distorting effect of deflections. He was very much in favor of getting more deflections, but warned us that the existing control stations on which to build the new work, might not be recoverable any more, and then one would have to start from scratch with more time. I talked this problem over with Rutscheidt and pointed out that twenty new but very likely unfinished stations, at the outer fringes of my scheme *nota bene*, would not do us any good, while these resources would do a lot of good, e.g., on the new road from Brasilia to Belem: here they would brace the eastern part of the country and start a check on the doubtful HIRAN net.

At this occasion I was told that someone in "P and P" (Plans and Production Office) was establishing a South American Datum plan independent from Byars' and my work, that the twenty stations were part of that plan, and that Mr. Glusic had been asked to prepare the plan and advise on its feasibility in the next several years. That certainly was a revelation and some puzzles fell into place. Andy Glusic, a friend and "honorary member of the Geoid Branch", had made cryptic comments to me, with his characteristic drawn-out accent (different from mine): "Mrs. Fischer, I don't like my job any more. They ask me for an opinion, without giving me the full story and context. They want to use my opinion to take someone else down. I am not allowed to tell you or anyone, so please don't ask me. But I don't like this kind of job. I am an old soldier, however, and I have got to obey orders." I was sure glad I was definitely not an old soldier, and I felt sorry and sad for Andy's anguish. But I had not taken it too seriously. Now, however, it dawned on me: Had Andy hinted that there was something afoot against Byars or me or both? But why?

I asked Rutscheidt directly and he said that this Mr. X in Staff had once worked for Byars and did not like him. Now in Staff, he was in a position to get back at him. "Are you serious?" I asked in amazement. Mr. X was two grades lower than Mr. Byars and no match at all in technical ability and competency. Rutscheidt replied, maybe in an attempt to ridicule the whole thing: "You know, Byars is a man who never smiles." I certainly could not agree with that from my own very pleasant working experience with Byars. I still have a little green routing slip that I wrote to Rutscheidt about some discussion with Byars, and it has this footnotes "P.S. He did smile." (The word "did" underlined twice.) I asked Glusic about his paper for Staff and he said that he was analyzing the accuracies of the triangulation network, but was not allowed to show me the paper. When I told him that we had a commitment to come up with a specific proposal for a new South American Datum at the PAIGH Assembly next year and that we had all the South American cooperation for it, he was very surprised. "Mrs. Fischer, they did not tell me a word of that. They said to make a plan for several years into the future." So I told him some more about our concrete plans and the strong momentum of our progress. Even a few weeks later when that paper was presumably finished, with my information probably included or at least considered, I was not allowed to see it. Rutscheidt acted surprised that I did not have a copy; of course, I should have one. I went immediately to Glusic to ask for my copy, but he said: "Mrs. Fischer, I would gladly give it to you. But Mr. Y was here just five minutes ago and forbade me to give it out under any circumstances, especially not to you. Mrs. Fischer, believe me, I don't like this kind of thing and I don't understand it." Neither did I. Half a year later, I was finally permitted to see a copy. It turned out to be a conventional analysis of the type of fieldwork needed for the (unrealistic) goal of bringing all triangulation including mountain routes and HIRAN up to topnotch standards. It specified in detail the number of field parties required, and assigned a standard time estimate of six months to each section of this plan, but wisely refrained from adding these up to an over-all time frame which would have run into many years. It made no mention of the practical improbability that such grandiose textbook plans would ever find concerted support in real life in the foreseeable future. It still seemed to assume that the priority of La Canoa and first-rate connections from there to the rest of the continent was the only way to salvation. Focusing on future fieldwork, there was no mention that its computation on the current faulty reference datum would introduce and perpetuate artificial and unnecessary distortions in a vicious circle. The uninitiated reader was left with the misleading impression that all this field work must be completed before a South American datum could possibly be designed, unaware of the fact that an unconventional, imaginative, yet realistic way to skin the cat from the other end and break out of the vicious circle was already successfully under way. Glusic was a competent geodesist of the old classical school of two-dimensional geodesy, an expert in the requirements of local triangulation; new, original ideas with a three-dimensional approach of continental scope were not his forte. Why his paper should have been kept from me for so many months was unclear. In its factual part, it was an unneeded duplication; and it contained nothing that helped or hurt our work. It did not even address the main purpose of our project: to remove the systematic errors and the inconsistencies of the currently used mathematical reference and replace it by one that would not distort otherwise good field work, new or old.

At the end of summer 1968, I was asked to say that I needed for my project a network of the so-called PC-1000 camera observations (the satellite technique promoted by the Air Force). This was an odd request to put it mildly. I had analyzed the PC-1000 results for the triangle of Curacao-Trinidad-Paramaribo, the only ones so far existing in all of South America, and had found them wanting. I was told that higher echelons needed a playground or practicing area for the PC-1000 to which they had been overcommitted, and South America would do nicely; but they needed a justification for such a project, and my South American datum project would serve just fine. Not only did I not need the PC-1000 and would put my dollar rather on any number of other data, but such a statement would scuttle my project there and then; it would mean a delay to a never-never due date, while in fact my due date was approaching very fast, and I meant to meet it. Next, Mr. X, apparently in a team with Mr. Y but most probably being used and directed by Mr. Y, began to question the validity of the mandate for the PAIGH Working Group, although that had been officially formulated and published in 1965. In October 1968, Mr. X wrote an official memorandum to the Commanding Officer, and Mr. Noell Fischer, then Chief of the Planning Division, sent it to the Department of Geodesy for an official response. It said that "…it has cone to the attention of members of the Planning Division that Mr. Byars…and Mrs. Fischer…feel that they have a mandate to recommend a datum for South America at the PAIGH meeting in May 1969…A study on the work needed for the establishment of a datum for South America was made by Research and Analysis Division as requested by Planning Division…was presented to higher headquarters, OCE and DIA, by AMS…calls for a volume of work that is impossible to accomplish…in the near future. To make a recommendation before work called for by the study is completed will probably result in another provisional datum as the recommendation would be based on insufficient information and…containing systematic error…It is felt that there now exist enough datums in South America without adding another interim datum…" Byars and I had to explain the ABCs of geodetic reference systems to an in-house conference, which had come to discredit us but behaved like Balaam who came to curse and instead blessed. Besides Culley (always supportive), Byars and myself, there were Rutscheidt, Bill Doxey, Bob Yater, Ora Smith (Coordinator for Latin American Mapping), Andy Glusic, Link Genoung, Noell Fischer, Rudi Salvermoser, and maybe others. Noell Fischer summed it up to me, "Now I understand what you are up to, and it makes a lot of sense to me now." To which I happily replied: "Thank you. And why don't you come up here more often to get stories straight instead of by detours." But that has become a more and more established government procedure in recent years: to listen to in-betweens instead of getting a story straight from the source. It is easier for in-betweens, if they wish to mislead and discredit, if a direct connection is out of bounds.

On the following day, Culley, Byars, and I, accompanied by some of yesterday's participants, had to go to DIA, who also had been alerted to interfere and played Balaam instead. The DIA people had a better geodetic understanding and that helped, of course. They also were familiar with the events within PAIGH and interested in maintaining U.S. prestige and reliability there.

When it was time to get my paper "The Geoid in South America Referred to Various Reference Systems" approved, cleared, and produced for distribution, curious obstacles were put in the way. Time was getting close and I wondered why Henrietta, an otherwise very efficient secretary, did not make any headway. "Oh, did you not know? I was told to stop everything, because 'downstairs' they are having some trouble with your paper and they are holding it. I was told you knew." Of course, I did not know. Holding my paper for two weeks by now and not telling me could mean only one thing: a maneuver to make me miss the clearance deadline. What should I do? When I walked to the parking lot that evening, I saw Mr. Culley getting into his car. Of course, I would appeal to him for help. I told Mr. Culley what I had heard and that time was getting short. Mr. Culley offered to call a conference to force things out into the open and see what was brewing. That was Friday afternoon. On the following Monday, he had called an urgent conference with Byars, Rutscheidt, Yater, Genoung, Doxey, myself and maybe others. It turned out that there was a problem with the bookkeeping, because they had justified sending astro parties to South America as being absolutely necessary for the South American Datum and here my paper had succeeded without them, that is, succeeded "in contradiction to TOPOCOM policies." Mr. X's planning had backfired. The following day there was another conference with Col. Conard who decided that the "P and P" Office had no right to hold up my or any other paper. It was to be sent out immediately for clearance. DIA cooperated with a quick clearance return.

The production of the paper for distribution, however, ran again into some odd little obstacles, and it was not always clear whether I should laugh or cry. Considering the extraordinary background of this work, I had written a prefatory page of acknowledgments: "Many people contributed to the assembling of various geodetic facts for the present integration and analysis. The Americas Division and the Geoid Branch of the U.S. Army Topographic Command cut through the red tape of organization barriers and worked in exemplary cooperation to meet the short deadline. The Geodetic Division within the Inter American Geodetic Survey has assisted very efficiently through its cooperative efforts with many South American countries. But most impressive was the extraordinary international cooperation within the Working Group on the South American Datum under PAIGH and within the individual countries. Rather than mention many individual names, I would like to emphasize that the possibility of this study stands as a shining example of what international good will can accomplish."

When the Publications Office (wouldn't you expect?) sent the paper up for proof reading, this page was missing. It was not an oversight: first they had not liked the phrase about the red tape although or because they were wrapped in streams of it as often demonstrated. And then they had not liked the whole thing. Maybe, they did not like the stress on extraordinary cooperation and good will, of which they did not have any. Whatever it was that irked them, they searched and found an Army Regulation disallowing acknowledgments on conference papers. I questioned that, of course, and appealed to Mr. J. R. Kephart, Technical Liaison Officer. That was not enough, so Culley, Kephart, and I appealed to Col. Conard for help. The winning argument was that Army regulations did not apply since DIA's clearance had stipulated identification as a PAIGH paper rather than a TOPOCOM paper, by putting under my name: "Member of the PAIGH Working Group". Furthermore, the Publications Office should not have the right to change anything against an author's wish after a paper had been approved and cleared. Hurrah, the acknowledgment page was to be in! But wait a minute. The Publications Office did not give in easily. They sent us 700 copies of that page, cut much too short, and told us to slip them ourselves, without staples, into the

700 already stapled papers sitting at the loading dock for shipment. Clearly, such little slips would flutter out easily and thus defeat the purpose. It was two days before the meeting. Several men of my branch and I, armed with staplers and staple removers, went to the loading dock to see what we could do. And what we saw was a clearance violation in the form of the unauthorized use of the TOPO-COM conference covers complete with official seal. What do we do now? There was just enough room under my name for the required identification line, but write it in by hand? 700 copies? I went to see someone in the Typography Branch and they said they could put that line on the bound copies, a small job that could be done over night. I got permission from Col. Conard and sighed with relief. But next morning the job was not done; it had been countermanded by the Publications Office refusing to allow any more money for this paper, a matter of twenty dollars. Mr. Byars spent all day trying to set this correction up as reimbursable from PAIGH, but he had to call DIA for reinforcement. Col. Knipling of DIA considered this clearance violation as serious and sent a directive to TOPOCOM that the paper had to stand as cleared. The covers had to be destroyed and replaced, and the Acknowledgments page included as originally requested.

The paper was not ready for the Working Group session the following day, which Byars had called for May 29-30, 1969, at our Agency, immediately prior to the PAIGH Assembly at the State Department. We had to apologize to our foreign visitors for the delay in our printing service. But otherwise, this session at the Ruth Building at TOPOCOM and also Byars' and my presentations to the wider audience at the Assembly at the State Department was a huge success, which resulted in the official PAIGH acceptance of the "South American Datum of 1969" (SAD 69) and a request to us to finalize it and to work out all the technical details needed for the users. All's well that ends well. I wished Mr. Mills and Mr. Fuller were still with us to enjoy the success of what they had started with so much enthusiasm. I wrote about it to Mr. Hough who had given us advice and encouragement.

We had a little celebration at our house with the Byars family, the Christensen family, and Ing. Eias. And Byars had a big reception at his house for the PAIGH officials and friends. There also was a reception hosted by the Argentinean delegation at the Mayflower Hotel, and General Podufaly, our new Commander, was there, very much interested in our work and beaming about the success.

Andy Glusic congratulated me warmly and sincerely on my "excellent paper" when it was finally released for distribution. The sulking Publications Office had refused to let the author have any copy of the 700-copy edition before the meeting. After the meeting, they graciously sent three (3) whole copies as "leftovers."

For my helpful assistants I had written up awards and/or promotions, and I wrote also a Letter of Commendation through channels to Mr. Kurt Streit in the Visual Communications Branch of the Department of Applied Cartography, who had given my many illustrations a professional polish. Byars special Commendation from General Podufaly has been mentioned earlier.

During the Working Group session I had a curious phone call from Joy, our division secretary. She wanted me to name all my publications up to date over the phone from memory. She said she had a list of so and so many but she knew there were more. She could not wait till I came back to my office in the afternoon, and she was not allowed to tell me more. When I got back to my office, Ray Shirley apologized for having gone through my personal file for some needed information but he was not allowed to tell me what for. A few days later, my friends in the Personnel Office told me that an invitation to nominate someone for the National Medal of Science had come in, and while some thought that TOPOCOM did not have that caliber of employees, others thought that since I had been a winner for the Agency several times before, they might as well try my name again. With the approval of the Commanding Officer, General Podufaly, the Geodesy Department had been asked to prepare my nomination. Had I not been asked to help with that? No, I had not. They showed me the qualification criteria, which seemed to give me an outside chance, considering the spectacular work with South America and the previous high awards. So that was why Joy and Ray had wanted to get more recent listings: good souls in Personnel told me later that a nomination copy, signed by Mr. Culley and General Podufaly, had been placed in my "201 file" and showed me the attached justification write-up. It read very nicely, more or less copied from the one for the DoD Distinguished Civilian Service Award two years earlier, with listings brought up to date. It was not geared, however, to the criteria given, and curiously missing was any mention of my significant work in the last two years with South America. Curiously also, my supervisor never told me about this nomination. Years later, at another occasion, it was discovered that the nomination copy was missing from my "201 file", upon which it was replaced by a xerox of the Division copy. Why would anyone want to quietly remove a signed nomination copy from another employee's folder?

Mr. X's plan of blocking us made a last flicker before vanishing into oblivion. I had been shown a copy of a TOPOCOM Operations Plan 69-0001, a Plan for a Datum and Unified Geodetic System for South America" already at the Work Group session. It was between official covers, without signature of the Commanding Officer, and thus unauthorized to be between official covers. A month later it came to my desk for routine evaluation. Needless to say, I tore it verbally into shreds. I pointed out the technical miscomprehension of the unnamed author, his misquoting of Glusic's paper, and his conspicuous unawareness of the current status. Mr. Y's part in this attempt at undermining our work was confirmed (if ever doubted) when years later he complained that DIA had not supported his managerial rank by refusing his (negative) advice. His arguments showed that he still did not understand the technical significance of the project.

Mr. X showed no personal hostility towards me. When a RIF (reduction in force) threatened, he confided his worries of losing his job. "Maybe, you could come into my branch and do some computing for me?"—"Oh, Mrs. Fischer, I would not be qualified to work for you."—But he had been considered qualified to interfere with Byars' and my work.

3. CONTACT WITH DIFFERENT GEODETIC GROUPS; PRAGUE 1969

The new South American Datum of 1969 was of immediate interest also to geodesists not usually involved in PAIGH affairs, because of its two-fold contribution to global geodesy: it produced a solid continental piece for the picture puzzle of an improved world datum, and it had given meaningful practical experience with several competing satellite techniques. Ray Shirley, Sandra Todd and I extended our analysis of the various satellite results to their networks stretching from South America across the Atlantic, Africa, and the Mediterranean to Iran, trying to identify possible systematic errors and listing the as yet unexplained discrepancies in order to caution against the premature over-exuberance of putting all geodetic eggs into one satellite basket. In a paper for the yearly DoD Geodetic Conference in October 1969 ("The Use and Limitations of Satellite Geodesy in Specific, Real-Life Geodetic Problems") we also mentioned our difficulties with incompletely recorded satellite data, and put in a plug for the DoD Satellite Record Center established at TOPOCOM under Byars. Responding to some confused geodetic statements by a couple of high-ranking geodesists who should know better, I inserted once again an explanation of the different purposes of a continental versus a global datum, and included the application to problems of delineating boundaries at sea for exploiting the resources of the continental shelves, and a brief discussion of the concepts and legal terms concerning the continental shelf and the Law of the Sea Conference.

Little administrative obstacles kept popping up on the way again, just to keep us in practice. The call for abstracts for this conference had reached me mysteriously just before its deadline it was dated three weeks earlier; and just as mysteriously, my abstract, which I wrote quickly, and still on time, had not reached its destination a week later. When the completed papers were circulated as usual in advance of the meeting, ours was not among them. It took again Mr. Culley's help to pry it loose from somebody's desk. There were more things that mysteriously went astray. Some mischievous gremlins seemed to be around. For example, Mr. Culley asked me to suggest a showpiece for General Podufaly's office to represent some TOPOCOM achievement. I thought of a relief model of my latest unclassified world geoid since the relief model of my geoid in North America had aroused so much interest.

Culley as well as the General liked the idea. I received a note to that effect from the General asking me to discuss it further with Col. F. J. Spacek (Commanding Officer of the Production Center to which our department belonged), particularly concerning the costs since funds were low. (Aren't they always?) I was ready to go see Col. Spacek and use all the powers of persuasion I might have, when Rutscheidt said he would rather send the message formally through channels. I never got a yes or no response, and when I quietly checked through friends, the note had not been seen anywhere. It all just went to teach us that "there is many a slip 'twixt cup and lip."

More important was the good reception of the "Uses and Limitations" paper at the DoD conference. There was an "Item of Interest" about it and a phone call from Mr. Gilchrist, DIA, asking for more details, as DIA was discussing its implications with a view to canceling PC-1000 plans and give more support to our Satellite Record Center. For the wider audience of the 1969 AGU Fall Meeting in San Francisco I had already submitted the abstract of the PAIGH paper, but now I distributed also an unclassified version of this DoD paper and combined the contents of both in my presentation there. It was published in *Surveying and Mapping* under an editorially slightly changed title, to reach a different audience again. A concise summary of the new datum and its whys and wherefores was formulated in "The Development of the South American Datum 1969" for the *Survey Review* in England (October 1970) for yet another audience. At the presentation to the 30th Annual Meeting of the American Congress on Surveying and Mapping in March 1970, I made the personal acquaintance of Mr. Bill Lynn, editor of the *Newsletter*, American Society of Photogrammetry, with whom I had exchanged some amusing correspondence in the past months. A year earlier, he had lampooned, in that *Newsletter*, a newly formed seven-member committee of the National Research Council charged with the task of determining a new North American datum, and had suggested instead that in the spirit of chivalry, "…we adopt Irene Fischer's (reference ellipsoid), whichever one she chooses to call best." I had written to him in the same light vein and we both had fun in the correspondence.

My letter files of worldwide IAG correspondence kept growing. When the organizers of the 1969 Prague Symposium on Physical Geodesy asked me for the names in my SSG V-29 for personal invitations, there were 27 foremost geodesists on my list: Brig. Guy Bomford (England), Don A. Rice (U.S.A.), E. Tengström (Sweden), E. E. Baglietto (Argentina), S. Bakkelid (Norway), Anthony G. Bomford (Australia), M. Bursa (C.S.R.), A. A. Cerrato (Argentina), G. A. Corcoran (Canada), Pablo Dragan (Argentina), H. M. Dufour (France), E. Fitschen (South Africa), C. Gemael (Brazil), V. K. Hristov (Bulgaria), B. M. Jones (South Africa), R. P. Lambert (Australia), J. J. Levallois (France), Rene de Mattos (Brazil), A. Muminagic (Yugoslavia), S. Perlman (Israel), M. Pick (C.S.R.), A. R. Robbins (England), L. Rodriguez (Argentina), Uri Shoshani (Israel), N. de Sousa Alfonso (Portugal), Helmut Wolf (Germany), A. Zivkovic (Yugoslavia). The Symposium in Prague (September 1969) attracted about 60 participants from 19 countries, and heard more than 30 papers and another lively discussion of the new IAG Reference Ellipsoid of 1967. The fact of the continuing discussion illuminated the interesting change in the history of this concept from a primary geodetic research goal, to being just a means to an end, a useful tool that could be slightly different for different purposes.

The time of the Symposium happened to be about a year after the Russian invasion into Czechoslovakia, the end of the "Prague spring", the destruction of Prime Minister Dubček's version of "communism with a human face." The embittered resistance of the population could be sensed everywhere. The German occupation was still in hated memory, but the Russian occupation was feared to be even worse; yet the outspoken Czechs dared to show their hostility. A young student, Jan Palach, had burned himself in protest on Wenceslav Square the previous year; and on a bus tour we suddenly made an unannounced detour to be shown his grave hidden under a mountain of flowers. Streets were torn open for repair work, which, so we were told, would take a long time as would also all the other construction work because of a general work slowdown. When we walked around the beautiful city for sightseeing, we asked our way first in German, thinking that would be more readily understood than English, but that person did not seen to understand us at all. My husband said to me in English, "Well, we'll go around that corner and see," whereupon that same person said in fluent

German: "Oh, you are Americans? You go straight here and then right..." And when Levallois wanted to show off his amazing multilingual talents and asked a cab driver in Russian to take us somewhere, that cab driver first refused to take us at all; he relented when sufficiently persuaded that Levallois was French and none of us Russian. Even during a technical discussion during the meeting, a point made in Russian and needing the translation of a specific term could mysteriously not be translated into any other language. My husband took advantage of a sight-seeing tour of the Ladies' Program (as the only "male wife") and observed more of the Czech courageous resistance.

We made a pilgrimage to the famous 14th century Jewish cemetery and the five once flourishing ancient synagogues comprising the Jewish State Museum, where the Nazis had collected the artistic ritual objects in silver and embroidery from all destroyed synagogues in Bohemia and Moravia. Only the 13th century "Old-New Synagogue" was still in use by a handful of old people who had sent their surviving children to better lands. Behind it there is the 13th century Jewish town hall with the tower exhibiting the large famous clock with Hebrew numerals that move from right to left. A memorial to the Nazi victims lists their names and says, "The names accuse." A small temporary exhibition shows children's anguished poems and drawings, found in concentration camps. We had seen a similar collection in Vienna two years earlier.

Our Czech hosts went out of their way to make us feel comfortable, show their desire to come closer to the West and put distance between themselves and the U.S.S.R. At a reception, Dr. Pellinen (U.S.S.R.) was constantly flanked by two people of his delegation whom nobody knew, and the organizers did not even know whether they were geodesists at all. The three kept to themselves in a corner of the reception room and no one approached then. At the banquet, however, they could be separated at least a little. My husband and I were seated next to Mr. and Mrs. Pick, the hosts, at the head table, and Dr. Pellinen between my husband and someone of the Polish delegation. Dr. Pellinen's two shadows were seated at another table across from us; they did not say a word to anyone and probably watched him all the time. Geodetic meetings usually succeeded in transcending the international tensions, but this time blood pressures were high from history and memories. Thus, the unexpected availability of pictures from Apollo 12 and the moon landing was a great event. Everybody crowded into a too small room to view man's new achievement together. When Neil Armstrong had set foot on the Moon on July 20, 1969, the first man ever to do so, we had not been shown that many pictures yet. But the unprecedented event had electrified the world. Like many others probably, I had been sent comments from colleagues abroad. One wrote: "Though you did not personally land on the Moon, may I felicitate a U.S. citizen of this splendid achievement which is the honor of mankind and uniquely of U.S." To which I naturally replied: "To my way of thinking it is just incidental that U.S. citizens were carrying out the actual performance. I wish that this incredible achievement would inspire mankind to attend to and honestly pursue its major assignment: to create conditions here on earth, where everyone may sit in peace under his fig tree and none shall be afraid." And here we were, not exactly under a fig tree, but people from many countries crowded in this room, in communist Prague of all places, united in awe to see a man walk on the moon. Just before we left, however, the Czechs were suddenly forbidden to leave their country, even on a short vacation. When we were accompanied to the airport, some said: "I wish I could go with you." Back home in Washington, we saw a little note in the newspaper: the headwaiter of the hotel where we had stayed in Prague had been arrested because his response to a Russian customer had been considered disrespectful.

4. SOUTH AFRICA, 1970

"So geographers, in Afric maps
With savage pictures fill their gaps,
And o'er uninhabitable downs
Place elephants for want of towns."

Thus Jonathan Swift (*Gulliver's Travels*) ridiculed the African maps of his time. Geographic maps have improved since, but geodetic charts of Africa have still plenty of space between the sparse geodetic controls. It is not fashionable now to fill the gaps with savage pictures; one just leaves them blank. Yet, ironically, it was Africa where geodesy started more than two millennia ago, and where milestone contributions to geodesy have been made since. To recapitulate briefly: the first determination of the size of the Earth by Eratosthenes (3rd cent. B. C.) from an arc between Alexandria and Aswan; the Earth's ellipsoidal shape derived from an arc in South Africa by the French Academy of Sciences (18th cent.) similar to the more famous derivations in Lapland and Peru; the farsighted start of a geodetic survey along the 30th meridian to connect the Royal Observatory at the Cape with that in Greenwich (end of last cent.); and the observation of the lunar parallax simultaneously at these two observatories (1905-1910). The last two endeavors were completed several decades later by the Army Map Service as told before. The completed 30th meridian arc had been crucial for deriving a new, smaller earth model in 1956 and for demonstrating the inappropriateness of extending the European Datum to Africa. In the east-west direction, there was triangulation from Morocco to Egypt, which I had used, in a geoidal loop around the Mediterranean. AMS was heavily involved in work along the 12th parallel north, and there existed some work along the 6th parallel south. Professor B.M. Jones of the University of Natal was working along the 30th parallel south and had let me have some results for inclusion into the "Picture Puzzle" for Lucerne. So it was with a feeling of considerable involvement and pleasure that I received, out of the blue, an invitation in January 1969 to give a paper at the Fourth South African National Survey Conference in Durban, July 1970, with a request for an early response.

As usual, I sent the letter through channels asking whether I should accept. The question went all the way to DIA and the State Department. Then came informal guidance from the Department Chief, Mr. Culley, and from Mr. McCall and Mr. Yater: I would represent DoD or IUGG, and they suggested topics of interest; TOPOCOM would accept the invitation, but for the time being I should proceed on a personal level, since the political situation was delicate and the date was too far away for official paper work; I should submit an abstract. In April, the abstract was cleared and I sent it on to Dr. B. M. Jones in Durban. The full paper, "Constructing a Geodetic Datum that Fits a Continent," was expected in Durban by January 1970 for preprinting in the *Proceedings*. That suited me fine as I was very busy with PAIGH just then and intended to draw on comparable geodetic problems in South America and Africa.

Sandra Todd and I had in fact already worked for quite a while on the African project prominently listed among the twenty Branch projects for FY 1969. We would follow the South American procedure in first demonstrating the strongly distorting effect of extending existing datums beyond the area of their original derivation, and then going back to the original field work in order to avoid these pitfalls. Access to the fieldwork of the crucial 30th meridian was arranged with the specialists of the Europe-African Branch in the Foreign Control Division who expertly readjusted it on my 1968 Modified Mercury Datum, shown to be non-distorting. Work on the 12th parallel north would be completed in 1970. Mr. John S. McCall, the TOPOCOM's "Chief Geodesist", was greatly involved in this work for which he received a specific Commendation from Lt. General D.V. Bennett, Director of DIA, and General W. C. Westmoreland, U.S. Army Chief of Staff, with cover letters from the Chief of Engineers and TOPOCOM's Commanding Officer, Brig. General E. T. O'Donnell. We had already analyzed the satellite nets across the northern part of Africa, as mentioned earlier. In the southern part there were also Doppler, BC-4, and Baker-Nunn stations, but some of their locations with respect to the local geodetic control were not quite clear: it would be part of my mission to South Africa to help clarify such questions for our Satellite Record Center.

Reorganization tremors causing mountains of paper work reached us in the form of a request for input information into a "Long Range Geodetic Plan". The request came in March 1969 from Col. D. N. Hutchison, Director of Operations, and Col. S. J. Hathorn, Director of Plans, Programs, and Requirements, to the Department of Geodesy, and Mr. Culley appointed a committee to serve under Mr. McCall as chairman. It was suggested that we might as well go all the way and dream up all the projects we might reasonably wish to do or see done; the committee would then organize it and trim the fat. Bob Yater and Marian Hardy gave very much of their time to designing a reasonable format, and we all spent time formulating, formatting, and reformatting our projects plus justifications. Some of our inputs got mysteriously lost on the way again and one had to watch the transmission lines. Marian wanted me to help speed up the organization of the material into the "Plan" but I was not permitted to help. What we did not know then was that this would be a recurring and repetitive exercise in the coming years. In fact, this was already a duplication of listing the twenty projects of June 1968 for my Branch alone, the same in content and different in format only. Fortunately, I always kept copies of whatever I submitted, so that I would not have to spend so much time again to say the same things in blue and then in green in another year. The aim and justification of long-range authorized research does not change so fast. It is one of the wonders of government that they hire you to help them achieve their objectives, authorize your projects leading to those objectives, and then require you to justify again and again why you do what you were hired to do in the first place.

More upheaval, excitement and intrigues were involved in attempts to completely reorganize the department by a committee that curiously excluded the department chief. I managed to keep away from the turmoil as long as my work and my branch were apparently not or not yet affected. But when one could sense the muddying of waters and watch the undermining of branches and branch chiefs, I was alarmed too. Some of my people began complaining that as soon as I left for lunch, a certain administrator would appear and ask questions of how I did one or another thing, how work was organized, how certain data were utilized for what projects, what methods were used and how and why, leaving the impression that a duplication elsewhere was contemplated with possible transfer of functions. They were particularly irritated by his refusal to ask me as the Chief directly, and by his demand that I should not be told of his inquiries while he should be kept informed confidentially of all happenings. Well, well, so the intrigues had reached our shores too, and here we had a sinister spy in administrative clothes! I began to return from lunch unexpectedly early, and sure enough, he was there. So I asked with my most courteous smiles "What can I do for you?" This happened quite often before he gave up bothering my people and bypassing me, but all of us became more wary. It certainly did not create much respect for some managers.

In November 1969, Mr. Culley gave up waiting for the appointment to the job he had been considered sufficiently qualified to fill for almost three years in an "acting" capacity. Mr. Harry Adams took over as "acting" chief and was treated similarly. His was a caretaker regime until they found an outsider. He was "acting" again to fill the gaps between the parade of one-year department chiefs, some of whom clearly took the job as a stepping stone for something better, and some were colonels parked here on a short-term tour of duty. On 20 January 1970, Andy Glusic suddenly died. He had been suffering from severe breathing difficulties, a memento from the concentration camp way back in Yugoslavia. He used a breathing pump every few hours and strong medication; and he was aware that some day he might go out suddenly. His great wish was to finish his manuscript on leveling before that happened and he was unhappy that his time was taken up instead with silly little requests. I was familiar with his careful manuscript and when he died, I offered to see it through to publication in his memory. I was not permitted to do so, and the manuscript passed through other hands and never made it to publication.

In February 1970, Mr. Charles H. Frey, a member of the DIA MC troika, became our department chief (to the disappointment of some crown pretenders), on his way from the soon to be dissolved DIA MC to the soon to be established headquarters of a new Defense Mapping Agency (DMA). While he was with us, he concentrated on a thorough reorganization of the greater part of the department to accommodate newer techniques and objectives. Aided by Bill Doxey, the new Assistant Chief, and Carl Born, the new Chief of the Geodetic Control Division, he could succeed with centralized authority in what was beyond the chances of a possibly (not assuredly) well-meaning, yet lower-rank committee under internecine circumstances. The reorganization upheaval spilled over into our "Research and Analysis Division" with time consuming paper work on drafting functional statements that were rejected for being too specific or too general, and re-drafted to say less with more words. Much more interesting was a new award, offered by the Assistant Secretary of the Army (Research and Development), for "Recognition for Outstanding Accomplishment in Science or Technology for Southeast Asia" which seemed to fit Thelma Robinson perfectly. Thelma had become a specialist in navigation systems with specific application to Southeast Asia, and I quickly and happily wrote and forwarded the supporting paper work. It was readily approved but changed along the way into a group award to include Jim Walker, Chief of the Datum Branch, who had worked also on these projects. There was a very nice ceremony at the Army Engineer Topographic Laboratories, Fort Belvoir, and also a "Special Act" Award for them at TOPOCOM.

The paper for South Africa had been cleared and mailed to Durban in January 1970. I remembered that Mr. Hough had visited in Cape Town way back in 1953 in connection with his leading involvement in completing the historic 30th meridian arc, and I wrote to consult him on my mission. My husband and I had visited him in Woodstock, Va., the previous fall, and he had taken great interest in the success of our PAIGH work. I was sure of his interest also in this South African opportunity. In a warm letter, he congratulated me on this invitation, and he named a number of people he knew at the "Trig Survey" of South Africa and the University at Cape Town who probably would be at the Durban meeting. I missed the guidance of Mr. Mills, the late department chief, who had led the field party then, and had been involved later in the problems of a unified African datum. Mary Slutsky and I had prepared for him some background work to take to a conference in Africa. He would have welcomed the opportunity of this trip to push for cooperation towards datum unification. I also consulted with our Satellite Record Center and with the African specialists in the Foreign Control Division about specific current problems where I might have a chance to bring back information. And I looked up the Proceedings of the previous conference a couple of years back, and was pleased to see that AMS had been represented there by Al Nowicki, the AMS Technical Advisor. So I was shocked when the new department chief suddenly cancelled the trip. He said he regretted to have to start out with a negative action like that but travel funds had been severely cut. Hopefully, the next fiscal year would be better when my trip to the IUGG Assembly in Moscow would come up. I felt it odd that after having been asked, since more than a year, to step forward personally for this international representation, which had aroused interest all around, the rug was suddenly being pulled out from under me. I knew there was still a lot of traveling going on. Maybe the new department chief was not aware of the history and commitment of this assignment? I had no way of challenging his decision in the formal setting which the new boss seemed to establish, and the uncertainty surrounding a change of regime, and thus I had to take the explanation at face value; and I wrote to Durban accordingly. My husband and I discussed the matter for a while and then decided to inquire about prices and sightseeing possibilities and make a once-in-a-lifetime vacation of it. I had a very good salary and was willing to financially support my Agency's interests and prestige at its time of ostensible near-bankruptcy. The possibilities of this mission were quite obviously enmeshed in the objectives of my Agency with which I identified in my own aspirations and whose leadership had been good to me. Dr. B. M. Jones and his organization committee came through most generously. They offered both of us free accommodation at the Conference and a land rover to roam the country for sightseeing as long as we wanted. We did not take the land rover but accepted gratefully his help in planning our sightseeing trip and making all the reservations, which seemed impossible to be made from here. In March 1970, I received for the second time the nomination by OCE and by the Secretary of the Army for the Federal Woman's Award, with a glowing cover letter from General Podufaly, covered in turn by Col. Hutchison and Mr. Frey. Mr. Frey also arranged for an "Outstanding" performance rating, suggested by Mr. Bill Lane of his office and first recommended by OCE when transmitting the award information. The warmth and appreciation of my Agency all around me made me feel good, particularly at this moment of disappointment about the unfair slicing of the travel fund pie. It also put into focus the distinction between the Agency and individual antagonistic officials. One such lovely official came to tell me that he had advised Mr. Frey to cancel my trip as "non-essential", since it was only a land surveyors' conference. Why would he want to tell me that? What was wrong with land surveyors who asked us for a geodesy lecture? Mr. Rutscheidt, the division chief, was Secretary of the International Federation of Surveyors (F.I.G.) and was slated to go to Wiesbaden for their conference next year. Also, two South African geodesists at the Conference were members of my International Study Group Mr. E. Fitschen and Dr. B. M. Jones. I showed this administrator the *Proceedings* of the last South African conference with Nowicki's long paper, and it made him gasp with unpleasant surprise. I could not suppress the question: "Should you not go and see Mr. Frey and tell him that you misinformed him?" I just got a certain look that might have meant: Are you crazy or do you think I am crazy?

Around that time, somebody from OCE asked me to suggest of a showpiece about an Army geodetic achievement to place into some bigwig's office. Had I not heard that one before? I told him of the idea of more than half a year ago to make a plastic relief model of my MMD 68 geoid, which General Podufaly and Mr. Culley had liked; but that the General's note had mysteriously fallen between the cracks. I gave him my xerox copy of that note: maybe he could use that to revive the project. And he did. The original could not be located anywhere. It was heart-warming to see how quickly things happened now. The cost estimate was kept low by utilizing trainees. Col. Hutchison's endorsement said: "Topocom is foremost in the field of geoid determinations, yet this fact is hardly ever publicized in the world of mapping and geodesy. The distribution of plastic relief geoid models to well known mapping agencies throughout the world would add prestige to the TOPOCOM's image in the field of geodesy." The General approved the project and may have wondered why it took more than half a year for that cost estimate. A hundred copies of a wall map about a meter wide were ordered by October 1970. Rutscheidt handed me the official PAI (Project Assignment Instruction) and asked me to write a legend and explanatory paragraph and to compile a distribution list. Mindful of DIA's attempt to suppress my world datum, I cautiously listed only about 50 geodesy offices throughout the world but that was enlarged to 200 for a starter by Mr. Frey who was now on our side of the fence. A copy was proudly displayed in the lobby as the "Map of the Week." Many more charts were made later and were received everywhere with pleasure, but also with amusement about the title. In an eagerness to use the chart as an advertiser for the new, unfamiliar name of the "U.S. Topographic Command" (what's that?), it had been printed in huge letters across the top as if that were the title. Only two years later, the hardly known TOPOCOM passed into complete oblivion when it changed into the Defense Mapping Agency; and a strip with the real title "The Shape of the GEOID" had to be pasted over that name until a new edition was made. I kept receiving requests by foreign geodesists who had seen it hanging in somebody's office and wanted one too. There had been a question whether my explanatory paragraph for this chart would fall under the FOUO restriction (For Official Use Only), since that had been clamped on TR-67 on the AmoMD—F68. Of course not, because the information in the paragraph was by now widely mentioned in the open literature. But the question gave the impetus to have that restriction removed from TR-67, and I wrote a memorandum to that effect. I cited the appropriate regulations and meant to include a copy of the official letter ordering the FOUO restriction in 1968 so that it could be revoked. Who had written that letter? General Ploger? No such letter could be found in the files of the Commanding Officer or anywhere else. Is it possible that there never was such a letter? Or had it been removed for fear of ridicule, or abuse of power?

In a pleasant surprise, Mr. Kephart, Chief of the Public Affairs Office, called to say that my paper for the South African Conference "Constructing a Geodetic

Datum that Fits a Continent" had been selected by DIA and by ACSI (Army Chief for Staff Intelligence) to be "submitted by the Government of the United States of America" (with my name) to the Sixth United Nations Regional Cartographic Conference for Asia and the Far East, in Teheran, October—November 1970. I was impressed. No way to go along, though.

As soon as the complicated schedule for our South African trip was settled with the help of Dr. B. M. Jones and the flight booked, I applied for annual leave and seven days administrative leave (for part of the conference including travel time). The routing slip which I kept as a souvenir is dated May 11, well in advance of the July 7 flight. Mr. Rutscheidt approved the request but did not pass it on to the department, saying it was too early. On June 30, I filled out the official leave slip, referring to Mr. Rutscheidt's 11 May approval. That same day, I recapitulated with Mr. Doxey, the Assistant Department Chief (Mr. Frey was on leave) what I should try to accomplish on this mission, as I had done with several other people a few days earlier. Then I told Mr. Doxey that my request for a few days' administrative leave would reach him soon. He thought that request was fair. I was stunned when on July 2 my request was refused; an accompanying note (which I also kept as a souvenir), written not by Mr. Doxey who but by that official who, by his own information, had persuaded Mr. Frey to cancel my trip, said I sympathize with you but "a phone call to OCE had been made" and it had been found that I would require country clearance to get administrative leave, even though I had a visa. The "sympathizing" of that official could only be meant mockingly, and Mr. Doxey had not mentioned any intent to call OCE. What was going on? The note sounded almost sinister to me: for once, if it were true that country clearance (military approval or what?) was needed, why did they not get it in the almost eight weeks since my 11 May leave application? Secondly, administrative leave could be given, as far as I knew, within the department. The secretary dug out the leave regulations to that effect and I went for help to Mr. Joe Bernard, then Chief of our department's Operation Division, to cry on his friendly broad shoulder. Mr. Doxey was there too and both looked at the regulations and discovered that they had been misled: there had been no need to bring OCE in, since the leave authority did rest with the department. But, they said, one cannot disregard OCE after they had been consulted, and that could not be undone now. "Tell you what," said Mr. Doxey, "when you bring back materials for us and write us a letter in lieu of an official report, then we can acknowledge your work in retrospect with administrative leave. That has been done before." The world looked bright and friendly again.

The day before I left, I meant to review with my people what should be done or might come up in the three weeks of my absence, but there was no time. There suddenly came an unexpected, urgent request to prepare, by the end of the day, a detailed list of all subprojects to be accomplished in the foreseeable future, complete with man-hours required for each phase. The essential information requested was available in the division anyway, which made one wonder about a few things. It seemed almost impossible to write out predictions of all the details of the future in the short hours left, but not to finish might conceivably have dire effects on projects as well as manpower. A colleague from another division had applied through channels for transfer into my branch, but received no answers. I had applied for his transfer from my end, to no avail so far; was this connected with today's chore? Particularly mystifying was the demand of the Assistant Division Chief to be sure to include all sub phases and man-hour needs of the WGS-project, since his superior had not told him what the WGS commitment involved (!). Whether this claim was true or not, either way one could draw some conclusions. All my people offered to help. Together we produced reasonable "guesstimates" that would be compatible with previous similar listings and show the need for at least one or two more people. Neither did this colleague succeed to transfer, nor was our urgently requested listing. ever used. "Keep smiling."

At last, on 7 July, we were on our way to see an exotic world. We were met at the Johannesburg airport by someone from the Wild Company, the South African subsidiary of the Swiss firm for precision instruments, who kindly played hosts and guides to us for the three days we had in Johannesburg. It was an intensive introductory course into the life and thoughts of another land, about which you may have read but without grasping what you read until confronted with the stark impact of reality. The people we met during these three weeks (naturally only English-speaking whites) were extremely friendly and outgoing—and eager to explain to us the policy of apartheid although we never asked. We only observed and listened intently; there was defensive approval in some and frustrated criticism in others, all preoccupied with a feeling of impending doom (maybe in a decade or so, they thought) and at a loss or with opposing views of how to avoid it. We were amazed at the open and strongly worded criticism in the English-language newspaper. A visit to a training center for black miners from neighboring countries, here on contracts limited to a half to two years' duration, gave us a deeply upsetting demonstration of the methods used to keep these laborers in a robot status (no common language other than "Fanilago", a collection of a few artificial words and phrases denoting mining instruments and activities, minimal pay and no possibility of advancement, no accompanying families) and the diabolic sounding ideology of justification. Did the director who showed us around really believe what he told us?

There was a little sightseeing of modern Johannesburg with the golden dumps from exhausted gold mines in the vicinity, a drive through a new, small game reserve with exotic animals roaming freely and playful little monkeys crawling all over our car, and a visit to "Mrs. Plessis", the skull of the oldest so far known female Paranthropus which had been found here in the Sterkfontein cave. This and several hours spent on the NASA premises at Hartebeesthoek where a BC-4 tracking station was located, filled our time very tightly. The station manager was extremely helpful, glad to have a visitor from so far away. He showed me the BC-4 location and went out of his way to find all the documentation for it in answer to my many questions. He let me have copies of a stack of background papers with diagrams and detailed computations which should clarify some question marks of our Satellite Record Center; he showed us around his NASA station and let us listen in to a telephone conversation with NASA at Greenbelt, Maryland, on the other side of the world, asking for the weather there, which is near my home town, Takoma Park. It was a thrill to experience that my far away home was literally as near as the telephone. He then offered to phone to the Baker-Nunn station in Olifantsfontein and to the Trig Survey Office in Cape Town for some more information. Only with the Doppler station he regretfully could not help me because that was all-American and I would have to go through the Embassy. Ironically, I could help my Agency with everything but our own station because someone back home had deliberately messed up my official status.

In Durban, there were about 300 people: geodesists, astronomers, oceanographers, photogrammetrists, cartographers, land surveyors, and instrument firms. People remembered Mr. Nowicki's participation at their previous meeting. Some of the old-timers remembered Mr. Hough and also Mr. Mussetter and Mr. Mills from the days of the 30th meridian work and cordially returned Mr. Hough's greetings. Some people were IUGG members and had been in Lucerne in 1967. Two land surveyors hoped to be sent to the FIG conference in Wiesbaden. The geodesists welcomed my talk and the following lively discussion as a support for their efforts to get more understanding and funds for their projects from the offices with the purse strings. It sounded familiar. All big office holders of the profession including those from Rhodesia and Malawi were there and extremely cordial and willing to cooperate with us. They had plans for a readjustment of the whole South African triangulation after re-observing several known weak parts and at the same time would get rid of the different measures in English feet, Cape feet, etc. and change to the metric system. But they wanted to take their time, no need for American hurry. In the exhibition there also was a Wang 700 computer like we had, but this one had already a printer, and I was presented with a print-out of Snoopy the dog as a memento. Get-togethers in the following days brought more discussions and offers of cooperation in the spirit of the 30th meridian work at Mr. Hough's time. The Figure of the Earth and a unified African datum were appealing topics. Maybe one could arrange an exchange of observations for assistance with our computer capability? There also were questions about new satellite tracking equipment, which I would try to get answered after my return. The group from Cape Town offered cordial invitations for our stay there ten days hence, and help in clearing up all my remaining questions about the other two tracking stations. Meanwhile, we were taken to beautiful game reserves in Zululand and to the fabulous bushmen paintings in the caves of the Drakensbergen, and then we took a seven-day bus tour to Cape Town along the scenic garden route. On the way, we came through the Transkei with its capital Umtata, the first "independent" Bantustan and another disturbing exhibition of the country's attitude to its race problems.

In Cape Town, we were given the red carpet treatment. Our bus was met by one of our new friends from the Conference, and they took turns in driving us around to show us Cape Town's exceptionally beautiful location. Imagine standing on the southernmost point of the continent and waving to the opposite coast in Antarctica; or watching the differently colored waters of the colder Atlantic and warmer Indian Oceans mingle. What a fascinating beautiful world we have, if only people would not make such a mess of living in it. A plaque on a house wall in downtown Cape Town commemorates the starting point of the 18th century arc to determine the shape of the Earth. We also visited the Royal Observatory in homage to Sir David Gill who thought up and started the 30th meridian project and thus provided me with a fascinating career. There, Dr. Churms kindly showed us around. He had recognized my name from the literature, having looked up my parallax article by coincidence a month before. I tried to interest him in solving the remaining part of the parallax mystery, but he said that all records would be in Herstmonceaux, in England. There was an interesting invitation to Stellenbosch with further insight into the country's problems. We spent two days at the Trig Survey Office, where my questions had been researched in the meantime, and all remaining answers had been provided, including even those concerning the Doppler station.

I had written to Mr. Doxey, Assistant Department Chief, four very long detailed trip reports from South Africa, which were widely circulated, including at DIA, and I was happy to bring home so much useful material and offers of future geodetic information exchanges. A whole stack of interesting conference papers were catalogued for the library. I gave several informal technical briefings for Doxey, the Satellite Record Center, the African specialists, Yater, McCall, Byars, and all other interested colleagues, and there was a long column in TOPOCOMMENT, the office newspaper. Our many pictures of this exotic land made the rounds, including the front office of the department, before landing in another of our many travel albums. These three weeks had been an extraordinary experience, on several levels. There were two ominous little sequels, however, alerting me to the fact that the mischievousness of my favorite gremlin who had tried to prevent my trip was still there and virulent. Mr. Frey had asked me whether I would be willing to give a more formal briefing on the trip at some later yet unknown date, and I gladly prepared myself to be ready on short notice. Mr. Mischief who knew of my plans for a specific week's vacation for which I had made reservations long in advance, sent me a message to be ready at 3:30 p.m. on some yet undetermined day of that week, implying that I had better cancel that vacation. A written appeal to Mr. Frey revealed, however, that the briefing was not even set for that week but delayed due to several other people being out on vacation too. "No need to change vacation plans" wrote Mr. Frey. Mr. Mischief also changed my "Item of Interest" (without even telling me) by deleting the information that the trip had been "self-financed—no expense to the government," and that both my husband and I had been guests of the South African Surveyors Conference during the week at Durbin. When that same information appeared in the TOPOCOMMENT, I was taken to task for giving out incorrect information, since I was given seven days administrative leave. Doxey's promise of retroactive leave had been sidetracked, however, and I was wondering whether it would not disappear into the cracks in these times of intrigues. The request had not even left the division several weeks after my return. Also, it did not do anything for the flight expenses. But I offered, if and when I would actually see the leave restored, I would put a factual correction into the TOPOCOMMENT—which drew a hasty protests: No! That would make it worse by drawing even more attention."—Why do you fear attention to these facts?—No answer.

11

THE WIDENING GAP BETWEEN RESEARCH AND MANAGEMENT (1970–mid 1972)

1. MANAGEMENT ON A RAMPAGE

> "We trained hard—but it seemed that every time we were beginning to form up into teams, we would be reorganized. I was to learn that later in life we tend to meet any new situation by reorganizing, and the wonderful method it can be for creating the illusion of progress while producing confusion, inefficiency and demoralization."—Petronius Arbiter, 66 A.D.

So there was nothing new under the sun. The above pin-up in my office provided some sort of solace and detachment when watching the government infatuation with almost continuous reorganization, which encouraged secondary dislocations and unrest, out of proportion or even counterproductive to the basic purpose. In April 1970, the reorganization upheaval had reached our division again. We had just had one the previous year under the previous department chief, so we stayed in practice; we also had saved ourselves the effort to memorize the new names of the division and branches, since they were now changed again: the good old Research and Analysis Division had first changed to the Geodetic and Geophysical Sciences Division and became now the Geosciences Division; three branches then called Astronomy and Astrodynamics, Mathematical Support, and Geodetic Applications, were simplified into only two, the Geodetic and Astrodynamics Branch and the Mathematics Branch. There also was an Engineering Applications Branch to give a home to some survivors of the dissolved Engineering Services and Support Division at the Herndon location from where the phased out SECOR field work had been monitored. We had to give a home also to its division chief as our assistant division chief, which aroused strong pro and con feelings from the people in the know. My Geoid Branch kept its well-known name and identity upon my urgent request, since I could not see any virtue in exchanging it for the non-descript name "Geodesy Branch"; this name had been proposed by someone with the grandiose claim that "geodesy was in the process of emerging from the age of the geoid," thereby broadcasting his misunderstanding of the centrality of the geoidal concept in theoretical geodesy.

The formulations of functions, comments to these, and reformulations kept us busy again, as did another series of repeated descriptions of our several projects as to their objectives, background, outline of approach, manpower required, machine time needed, PERT charts (Program Evaluation Review Techniques) quality control, priority status, relationship to other projects and relationship to the department's master plan (which, however, was not available yet.) Fancy, time-consuming flow charts were required, parading as modern business tools supposedly for our own good, while so obviously they were a means by some administrators to get a superficial hold over our activities without having to really understand them. I guess we were all a bit irritated at the wasteful repetitiveness, which so clearly stemmed from lack of understanding and preplanning. The person in charge of getting all this information into an automated accounting system was not a geodesist and had a hard time, I am sure, but it did not exactly help matters when he countered with the preposterous statement that it was the researchers' job to guess what he needed and in what form, and that his rejections and our guess at re-rewritings would get us there eventually. This imperial statement was underlined by an insistence that certain listings and job definitions were needed now—or else"; and we scrambled all over ourselves to provide them, only to find out weeks later that nothing had been done with them and that they were going to come back for another rewrite in a different format. After a while, our initial good will for a new effort toward sorely missing orderliness was being used up and we realized that this show of imperial activity was just camouflage for inefficiency and frustration, coupled with an arrogant lack of skill in dealing with people. I managed to spend a minimum of time on each repetitive round, but some of my people worried about a spill-over to them: "Whenever you are not here, Mrs. Fischer, something is bound to come up." And a note from one of them, in charge of the branch during my vacation, described drawn-out discussions lasting five whole hours as completely useless: he had been required to immediately rewrite some of my project definitions with much more specific, yet non-existing information, which supposedly could not await my return lest the authorizing job numbers would be cancelled. His protest was countered by the peculiar guidance that he do the best he could and that what he wrote would not be used until I approved it upon my return (!). It all added up to a much increased appreciation and respect for the previous, lower-grade job number jugglers in Staff throughout the years who had been capable and willing to take basic information, put it into the correct format for their needs, and come back for checking against possible misunderstanding. The new automation angle had really nothing to do with that. Hurrah for the unsung heroes of the past.

We also heard with amazement that the celebrated, yet elusive, department master plan was a "first" in the making: the department never had had one before. Really? Did we old-timers not remember having been asked periodically for input and justification? Somebody looked for and found a master plan of Mr. Hough's time: the department then was about the same size, about the same number of employees as now. It not only functioned smoothly but managed to get all the projects approved that the department chief considered important; and he succeeded to build up a vibrant organization of international repute, all without any of the present gyrations. Hurrah for the old system. Administrative paper work grew like Topsy. It just grew. And it seemed to take priority over any technical work, which, in my naive opinion, was the purpose of the whole establishment. If we were not given enough time to produce, what would there be left to administer? Time just did not seem to count when applied to administrative work, while it figured as a major item in juggling manpower resources for production jobs and adjusting time-scale graphs for slippages of due dates. As a manager myself, I was involved in all this, but as a researcher at the same time, I could not help seeing the disproportion in efforts and the double standard in accounting for time spent. The weekly and then bi-weekly time sheets grew ever more complicated and time-consuming in themselves by requiring to peg sub-phases of jobs in units of an hour or half hour to different job numbers from a long list of possible job numbers, while all of an administrator's time was covered by one overhead number, irrespective of whether he accomplished whatever he was supposed to accomplish, or anything else or daydreamed, helped or hampered the production effort. Wasteful overhead unfairly raises the cost of the whole enterprise. In my mind, a practical solution was rather obvious why don't these "administrators only" from throughout the plant come and do some technical work for a. change, and thus help increase production in my (or another) branch; we'll teach them how to compute or do some other useful work at least for part of their time. They might at the same time learn to appreciate the value of time for researchers and refrain from bothering them for repetitive administrative information already accumulated in the files. They might learn to do more in less time by using common sense (if that information is needed in a different form, shape, or color) to make these earth-shaking adjustments themselves in short time. Of course, you can't make such obvious suggestions other than jokingly; but I applied them to myself in proportioning my time. More disturbing were peculiar difficulties in getting and transmitting information which seemed to fall into black holes on the way. Upon complaint, such happenings were acknowledged as just communication gaps", a label in lieu of a solution. I formed a habit of keeping a dated carbon copy of everything I sent out, even of the informal, handwritten routing slips, and letting someone know that the information was on the way. Information in the other direction was harder to monitor. A note that I should attend an informative meeting was received after the meeting. An assignment to brief the Commanding Colonel was received on the day before although the date of the request was several weeks earlier. The only representative to an important technical meeting out of town could tell us about the snowstorm there and with whom he went for a beer, but not the content of the meetings: "I can't remember thy details of a five-hour discussion."

Somebody who had a closer ring seat to observe events than I did, was worried about certain sinister aspects such as undermining people not present by insinuations from half truths, and arbitrary decisions in an abuse of power. The phenomenon was described to me as comparable to the infatuation with power, which a child might feel when left alone in a room full of new toys and experimenting how far his new power might go unchecked. Something like that may have applied to the experience of another employee at the time of the annual job performance appraisal, when the conversation went something like this: Supervisor: "Things are different now and tougher. I am going to downgrade your performance because you did not meet all the due dates." Employee: "But I did meet all due dates. I make it a point to watch all my due dates. Can you tell me which I missed?" Supervisors "I have been downgraded myself on some of those many due dates that I did not think were that important; and also on my files. Nobody has files in perfect order; so I will downgrade you on your files." Employee: "But my files are in order. Remember, you sometimes come to me when you can't find something." The employee eventually got out of a tough situation by indicating that he would quote this conversation in the space marked "Employee's Remarks" on the appraisal form. Stories like that did not help to enhance respect and trust or to fight general cynicism about management. Petronius Arbiter looked down on us from his perch on the door of my cabinet to remind us that it has been thus already 1900 years ago. The long view had a soothing, philosophical effect, but that alone was not enough. To calm down the waters to a state of detached serenity was one good thing, but a force was needed also to perk them up again for meaningful activity. I had a second pin-up on the wall of my file cabinet next to my desk, a picture of Galileo Galilei that I had found somewhere. He had a pensive look into the far distance; and the impact of the inescapable contrast between his probable thoughts and the paper madness around us was refreshing. His presence reinforced awareness of a clear distinction between real work and busy-work in a researcher's value system, and helped allocate precious time and energies accordingly. I had to defend his presence repeatedly against those who saw the purpose of a government office in a drab "business-like" appearance and not in inspiring creativity.

In the middle drawer of my desk where I kept pencils, paper clips, and little notes of short-range unfinished business etc., there also lay a small piece of paper with Lord Acton's famous: "Power corrupts, and absolute power corrupts absolutely." I don't remember what occasion made me put it there, whether in these trying years or later, but I kept it there all the later years until my retirement. It served as a sort of antidote or immunization against the bug of managerial infatuation with, if not abuse of, power that could be observed traipsing around the neighborhood now and then, making the bitten patient look like a fool. Whenever I needed a pencil or such, I could not help seeing that message to myself, which had a sobering as well as a detaching effect, a source of strength. Mr. Frey's regime brought some welcome improvements. Weekly branch chiefs meetings were started to follow the weekly division chiefs meetings to follow the weekly department chiefs meetings, allowing some information to filter through. Upon urgent invitation, Mr. Frey and Mr. Doxey agreed to visit the branches directly in order to prevent too much garbling of information in either direction. At one such occasion, Mr. Frey realized the need to establish more clearly his intended "open door" policy, thus correcting previous deliberate intimations to the contrary. He opened up again free communication with elements outside the department, which had been hampered by restrictions of an insecure management. In other words, we were again allowed to behave like real people—at least for a while.

During this reorganization, some people transferred or were transferred to other units.

Mr. Byars transferred to a planning position in Staff. Bill Veigl, and later Marvel Warden, joined my branch, and Thelma Robinson went to the new Math Branch. Marian Hardy, whose branch had been abolished, went into the recently established Operations Division where she served as the Chief and valiantly fought an up-hill battle against managerial capriciousness, the ubiquitous deliberate communications gap, and the unreliability of transmittals. She set up a system to clarify assignments and due dates, and initiated a mechanism to apply for approval of new projects. She was a symbol of helpfulness in a widespread sea of distrust and unhelpfulness.

Considering the many new faces in the higher management layers, particularly the unaccustomed abundance of military personnel in authoritative positions and/or on short-term stays, briefings became more frequent, more formal, more repetitive, limited to so and so many minutes of duration, and were sometimes cancelled in the last moment or even after an unannounced delay. Mr. Frey collected a set of standardized talks and visual aids for his own briefing needs and invited contributions. Under the project "Visibility", we designed beautiful briefing boards, viewgraphs, slides, and exhibition pieces, which could be displayed on short notice to suit every occasion. It was a rather time-consuming, if enjoyable activity, but necessary for survival, since unimpressed bosses might pull the supporting purse strings shut. Convinced that a picture was worth a thousand words, I had used carefully designed illustrations in almost all my publications and presentations for technical audiences, visitors, and newcomers to the field, but not for a parade of changing and possibly preoccupied bosses. My collection needed to be enlarged for the purpose and spruced up in colors. Here was a curiously inverted situation: explanations, overviews, outlook, inspiration, teaching used to come down to me from my superiors; now I used colorful teaching aids in the opposite direction, to arouse the interest of my superiors and inspire them into supporting my projects as fascinating and important endeavors. Having been a teacher all my life, I rather enjoyed this task. My branch exhibitions were really quite impressive: large colorful briefing boards for the background and products of major projects; the large relief model of the North American geoid with the story of the ice-age depression in the Hudson Bay, which had been a conversation piece in the Commanding Officer's room; the new plastic relief model of my MMD 68 which was being sent out as a much appreciated TOPOCOM goodwill gift all over the world a relief map of South America to explain the idea of the new South American Datum of 19691 a graphical history of Special Purpose Mapping in Geoid Branch"; several geoid maps and a mountain of publications. I could quickly tell from the face of the visitor whether he was interested and in what, and then elaborate on the appropriate items in my many-sided collection.

At the occasion of a briefing for someone from the Civil Service Commission for the purpose of reviewing and updating Civil Service standards for the geodesy series, I suggested to Mr. Frey and Mr. Doxey that I would use the experience of my many briefings on basic geodetic concepts to devise a standard talk for their collection which could be adapted to various length. They liked the samples of what I had in mind and Mr. Doxey suggested a grouping around the three major topics: the shape and size of the earth, the gravity field of the earth, and point positioning. And so I started to design one of my "best sellers", called "Basic Geodesy, An Initiation into the Mysteries of Geodetic Concepts." It was to be a series of colored slides with a couple of brief paragraphs accompanying each, for the use of the Department Chief and for my own briefings; but its instant success turned it also into a taped narration for the use of the TOPOCOM Training Center in its introductory courses for technical and administrative employees, and into Student Pamphlets of the U.S. Army Engineer School at Fort Belvoir. Mass production there had been arranged by Col. R. A. Whelan (Topographic Production Center), Col. S. Hathorn (Chief of Operations), and Col. E. Kurtz (Chief, Department of Topography, at the school). I mentioned already earlier that I had saved, in 1959, a delightful "PEANUTS" cartoon about Dr. O'Keefe's pear-shaped Earth, which I wanted to include but was refused copyright permission by the United Feature Syndicate, Inc. in New York. Another disappointment was the refusal of the Army to accept the subtitle for the cover and some choice stories. The pamphlet was available in 1972 with colored illustrations, and distributed all over the (geodetic) world in several thousands of copies. It was even translated into Spanish and French, and had to be reprinted to fill the many requests.

The ascendancy of management could be witnessed from the rash of managerial courses all supervisors were required to have all of a sudden. The idea was probably well meant, though we could not see any beneficial effect on poor supervisors, and marveled about the amount of time these high-paid employees (including myself) were required to spend. There were in-house courses at our own Training Center, where you could still squeeze in some business in part of the day; and there were fancy outside courses, where you were stuck for two solid weeks, though rewarded with the opportunity to listen to a sequence of high placed, knowledgeable personages in government and private industry. The in-house courses had the advantage of being more applicable to in-house situations. Mr. Gil Monck, Chief of the Training Center, showed a rare talent of monitoring the discussions around management films and reading material in a relaxed atmosphere of general trust and openness. The freedom to gripe gave at least the solace of common misery, and thus a better chance of pertinent analysis and possible remedies. Mr. Bill Evans, instructor at the Training Center, turned out to be a fantastic actor. I still remember with admiration his demonstration of "how to tie your shoe" and how to put on a coat" when following exactly a sequence of orders supposed to fully describe the task. It splendidly drove home the point of the difference between what we think we are saying and what the listener may hear and understand. The outside courses had other advantages. First, they provided the stamp of validity and authority to the concerns of these courses as dealing with really, generally and seriously discussed problems, which seemed trivial to some of us. They also provided the opportunity of studying the performance of some very polished or not so polished speakers, both in delivery and in logical development of their messages. That message reflected, of course, the viewpoint of the speaker and differed interestingly from that of another speaker. Different viewpoints were also in the pertinent literature; and I naturally took sides when these supported or contradicted my own experiences as a long-time teacher and supervisor and also as an involved observer of all types of supervisors. Some of the old AMS managers appeared to have been at the forefront of modernity, while the current trend seemed to point in the opposite direction despite the high investment in these Army-sponsored courses. As a doer, I attempted to find ways to make this investment relevant to my agency and proposed a committee of interested first-line supervisors for a start. The suggestion program was being pushed at that time by the requirement of monthly submissions, and the "Resources Conservation Program (RECON)" was a buzzword. So I tried to catch three flies with one stroke and submitted a suggestion called "Motivating People at Work, a RECON Study." It was disapproved, of course. One of the anonymous evaluators (rejectors) wrote: "...It is obvious that...feels that the concepts taught in the various personnel management courses would be beneficial to TOPOCOM if they were applied...Formal application of personnel management concepts on a Command-wide basis is much more difficult to achieve...The suggester seems to favor a participative form of management...Possibly this sort of management could benefit the Command. However, it is highly questionable whether a TOPOCOM study on how to motivate people at work would result in any tangible benefits..." Well, I tried. I must have expected a rejection, as my suggestion carried an intriguing motto, copied from one hanging in a big frame in the Ruth Building lobby "Only he who attempts the ridiculous can achieve the impossible." (Will Henry)

2. PREPARATION FOR THE IUGG ASSEMBLY, MOSCOW 1971

The first reminder of the approaching 15th General Assembly of the IUGG in Moscow, August 1971, came as usual in the form of a request for input into the U.S. National Report, in August 1970. It was time to recapitulate what had been done since the last General Assembly in Lucerne in 1967, and these recapitulations would take several forms: the national reports, the reports of the section presidents, the section secretaries, the study groups, the commissions, and individual papers. While this arrangement included some duplication, it provided a convenient grouping of the huge amount of material, and authority to request information with the certainty of all around cooperation. My own obligations included writing an input about my own work for the U.S. National Report, writing a Secretary's Report about activities in the topics of Section V with an appropriate bibliography, write a Report on the activities of my Special Study Group V-19, and assist the Section President, Dr. Tengström, in collecting information, in setting up the program for the allotted sessions, and in helping him at the Assembly in every way he would suggest.

My input into the U.S. National Report described the background of four major projects the revised geoid chart of North America in three map sheets which were already widely used; the geoid chart Of Australia; the South American Datum of 1969; and the Modified Mercury Datum of 1968. The latter had gained wide recognition as the only available unclassified, up-to-date world datum, and requests for information and reprints came frequently. The DIA must have had a change of heart as there was no difficulty anymore from that side, neither against advertising the MMD 68 through the worldwide distribution of the plastic relief model not through filling frequent requests for TR-67. Despite its restrictive marking, this Report was released even to foreign countries, as to the Greenwich Observatory upon request by the U.S. Naval Observatory for use in a common project. A beautiful letter of commendation for me came from the Air Force Systems Command at Andrews Air Force Base to Col. John R. Oswalt, then our Commanding Officer, telling him that my MMD 68 was being used in their advanced systems and strategic development analysis and in the joint DoD/NASA space transportation system. It was covered by a pleased Commanding Officer's letter of appreciation, and in turn with one by Col. D. N. Hutchison, CO of the Topographic Production Center, the next management layer above our Department of Geodesy and Geophysics.

The Report of the Special Study Group was, of course, based on the communications of the members, which arrived promptly in response to my circular. With several members I had been in correspondence all along, since exchange of ideas, results, and sometimes also data was the purpose of forming a study group. Dr. B. M. Jones and Mr. E. Fitschen of South Africa were members and I just had had a closer insight into their work on my South African trip. It was possible, with Mr. Doxey's help, to send them literature about a new geodetic tool, which they were interested in, and I also searched, upon Mr. Fitschen's request, our archives and Mr. Hough's memory for some documentation in connection with their work on the 30th meridian. Correspondence with Brig. Bomford kept me informed about the revisions and improvements of his geoid chart of Europe, and I succeeded to persuade him to issue a new chart at Moscow, for the benefit of all of us. He did so in an appendix to my report. President Tengström wrote about his plans for the program. He as well as Bomford and Levallois were always aware of what I was doing. Hristv continued discussions of the new IAG Reference Ellipsoid. Mr. Muminagic, President of the Yugoslav Committee of Geodesy and Geophysics, sent a new geoid chart of Yugoslavia. The Turkish Society of Geodesy sent the first issue of their new journal with an invitation for me to publish there. George Mourad, Battelle Memorial Laboratories, kept me informed of their great achievement of placing a geodetic marker on the ocean floor in more than 6000 ft of water, and he sent a program for a Marine Geodesy Symposium at Moscow. For the Secretary's Report I checked the literature of the past four years and sent another circular to colleagues in other countries who were themselves involved in collecting information in their country for their own reports. Receiving and giving such almost simultaneously collected information depended on timing, but worked out all right.

At that time a letter came from Dr. A. Barvir, Vienna, inviting me to participate in a "Festschrift" by the Österreichische Verein für Vermessungswesen in honor of Dr. Karl Ledersteger at his 70th birthday. Time was short and clearance channels were long, so I decided to write an abbreviated take-off on my own parallax paper, given as a successful after-dinner entertainment at the Washington Philosophical Society several years before. While my parallax studies were well known in the literature, the *Bulletin* of this Society was hardly circulating abroad; and this talk might entertain Dr. Ledersteger, particularly if spruced up a little in a German translation to suit the occasion and to soothe his ruffled feelings about my using English with him. So after securing the copyright permission from the Society, I went to work with gusto. First, I changed the sober title "The Distance to the Moon" to the mood-setter "Über die Schwierigkeiten in einfachen Problemen, gezeigt an der Berechnung der Monddistanz" (On the Difficulties in Simple Problems, Shown in the Derivation of the Distance to the Moon). After another mood-setter in an introductory paragraph of birthday congratulations, I managed the German translation (my first and last geodetic paper in German) in a light vein, even using German language associations here and there. I had even more fun when someone in the clearance channel who did not know German, insisted that I list and explain all places that differed from the English original. Did he think that in freely quoting Wilhelm Bush or Friedrich Schiller in a fun paper one might transmit classified defense secrets? The yearly DoD Geodetic Conference this year was postponed from October to April, and abstracts were requested. My branch had two appropriate candidates: one was a heap of numerical comparisons between results derived from the various competing satellite techniques and terrestrial results, all in connection with the ongoing activities of the three-service WGS-committee. Ray Shirley who continuously worked on these, with off and on support from others in the branch, would put his experience and conclusions into a systematic form and present it to the classified DoD meeting. The other was an unclassified analysis of the HIRAN net in the northeast South America, whose frustrating shortcomings had not let us sleep. Sandy Todd and I tried to pin down the villain by securing the original observations and duplicating the derivation procedures, where we discovered room for improvement. A discussion and demonstration of the effects of several refinements to the standard procedure was given in our co-authored paper "A Refined Procedure for Computing Geodetic Distances from HIRAN Observations" (*Surveying and Mapping*, No.1, 1972), which Sandy presented to the AGU meeting in April 1971. Our findings were endorsed as "significant results" by DIA and passed on to Air Force units involved in HIRAN reduction problems, which made us feel very good, naturally.

In preparation for Moscow, Professor H. Draheim, Karlsruhe, sent a questionnaire to a number of geodesists for their views of the current status of geodesy and the most significant developments in the last twenty years. The deadline for the answers to four key questions was short, but the answers were supposed to be brief and to the point; and I managed on time. Draheim published this collection in an interesting article in the AVN (July 1971). Everything pointed to the approaching Assembly in Moscow, in a hightened pace of activities, but now there appeared a sinister cloud in the sky. There had been an alarming incident in Moscow of harassing an American Jewish tourist. I don't remember the details, but I remember that many people asked the AGU to withdraw from the Assembly in Moscow and/or insist on a change of location. It was the first time ever that the IUGG General Assembly would take place in the U.S.S.R.; and the U.S.S.R. Academy of Sciences had already invested so much in the preparation and prestige of hosting the IUGG, that they were eager and successful in transmitting a guarantee to the IUGG that no one of the delegations would be exposed to any annoyance. After some uncertainty, the IUGG Secretary General, Dr. G. Garland, sent word to us that the guarantee had been accepted and we should feel safe to come. Thus I was greatly disturbed when I was told at TOPOCOM that I could not go to Moscow "because the Government could not protect my safety, since I had not come with the Mayflower. I was not to go on my own money either" (clearly referring to the South African trip I had paid for myself). I knew that Jewish colleagues of other agencies were going. That phrase about the Mayflower obviously meant that I was a naturalized, not a native citizen; but were naturalized citizens second-class citizens? Except for becoming President of the United States, in which I was not at all interested, there should not be any difference. A colleague of another agency commented that this was a symptom that the mighty United States of America did not feel as mighty anymore as it used to. But what about the guarantee? I checked with Charles Whitten, the leader of our delegation, and with Don Rice. Both were very surprised and confirmed the validity of the guarantees they could not interfere with decisions of another agency, however. I felt disturbed at being placed again in a position of having to back out of a TOPOCOM commitment (toward the IAG in my two officers' roles) which my agency had wanted me to accept on a personal level in their own interest all along. I had also the uneasy feeling that, considering the guarantee, there must be another reason. But which? The department chief had called several key people of the department's hierarchy to listen to his canceling my trip. Seeing all these people assembled I had expected some important business agenda involving us all, but nothing else came up. Odd—why were these (supposedly busy) people asked to attend? To impress me with the authority and infallibility of the hierarchy? I would have been the wrong party for that. It struck me sort of funny: several tall silent men and one small woman; did they fear I might pull a gun out of my pocket or what was the point? Or was the show meant to hide the hunch of another unfair deal? They must have known about the guarantee as well as I did. When we all left the room, I heard a well-known voice behind me ask the department chief: "May I go in her place?" That gave it another angle: Did someone just want to keep me from taking another trip? I had heard comments repeatedly in previous years (though not referring to me) that conference trips were vacations and not work times. These may have been confessions by whoever made such comments. A substitute for my place on the travel schedule would hardly fulfill my commitments at the Assembly nor bring home any pertinent technical information. Well, when I sent advance copies of my two reports to the members of my Study Group, to the President, and the Secretary General as required, I added the information that I could not come to Moscow. Back came a surprised and concerned letter from Levallois, the IAG General Secretary, deploring this decision which he could not understand, and offering, if the reason were money, to send immediately a grant by the Association for my expenses. Anxious that Levallois should not think that I was backing out by my own free will, I asked Mr. Frey what I was to give as a reason. Frey said to give the reason I had been given, but the channels wanted to see my letter. When I submitted the letter, however, it was changed (although it was a personal letter) and the (real?) reason taken out. It now said, correctly, that only one representative of TOPOCOM was permitted to attend and the department chief had been selected. So maybe that was the real reason? It did not explain why I could not go on my own or on IAG money. But it meant that those management elements had gained strength who approved scientific conference attendance only for the smallest number and preferably for administrators, and not for those involved in the subject matter. In fact, this trend grew stronger in later years when administrators are sent who do not contribute anything to the scientific discussions at the meeting nor bring back any worthwhile insight for the researchers who would have profited from attending for the benefit of the agency. One such administrator observed: "I am bored stiff by these technical sessions." Others were not seen anywhere for a couple of days. Another reported of a high-spirited, agitated discussion" that would have interested you" but was unable to tell the points of the argument.

Levallois' concern may have reflected yet another aspect my absence from Moscow upset the IAG applecart for the customary succession of offices. As the first secretary, I was expected to take over the chores of the section presidency from Tengström—if I were there. Could it be that someone in my agency had wanted to prevent me from becoming section president, or prevent a woman for that matter? None of these possible reasons had anything to do with a safety risk of going to Moscow, which was never mentioned again. Had someone in my agency used this idea as a pretext to muddy the waters? In Moscow, my Secretary's Report was presented by Milan Bursa (Prague) and my SSG-Report by Don Rice (USA). Mr. U. Uotila (USA) became president of Section V, L. P. Pellinen (USSR) moved up to First Secretary and R. S. Mather (Australia) joined as the Second Secretary. My SSG V-29 was continued under my chairmanship with the new title "Astrogravimetric methods for determining the shape of the geoid."

I was surprised and felt very honored when I received a telegram of greetings from the General Assembly. I also received from Paris a package of all the national reports presented at Moscow. My husband and I went on vacation to Europe that summer, and it felt very odd to have been actually forbidden to travel a little further and join the Association to do my job there. I did not hear of any other case like that.

When I returned from vacation, there was Professor Mather visiting, and he and Byars (then in DIA) and Rice (whom I had called to thank him for presenting my paper) told me about Moscow and conveyed many personal greetings from there. There also was a letter from the Bulgarian Academy of Sciences, Professor E. Ivanov, inviting me to participate with a paper in a "Festschrift" for Professor Hristov in honor of his 70th birthday. In due time I sent an article entitled "The role of Africa in the history of geodetic concepts."

3. A DOLLAR A DAY

A new assignment to study the effect of refined gravity information on missile trajectories within the TOPOCOM mission made me look up my notes from days of yore when I had assisted Chovitz in his TR-16 study and computed for him the effect which deflections of the vertical might have on a hypothetical flight path across the United States, or rather the error incurred by neglecting these deflections. Good to have a habit of making and keeping notes all along. I had to defend my growing files repeatedly against bureaucratic demands for destruction after two years or, what is practically the same thing, for transmitting them to the archives (read "morgue"). Each time, I had managed to persuade the various messengers of destruction that the work of a researcher is never really finished and that these notes were actually extensions of my personal memory of thought which may be applicable to future work. When a certain volume of transmittals was enforced periodically (by the foot of file space), there were two ways to circumvent that: (1) to send material that otherwise would have gone into the wastepaper basket anyway, but had been saved for just the occasion of such enforcements, and (2) to xerox material worth keeping. My files came in handy many times and proved great time and energy savers. In filling administrative requests for justifications, for instance, I could pick up a similar request of years ago and update it in a few minutes.

In this particular case of trajectories, my old notes brought me quickly back into the subject matter. Our rivals at ACIC, however, had taken over this whole complex of problems in the meantime, with the justification that the Air Force had more immediate need for such service, and that ACIC operated the official depository of accumulated gravity data. They had set up an entire section for continuous research on all aspects so that, with the help of authorized contracts with private industry, they could answer any specialized question at the drop of a hat, push-button like. This capability made our assignment look rather strange; why build up again a similar capability at TOPOCOM? To see whether there were any significant differences for TOPOCOM, that is, for the Army, we had to find out about specific Army mission plans for the foreseeable future; and we had to find and inspect existing programming packages for possible adaptation. Both types of explorations were not so simple, with resistance and incomplete information along the way. Marvel Warden and Louis Jones primarily, but also Phil Wyatt and I were involved in that assignment. ACIC understandably tried to defend their turf. They could not refuse an official request for the program packages but they could drag out delivery, and not send along the required explanations and test cases until repeatedly requested. The trouble with these involved programs written by someone else is that, in the absence of the explanations, it takes a while to see how that program functions and where the places are for changing it and adapting it for your own purposes. There was no time to write our own program, but we managed to do what we needed. The other hunt (for future Army plans) was also difficult, first to find the right source, and then to find that certain numerical information was not available yet. So we went as far as we could and produced a number of hypothetical examples, to be replaced when more information became available.

For the typical subject matter of our own usual research, the programs were home-made from scratch, of course. Phil Wyatt, the senior mathematician and programmer of the Geoid Branch, was involved in most of our projects, and he also helped other employees along. For a while, he conducted a little school, a series of regularly scheduled training sessions in programming skills for our more recent members, and some people from other branches joined appreciatively. Phil also repeatedly advised the computer section at the White Sands Missile Range in their efforts to compute a very detailed geoid there, which in due time earned him a gracious Letter of Commendation from the Missile Range, endorsed by our then department chief. One of his programs was lovingly named "Adam", in honor of his little son.

Ray Shirley spent some time to devise an automatic procedure-for our specific needs of plotting and contouring deflection values. These values are usually very irregularly distributed, but important to preserve as spot values. The usual run-of-the-mill contouring programs, by contrast, started with evenly distributed values and the original data were usually lost in machine averaging. Programming up to then was considered, at least in my branch, as a means to do a geodetic job, not as a job in itself. But with increasing complexity of programming skills and experience, programs were written up formally for the use of others too. Bill Veigl was interested in expanding plotting and contouring programs to cover special requirements such as overlays for several specific map projections, and different degrees of smoothing. He was an independent worker with enough initiative and enthusiasm to go out and find customers for his wares. He volunteered to stay evenings at work in order to get access to the automatic plotter. In time, he wrote some useful *Geodetic Memoranda* presenting and documenting his programs with a number of special options to meet the various needs of the customers. He earned a Special Act Award for his energetic efforts.

Briefing the brass on the Figure of the Earth and related topics was one among my several simultaneous chores. In 1970, General Podufaly had retired, and Col. John R. Oswalt filled in temporarily for a few months until the new Commanding Officer, a very different type, came in. It was Brig. General Edwin T. O'Donnell, and he needed to be briefed with stripes and flourishes. He was a very young looking, short and slender fellow, making up for that in snappy, military behavior. As mostly when facing the powerful Unknown, some administrators were jittery and circulated stories of how the new CO was asking specialized technical questions to test and/or embarrass you; and they insisted on dry runs for the briefings. Dry runs were relatively new to me and I disliked them as a waste of time and a likely way to produce dull performances. The spontaneity and enthusiasm of a free informal briefing would get lost, I felt, in a full-scale repetition. So I skimped on the dry run and dramatized the real performance. The content, in this and similar cases, was the general work of the branch, specifically showing off the plastic relief model of the MMD 68 and other show pieces. Col. O'Donnell really did ask some detailed numerical questions, which, since it was my own work, I could easily answer, pleased with his interest whatever, his motives (real or show business interest). An administrator who had wanted to replace me for that briefing, felt relieved because, as he admitted, he could not have answered those questions; and you could not have anticipated them in a dry-run. I could talk about my work in my sleep, without needing any extra effort or preparation. The interest for me lay in discerning the reaction of my visitor and eliciting his interest in turn. You cannot do that in a dry run either: that runs dry.

A requested briefing for Col. R. A. Whelan on the trajectory study was of a different type: he really wanted to know something, namely what I planned to do in this study and why; and I wanted to use that occasion to interest him in my new creation of the *Basic Geodesy* and its newly designed illustrations. Appropriately, I presented the second parts "The Gravity Field of the Earth". It starts out with the story of the apple that fell on Newton and started him thinking about gravity which causes things to fall; and it proceeds to the role of gravity in both helping and hindering geodesy, and in fooling us when we try to predict satellite orbits or missile trajectories. And that is why we have to watch that little villain. As mentioned earlier, Col. Whelan liked this part, wanted to hear and see the other two parts to my delight, and subsequently arranged for the publication. Bill Doxey and Joe Bernard helped me to sell this trilogy to the colonels, who helped smooth the way.

As the trajectory study proceeded, we were asked for a technical briefing on its status, which we then documented in a written Interim Report". Apparently, it was considered interesting despite its interim, unfinished nature, and I was scheduled for another briefing at the Command and Staff Conference. At that time, these periodic conferences were preceded by a roughly ten minute formal presentation on some important project, to keep the bigwigs informed on what we were doing. We called it irreverently the "show and tell" performances. General O'Donnell was about to leave after a little over a year with us, but his successor, Col. Edward C. Anderson, Jr., was there. I had eight slides for the ten minutes, but they were simple, beautiful, and to the point. I concentrated on the basic information of why, what, and what mores why we were doing this project, what we had done so far, and what more we planned to do. It was a big success. Some of the people there said: "I had no idea that you were doing such interesting things." *Dé jà vu* ; had I not heard something like that, years before?

A chore that was thoroughly delightful, was taking a course "Effective Speaking" from Mr. Joseph P. LePire, of our Department of Graphic Arts, in February and March 1971. At first, I had thought that I did not need such a course, but after the first session I was glad I had come. There was a lot that I learned from Joe. It was not only the well-known guideline that was forcefully brought home to me: "First, I tell them what I'm going to tell them, then I tell them, and then I tell them what I told them", but also the actual training in performance, enforced by obligatory, mutual, constructive criticism of the class to observe specific do's and don'ts. For instance, the appointed critic had to count the performer's fill-in "r—r"s, and I have become sensitized to such nuisance in myself and any speaker ever since. It still strikes me as inappropriate to hear a speaker end with the bland routine phrase "thank you very much", remembering Joe's comment that it is the audience and not the speaker who owes thanks. The speaker might say: "Thank you for your attention" or "for your time" or something similar, but not an empty thank you"; and I trained myself to follow Joe's advice.

There was homework to prepare a speech for each session, limited to the length of a specified number of minutes (stop watched), on topics characterized as descriptions (what it is), explanations (how it works), entertainment (with a central thought, not just jokes), persuasion (to act or support, sales talk), opposition (try to convince by reason, facts, hopes to help solve a problem, ideals), or manuscript (well-reasoned discourse). Formulate an appropriate purpose-sentence for your own guidance for which of these you want to achieve and how. Communicate with the audience, don't mumble monologues to yourself or to the slides. Use eye contact and pauses to involve the audience, to get and hold their attention. Impress the purpose and main points of your speech without distracting details. Watch your posture, voice volume, and phrasing. Use gestures for emphasis but don't overdo that. Don't fiddle with your notes; handle them unobtrusively, out of sight.

The organized criticism of the class had to judge you on each and everyone of the above points and more. Joe was a superb teacher.

In fall 1971, Charlie Frey left on detail to help organize the Defense Mapping Agency. Bill Doxey, his Assistant, had been named Chairman of the Tri-Service Implementation Committee already the year before. Frey's accomplishments in the reorganization of our department had been recognized with an Outstanding Performance Award and the Army Meritorious Civilian Service Medal. Before he left, he finally straightened out the peculiar difficulties about my promotion, first launched in March 1967. Since that time, the periodic Position and Pay Management surveys had come and gone, helping me to get promotions for my people and promoting many others, but Mr. Gensch's valiant efforts to turn the rising stack of my recognized contributions into a promotion for me had peculiarly run into periods of hiring freezes or restrictions of high grades or overload of priority business, all of which were just as peculiarly avoided when it came to promotions for men. My qualifications were never challenged in view of the Civil Service Standards, but at one time I was told that action could not be taken because it had not been decided yet what the name of my branch (Geoid Branch or Geodesy Branch) should be in the reorganization; this clearly violated the intent of my individual researcher title, which had nothing to do with a branch. At another time, Mr. Gensch was told to wait and see whether I would get the Federal Woman's Award nomination. I had already received that nomination by the Secretary of the Army years ago and was now being proposed for the second time.

Which man's promotion ever depended on getting a Federal Woman's Award nomination?

Mr. H. Dunning, Chief of Personnel, came to my assistance. He explained to the new department chief the history of that blocked promotion and arranged an interview for me, since my direct approach was prohibited by a deliberately incorrect announcement of a "closed door" policy. Mr. Frey first corrected this erroneous policy statement and then familiarized himself with the problem. It was said that the formulation of the supervisory control in my job description had been questioned and should have been but was not decided by authority of the previous department chief. It was the first time that I heard of any question about a specific formulation in my job description and I could see now how the lack of an appointed department chief in the period between Mr. Mill's passing and Mr. Frey's appointment had affected even my case through the intrigues surrounding the prolonged makeshift situation and the resulting insecurity of the only "detailed" Acting Chiefs. Mr. Frey updated my job description; but there was no easy sailing yet and things dragged on again. Yet, there was a celebrated promotion to the first GS-16 at TOPOCOM for the Technical Director, and there were GS-15 promotions for two men detailed to our department who had threatened to return to their previous positions elsewhere in the agency. Another GS-15 in our department had been detailed out of the agency since years, practically for good, and his slot was unused. Then one day, the division chief informed me that my promotion papers would not even be accepted by OCE at this time because of the pending reorganization. *Dé jà vu!* My promotion as an individual research geodesist was again being tied to an overall, slow-moving structural reorganization, contrary to the intent of the researcher title. I went again for help to the Chief of Personnel. Mr. Dunning had left due to his own problems. The new chief, Mr. B. E. Stevenson, and his assistant, Mrs. Mary Herring, listened with surprise and succeeded shortly in sending my promotion papers out to OCE, who not only accepted them right away but also returned their approval in a record time of a few weeks. The Department of the Army approved the promotion in February 1972 despite the current upgrading freeze: an exception was specifically authorized for me as permitted in cases of rectifying an improper classification. But don't congratulate me yet. When I was called into Col. Anderson's Executive Office where several dignitaries were assembled, including Mr. Frey who had kindly come over from DMA for the purpose, Col. Anderson began with a lengthy speech about the dire financial circumstances of the agency and my unique personal contributions to its prestige, and I thought: here we go again, they'll ask me to support their finances by foregoing my promotion, out of loyalty. But it was not quite that bad. He told me that my promotion had been granted (so far so good), but he then asked my consent to delay its effective date to the very last Monday before the end of the fiscal year. My consent? What choice did I have? What else could I say but that I had waited for this promotion five years by now, and a few more weeks would not make that much difference if I could rely on it. He promised that I could rely on it and he set the date for the ceremony. The date was important because with the end of the fiscal year the authorization for my promotion through OCE would expire, due to the reorganization into DMA. There was a congratulatory letter from General H. W. Penney, Director of DIA.

I could not help wondering, and I certainly had not heard, that the three men recently promoted to GS-15 and one even to GS-16 had also been asked to support individually the company finances by delaying their approved promotions. But I could see, like almost everyone else, the administrative waste that cost far more than the pay differential of my promotion. That money came from a different pocket, however.

Talking of financial straits that were noticeable even in the restriction of office supplies, many of us wondered again and again about a different area of waste that never seemed to go away: the administrative waste of time and people's energies, an ever recurring theme; but that money comes again from a different pocket and accounting procedure. Should one get used to that and agree to see it as normal? Administrative chores are a normal, common sense, integral, yet subordinate part of life in a research office, just as brushing teeth belongs to living. But in this age of managerial self-aggrandizement, brushing your teeth all day long seemed more important than eating or living itself. (That is obvious, if the toothpaste manufacturers are the lobbyists.) In July 1971, we had been asked to list again our major projects, as if nobody had ever heard of them before: Project definitions, objectives, background, scope, resources, approach, and time-graphs for each phase. Then we were to search for or request justification documents for these projects. Did the authorizers in Staff not have them in their files, and should they not be in the files of the division or department? If they were not, how come we got paid all this time and produced reports on their progress? Fortunately, I had duplicates in my own files, although these were not the official administrative files.

In November, we were back at it again: different format for the same information; with the headings: Project Title, Submitting Organization, Prepared by..., Approved by..., In Process (give identifying code numbers) or New Job (give suggested code numbers), Objectives (a, b, c,...), Background, Work to be

Accomplished (a l, 2, 3,..., b 1, 2, 3,..., c, d)...In December 1971, we were back at it again. In February 1972, the then new CO, General O'Donnell, wanted to know how many people in each branch were involved in which project. Rather than handing him an appropriate printout from the automated accounting system, the branch chiefs had to list that information again, the division chiefs collected it for the department chiefs, who briefed the Commanding Colonels, who in turn briefed the General. The numbers were changed on the way to make them look prettier. Some people said: what do you worry about? They pay us well and that's what they want us to do. Do "they"? Is that why the taxpayers pay me a good salary, or is it because somebody does not do his own job with appropriate planning, does not want to look into his files, can't or won't do the adjustments himself, and rather passes chores on? Maybe, it was not my business to question these wasteful arrangements; or was it? I was a taxpayer too, and I paid very high taxes, from which in turn my and their salary was paid. An intriguing viewpoint, indeed: So I had a right as a taxpayer to question what I as a civil servant was asked to do for the salary I paid him/her. In my case, the taxpayer got his money's worth. I was so involved in my various research projects that I did not let any shenanigans steal much of my time or I made up for it with work at home. Since I did not take these write-ups very seriously (knowing they would come back anyway in a change of plans), Ray Shirley and I decided very quickly on the sensible minimum and went back to more important business. But when there came forms to be filled out monthly about who was taking what technical or managerial courses, when and where, I refused: nobody could take any courses without previous elaborate permission procedures; everyone along the administrative lines knew, or had stored in their files, who was permitted to go to what courses. Whether these forms were dropped, or whether the division secretary filled them out, I don't know and I did not care.

Also, the suggestion and cost reduction programs took on absurd forms. A directive came down the chute decreeing that every supervisory level had to submit five (5) suggestions every month. The first time around that was easy: we submitted all the sensible suggestions that had been made before and disregarded. (1) Save money by buying from a cheaper source. (Sandy Todd got an award for suggesting to buy magnetic tape cassettes for storing computer programs from the drugstore instead of the GSA, since the items were identical.) (2) Save human resources by using people according to their skills. (A capable researcher had been waiting in vain for three weeks for work to be assigned. Another researcher gave up waiting and resigned in protest, writing on the resignation forms: nothing to do. This embarrassed the supervisors, of course, and the employee was coaxed back, given work, sent to school, etc. All this while, I had been crying for more people for my branch; even the machine printout showing the overload in my branch versus the underemployment in the other was disregarded.) (3) Shorter channels; what one person could do in an hour, now takes several people two weeks (arithmetic of Parkinson's Law.) (4) Establish a research publication series within the geodesy department instead of the Publications Office; the geodesy department did it before, and much cheaper and better. (A recurring suggestion, recurringly disregarded despite tons of evidence of inefficiency and even international embarrassment.) (5) Buy a machine to project book pages and drawings directly for one-time briefings on short notice; this would save taxing other departments for urgent slide services.

The "Resources Conservation" Program (RECON) had its own logic. Complicated forms had to be filled out for computing an estimated saving for the first and five more years. We tried to be cooperative and submitted Bill Veigl's automatic plotting and contouring programs in January 197, showing the estimated savings for users in terms of man hours. It was sent back: the estimate for the next five years should be computed differently. In March, it had to be redone for a different way of computing. In May, it was back again for recomputing in a different fashion yet. In June, there was another change. By that time, so much time had been spent by high-grade employees on this silly form that you wondered about any savings left. But, of course, time of managers does not count. On the other hand, somebody gets a paycheck for his initiative and imagination in thinking up such forms with elaborate computations of senseless five-year estimates.

Rumors of an impending RIF (Reduction in Force) made people fearful for their livelihood and the uncertainties racked the nerves. But when the decisions announced in February 1972, they went beyond the worst expectations: true, only a handful employees were dismissed directly, but there were a number of unbelievable demotions and corresponding reassignments. Some GS-11s and GS-12s were demoted to GS-4s or GS-5s. The salary protection clause did not apply to such severe demotions (where it would be needed most), even though they had nothing to do with performance. Some such cases were revoked or softened in the subsequent turmoil of bumping rights, others seemed to be meant as poorly disguised dismissals. Some people left, some people fought.

4. SUSTENANCE IN EXTRACURRICULAR ACTIVITIES

In contrast to the depressing atmosphere at TOPOCOM in that period, my "extracurricular" activities at home entangled me in an absorbing analysis of my own attitudes to the perennial controversy of science versus religion. In response to several contemporary books, lectures, and discussions, where I found myself in lively agreement or disagreement, I felt the challenge of clarifying to myself my own viewpoint and formulating it in a coherent manner. The result was an article called "A Layman's Search for Understanding," published in the quarterly journal *Judaism* (vol.21,no.2, Spring Issue 1972). I had listened to several young men (one, e.g., a high school senior convinced that modern science makes religion obsolete and irrelevant, another a prospective draftee into the Vietnam war, whose defeatist mood made him confuse God's will with human politics) and some not so young people (who had developed a sophisticated understanding of the secular scene while they were calcified at a primitive fundamentalist level of kindergarten Sunday School), and had noticed the different attitudes even of astronauts (one reporting that he had failed to find any white-haired, long-bearded gentleman sitting out there in the universe, and another quoting psalm 8: "…when I consider Thy heavens, the work of Thy fingers, the moon and the stars which Thou hast ordained, what is man…"; there even had been a postal stamp in honor of Apollo 8, with the motto: "In the beginning God…"). It sounded like the tower of Babel: people used similar words for different thoughts, and also different words for kindred thoughts. And I tried to go beneath the surface of words and empathize with the primary response to the eternal search for understanding life's puzzle. The key question seemed to be: "what do you really mean when you say…",—a key approach to that I had acquired from the "Wiener Kreis" and my mathematical training in my student days, and which had served me very well throughout my career. It provides, as a first step, a criterion to distinguish a verifiable statement of fact from an expression of feeling. For example, what do you mean when you talk about the size of the world? Do you mean a geodetic statement, which you can prove or disprove by certain specific measurements? Or do you mean an expression of awe such as Isaiah's "The heaven is My throne and the earth is My footstool"?, which does not have a factual but a strong, emotional content. Applied to the creation story in Genesis, you can tie yourself into insoluble knots if you insist on taking it as a competition to a modern textbook on geology, but you can get an inspiration from it when you compare it with other cosmologies of that time, and try to understand it as the moral and spiritual response to life's overwhelming eternal questions by an ancient, pre-scientific people.

Pursuing the meaning of a religious (versus scientific) sentence to its possible existential source, and trying to peal away the confines of stereotyped phraseology and contradictory tacit assumptions, can be a rewarding exercise. It can forge a glimpse of awareness of a kindred human condition across numerous generations, and one can find wisdom and guidance in an attempt to rephrase ancient insights in contemporary language about our current perplexities, and to affirm our basic values.

The confusion of the different realms of science and religion arising from the ambiguities and limitations of language should not carry over, one would think, to the distinction between scientific/technical exploration and moral imperatives of essentially religious provenience. The high school kid who proclaimed that modern physics makes religion superfluous, was obviously too young to understand the basic role of values governing people's actions, which is not even touched by physics: choosing between right and wrong, and accepting personal responsibility for that choice. The often-cited incompatibility of the concepts of free will versus limitations of our physical nature is no excuse to refuse personal responsibility. Rabbi Akiba's famous "All is foreseen, yet free will is given" is usually taken as formulating the paradox of free will versus God's omniscience and omnipotence. That never made any sense to me whatsoever; and its continuation "and the world is judged by goodness, and all is according to the amount of work" does not seem to be much help in clearing up the meaning. I was delighted, therefore, to discover in the course of working on this paper, that the original Hebrew text permits a much more sensible, even beautifully coherent translation, by referring the word "foreseen" not to an omniscient God (whatever that should mean) but to the choosing person. Translated freely, it means: Having searched all our options (considering also our human limitations), we are permitted to choose. (Consider thereby that) it is goodness that gives worth and meaning to the world, and all depends on our effort (in that direction).

12

ONCE MORE A RELEVANT LEADER (mid 1972–mid 1973)

1. COL.MAXWELL V. JONAH, OUR NEW DEPARTMENT CHIEF

In July 1972, General Penney announced the activation of the Defense Mapping Agency, merging certain elements of the Army, Navy, Air Force and Defense Intelligence Agency. Our name changed accordingly from Army Topographic Command to Defense Mapping Agency Topographic Center, or, to save your breath, from TOPOCOM to IMATC. Only last fall, a new gatepost displaying the then new name "Topographic Command, Corps of Engineers" had been built and dedicated by General O'Donnell in a ceremonious ribbon cutting. The name would be changed now, but at least we had a new gatepost. Also the name above the entrance to the buildings would be changed, but at least our newspaper kept its names TOPOCOMMENT.

Of more immediate impact was the sudden appointment of Col. Maxwell V. Jonah as our new department chief. He said that when he walked in on Monday morning, 3 July 1972, on some business, he had had no inkling that he was going to be assigned this job, starting that very morning: that's the Army. He came to us like a sudden gush of fresh air. Mr. Frey's attempt to reopen closed doors and reestablish trust had not survived his physical absence during his detail to DMA. Now, Col. Jonah's energetic presence opened doors and windows to a revitalization of individual initiative, encouraging and backing individual contact with sources of technical or planning information outside the department or the agency. He started right away with a meeting for all division and branch chiefs, followed by visits to the branches, in order to explain the principles of his working philosophies: everyone should familiarize himself with the overall objectives and plans; we should develop a confidence image of what we can do, contribute our ideas as the experts, point out what needed to be done and why, and not wait for "them" to tell us. He would keep communications open and he expected to hear of all problems, even small ones. Remember it is *our ideas* that are wanted.

It sounded almost like the good old days of the Army Map Service. The important difference, however, was in the expanding research economy then versus the shrinking defense budget now. The fight for survival of a small research unit within a map factory was perennial, though, and as old as the hills. Col. Jonah's common-sense evaluation of the reorganization trend was quite helpful and educational. The purpose of the reorganization, he warned, was bound to be cuts all around, painful in their gradually increasing severity and the concomitant fierce competition between the merging elements of who would get which part of the shrinking pie. Our perennial rival, ACIC, displayed a much better "confidence image," so that they hold the confidence of the bosses, with the effect that jobs go their way. We should be able to do that too. Agency cocktail parties are not the happy brotherly affairs they seem, he said, but opportunities to hear some information and to evaluate your chances of survival; while you lift your glasses, you know: it will be either him or me, it cannot be both. Thus we ought to look critically at our various jobs and decide which were worth fighting for with some prospect of success. No point in squandering your energies running after everything; you can't win them all. Better to make an intelligent choice for the fights that you can reasonably win and put all your efforts there. With his boundless energy and initiative, Col. Jonah really tried very hard to get us a place in the sun, to establish us as *the* geodetic center within DMA in this competition for survival, but he was not with us long enough. Maybe, if he had been permitted to stay, we might have made it.

I remembered the saying about the different responses by the three Forces to a requests the Navy does what it thinks ought to be done; the Army does exactly what was requested, not more and not less; the Air Force does twice as much to show what it can do. Col. Jonah was so right: not only for the past when we complained about ACIC's padding and duplicating our jobs, but also for now, and, as it turned out in later years, for the future. There won't be two research units, he predicted, it will be either here or there, if at all. And so it was: the reorganization after my retirement abolished the once famous AMS research unit.

When visiting my branch and reviewing our projects, Col. Jonah agreed with my feeling that the trajectory study for the Army was not worth fighting for; it might more economically revert to ACIC under DMA in the future. Since the Army had not yet decided anyway on their specifications to be used in this study, he advised to phase it out in the form of a "Threshold Report" which could be continued if and when further developments so indicated. I was asked to give a formal briefing about this to Staff people, the following week. On the other hand, Col. Jonah wanted to hear all about SAD 69 and my new involvement with deflections at sea. (More about this in the next chapter.) He liked my *Basic Geodesy* pamphlet, which just had come off the press, and only complained that his favorite song about isostasy was not in it. So I made it a point to find that song and include it into a second pamphlet I was working on, called: *The Geoid—What's That?—All You Ever Wanted To Know (and not more)*. It was going to be another bestseller. It was published by the Defense Mapping School a year later, with the support of Col. Kurtz of the School and Col. Cordova (then our Deputy Director).

Energetic Col. Jonah could be heard complaining that his requests sometimes seemed to disappear "into a bottomless pit without even an echo". (You too had that experience?) But even if I habitually responded as soon as possible, I may have appeared guilty just the same: when a request for a list of products by Geoid Branch (made quickly thanks to my good files) had not been forwarded after six weeks despite the secretary's repeated reminders, I decided to take a xerox copy with the original date directly to his office. He merely said: "This list is exactly what I wanted."

Common sense Col. Jonah also said that researchers should not be bothered with perennial rewritings of function statements; their time is needed elsewhere. We have administrators to do that kind of writing; just give them the basic information and they should take it from there. (Believe it or not!.) He also complained that he could not get to the bottom of why we still spent so much time on this never-ending, long overdue WGS project of the Three-Forces Committee. What significantly more could we expect to add? Get done and get out and do something really useful. He could not have failed to sense the padding and dragging out of this project, out of a desire to hang on to an already authorized major assignment, in the face of dwindling jobs and funds. Most of the many interesting jobs of Henriksen's times in our own agency had been lost through lack of forceful defenders. Aware of this ominous trend, I had been watchful to defend my turf from a similar fate by documenting the recognized usefulness and actual utilization of my products. Now, Col. Jonah asked the branch chiefs for memos on what worthwhile projects they thought their branches should pursue, and he himself periodically wrote memos about the status of the department problems and proposals about the next step to be taken; and he invited and expected comments. Someone started to call these periodic issues the "Jonah grams". In one of these Jonah grams, he suggested that my Geoid Branch might more economically be placed into another division to be able to draw more directly on the pertinent data and the manpower processing them, but he backed down when confronted with opposition. My group and I were all for it, as we had discussed the same idea a couple of years earlier and suggested it to Mr. Frey at that time, albeit without success. The division chief did not want to lose our high-level research functions, not then and not now. When the chief o the other division came to discuss with me a possible transfer in response to the Jonah gram, I watched my division chief stop him at the door and talk to him a while. When the visitor finally reached my desk, he told me that he felt it "unethical" to take these functions away from a colleague. This peculiar judgment apparently did not consider as ethical the support of the agency in streamlining procedures. When I asked my division chief what his answers to yesterday's Jonah gram had been, he refused to discuss it and said, he could not remember.

Col. Jonah's ideas about personnel management were clear from his support of initiative and free communication all around. He recommended "to put the personal back into personnel" and "to encourage participation in management thinking." He talked about the difference between participative management and the military "stovepipe" system where one could only communicate through single-line channels up and down the chimney; and the pro's and con's of the systems. Among many other things, he told me once about an attempt, in the apparent safety of the stovepipe system, to malign an employee in the department by low-key insinuations. He said that he was aware of the fact that a new chief could be told something at a deliberately oblique angle where it was difficult to distinguish between smoke from a real fire and an intentional smokescreen. To find out, he had told that employee the specific statements and received in return a "fact sheet" which listed documented facts in point by point refutation of these absurd intimations. There was a bestseller around then: Dr. T. Harris' *I'm O.K., You're O.K.*, which Col. Jonah liked and recommended. The title was used as a motto for that fact sheet; the employee may have meant to imply that the originator of the smear campaign was "no k.".

Stories like that gave you the creeps. It was an attempt at a neat little character assassination. There had been smear campaigns in the past, but feedback possibilities in an open system had led to a closing of ranks around the victim against the maligner. But if the trend towards a stovepipe system grew stronger (and it did just that in the later years) then a couple of rotten apples could easily spoil a whole bushel of healthy apples, and casual insinuations would be ingested almost unconsciously like a slow poison. Character assassinations would be easy then. In the marvelously witty series of mystery stories by Harry Kemelman, there is one that paints the extreme of a stovepipe management. In *Saturday the Rabbi Went Hungry*, a scientist supervisor uses the work of his talented subordinate as his own when reporting to his administrative superior, blaming occasional errors of his own on the subordinate. He usually finds condoning words for such errors, in order to appear as the lenient supervisor, while continuing the useful arrangement. The situation blows up when a gross error, fateful to the company, cannot be covered up and the administrator decides to fire the employee. There were only two occasions when the administrator, an adherent of the stovepipe management system, would talk to an employee when he hired him and when he fired him after telling him the reason: otherwise everything went strictly through channels. Terrified that the looming firing interview would reveal the real situation, the supervisor uses the circumstances of finding the alcoholic employee that evening parked along the roadside and passed out drunk, drives the employee's car home into the garage, closes the garage door leaving the motor running, and walks away to his own home. It seems to be the perfect, safest crime, as it looks like a suicide in despondency over losing the job. Fortunately, the Rabbi reasoned things out. Col. Jonah did too. But would others, who are less smart? Or fearful to get involved. By stepping forward to oppose the poison?

2. THE COMPLETION OF SAD 69

The SAD 69 project neared completion. Chairman Byars, now at DMA, was invited to the "CACTAL" Conference (Inter-American Conference on the Application of Science and Technology to the Development of Latin America) in Brasilia, May 1972, where he submitted a paper on the significance of our work, under the title, "Some Practical Applications from Recent Geodetic Acevements in South America." The Americas Section (Frank Reynolds and Lloyd Slama, then Chief of the Section) prepared separate national booklets for each South American state with the technical details of the national part of the continental adjustment; and I prepared a paper "The Basic Framework of the South American Datum of 1969", to be included in each booklet, about the historical and technical background, the procedures and results of SAD 69, with emphasis on practical information for the users. DIA had already recognized the merits of SAD 69 over the deficiencies of PSAD 56, and officially designated it in 1971 as the preferred datum for all of South America. NASA requested SAD 69 coordinates for their purposes. I was asked to prepare another "Show and Tell" briefing for the Command and Staff Conference, July 7, 1972. The briefing was cancelled on the day of the presentation, however, due to preparations for Col. Anderson's impending retirement. In his nostalgic mood of leaving, Col. Anderson sent me a lovely note with greetings for my birthday (I was touched that he took time to find out about that), and he also acceded to the pleas of two men for promotions that had been denied them before, for specific reasons. Since July 1972, the new DMATC had the power of promotion, independent of OCE. One of these men told me himself: "You know, they can do now what they please. It was simply called an "evolution of the job" and that was that. The only "evolution" one could see was the fact that they had a high-grade employee in their unit, a fact they could exploit without merit of their own. The simplified personnel action drew wide criticism, however, and an embarrassing note appeared in one of the Federal newspaper: columns and was circulated.

The TOPOCOMMENT as well as an "Item of Interest" described the SAD project as "one of the major accomplishments of the Department of Geodesy and Geophysics, representing a successful combination of serving the Department of Defense and assisting our South American neighbors to meet their needs on a continental scope." I was asked to write an article about its significance for publication in *The Military Engineer* (March-April 1973). Also DMA Headquarters showed heightened interest, as I heard from Byars and also from Frey who visited TC in September 1972. I happened to be present at this visit; in fact I had excused myself to leave, but Mr. Frey asked me to stay, to hear his suggestion of putting an SAD briefing for General Penney on the schedule through channels (that is, through our Commanding Officer),and also his comment that concept explanations on Doppler of the type of my *Basic Geodesy* might be welcome there. "And I just know who could do that very well" he smiled at me. Both suggestions were squashed as soon as Frey left, and were omitted from the required memo about the visit. The Doppler explanation was given to someone without teaching ability or sense of humor, with the predictable result of repeated rejections. To keep the second, more important suggestion from going down the drain, I sent a formal memo through channels to put the SAD briefing on the agenda for our CO, referring to Frey's visit and to the already prepared, then cancelled briefing for the Command and Staff Conference. That memo never made it; I never received the telltale yellow carbon copy back. Color-coded carbons were supposed to tell when a memo had left which office and who had seen it on its journey. When I inquired, I was told that Col. Jonah approved the idea, but, it was intimated, Col. Whelan (who had succeeded Col. Anderson as our CO) was not interested in that project and that the mere mention of it upsets him" and therefore I should not press the subject. It was suggested by someone who certainly knew better, that I myself could send the offer to brief the General directly to

Headquarters. That was a deliberately incorrect guidance: it would be a silly violation of protocol. The channels went from me through the division chief to the department chief or his Assistant, from there to the CO, and from there to Headquarters.

But then something happened independently: Dr. O'Keefe visited to borrow some slides from the abundant TC slide collection. When I accompanied him to the lobby for checking out of the building, Col. Whelan happened to come by and I introduced my former and my present boss to one another. I did not believe my ears when I heard Col. Whelan brag to Dr. O'Keefe about my South American Datum project. Obviously, something funny was going on in the stovepipe!

"I would love to brief you about the final results, if I may," I ventured to help my luck along a little. "Well, put it on the briefing schedule." "I did already several times, but it does not seem to work."—"You want me to ask for it?"—"If you would, sir; that'll be great."—When Rutscheidt brought me the scheduled date for the briefing, he said: "I forwarded your request only yesterday and here it is approved today already. It is set for next week and Col. Jonah wants a dry-run two days before that. It never happened that fast before."

The PAIGH Assembly took place in Panama City, April 1973. Most of the U.S. delegation went there on the same flight and we were met at the airport by the Director of the Panama National Geographical Institute and several people from IAGS who pleasantly relieved us from all passport and luggage chores. It was a hot and muggy evening and, after the tiring flight, we were grateful for the leisurely wait in the shade with cool drinks. There was a curious sequel to this helpfulness, though, when on our individual flights back home, I almost was not permitted to leave Panama: I did not have an entry stamp, so how could they let me get out? It took some anxious minutes, worried that we might miss the flight, before they graciously let me fly home. Arriving at the Miami customs entry into the U.S., the customs officer seemed puzzled by the heaps of papers from the PAIGH meeting in my luggage, instead of the liquor he was looking for. "What meeting is that?"—The Pan American Institute…"You work for the government?"—Yes—"Geodesy?" "Yes." "What's that, geodesy?" "Well, they compute the size of the Earth…" The officer interrupted with an amused and deprecatory shrug. "What a nut" was written all over his face. "You did not need to open your bags," he said and moved us on.

The U.S. delegation had eleven panels in the exhibition at the PAIGH, one of which told my story of the new South American Datum with several illustrations and bilingual explanations. This panel later became an important piece of the standing exhibition in my branch, much appreciated by Spanish speaking visitors. The format of the PAIGH Assemblies differs from most other scientific conferences. Individual papers of interest to Latin America, even if already presented elsewhere, are very welcome for distribution in several hundreds of copies, but they are not orally presented. Presentations and discussions in the sessions are concerned with commission or working group projects such as our SAD 69, which chairman Byars presented as completed. We answered several questions, and the work was unanimously accepted with congratulations and formal appreciations. Embarrassingly, my paper with the summarizing explanations had not arrived in time for that session. One of the South American delegates offered a friendly advice: "You must talk to or complain about your Publications Office.

It is important to have a paper on time." Didn't I several times? Many other authors at TOPOCOM have also complained again and again, but without success. I had tried to prevent any difficulties by submitting my four contributions (one on the HIRAN study in South America, one on satellite derived results, my new *Basic Geodesy*, and most importantly the one on SAD 69) in finished form, complete with publication clearances, four months in advance to allow plenty of time for producing the required 400 copies each under the required yellow covers, and transmitting them on time. But sometimes you just can't win. One colleague, citing his past experiences, gave up hoping to find his paper on the distribution tables in Panama before everybody left. But I did not give up hope so quickly. Of course, one has to help the hope along a little. On the advice of the chairman of the U.S. delegation, Dr. Arthur L. Burt, Department of State, I appealed to friendly Col. Ruthe, IAGS, for help. "...It was marvelous to watch an omnipotent U.S. Colonel in action," I gratefully recorded in my official Trip Report. "Telephones buzzed to IAGS in Washington and to Brainiff Airlines, and put everyone on a treasure hunt. In the evening, I was told that Col. Ruthe personally had trucked my papers from Brainiff Airlines to the distribution room at the Legislative Palace. I felt good; but next morning there were only two out of my four papers laid out on the distribution table, and my SAD paper was not one of them. There was a message that the other two had not yet left Washington. Another S.O.S. and repeated telephone performance. By that time, all of IAGS and Brainiff and the PAIGH must have heard of and took interest in my problem,...Late evening on Wednesday, I received happy messages from several sides that two big boxes had just arrived, so "all is well". Next morning, there was my third paper, but not yet the principal SAD paper. Repeat telephone performances once more. In the meantime, the Resolutions Committee needed that paper and there was no copy. I had made a magnificent present of my one and only personal copy to the Brazilian delegate and had to ask my gift back. The paper finally arrived at the end of the whole geodesy program, hand-carried to me by the lady supervising the Distribution Room. Other TC papers packed with mine profited from the intervention of Col. Ruthe, my deus ex machina. But I know of at least one other TC paper that had not arrived even then. Since this is by no means an isolated experience, it might be appropriate to request again that the publication of technical papers be disassociated from the publication of other DMA business. No doubt, the Publications Office was swamped with higher priority business (such has always been the explanation), but this is even more reason to permit the departments to take care of their own, as they used to do years ago—efficiently, on time, and more responsive to the author's manuscript...."

My *Basic Geodesy* pamphlet had been a huge success at home with repeated requests for twenty or fifty copies at a time; and it was a hit in Panama too. For Colombia, a Spanish translation, *Geodesia Basica*, had already been made by Eng. Rodolfo Llinas Rivera, President of the Cartographic Society of Colombia in Bogota, and published in their *Boletin*. Other South American countries were also interested in a Spanish version. Col. L. M. Sebert, Topographical Survey Directorate in Ottawa, Canada, surprised me later with a French translation, *Géodéésie Fondamentale*, which the Directorate planned to use in their courses.

Of the many requests for my other very popular papers, those for the useful North American geoid charts needed special acrobatics to be filled. They were contained in TR-62; and all TRs, even unclassified ones, had recently been adorned by our eager Publications Office with a caveat on the cover requiring the permission of that office for each transmittal. Not a security check of the recipient, since that was done very quickly by the Security Office. That caveat created very awkward situations, indeed. For example, Professor Ron Mather, geodesist in Australia and IAG Secretary of Section V, visited TOPOCOM in summer 1971 and requested a copy of TR-62 for use in his research. We could show him these unclassified charts but we were not permitted to let him have them because of the caveat. Mather was told the permission was just a formality of a few days and the charts would be mailed to him. Believe it or not, it took half a year before the Publications Office finally mailed them.

It is mysteriously easy to clamp some unwarranted restriction on government documents when no one knowledgeable is looking, but it is nearly impossible to get it off again even though it is acknowledged as inappropriate. In the case of the much-requested TR-62, a way around the obstacle was found: first, Col. Hathorn ruled that the charts could be given out independently since they did not carry the caveat. Hurrah: that was what the customers needed, since the content of the text was essentially known from my other publications anyway. How ironic to keep that text restricted by bureaucracy! Next I submitted that same text for clearance as an article for *Surveying and Mapping* where it was printed very soon. Now, who needed TR-62 when I could freely give out the reprints from that journal in its place?

Mr. James R. Kephart, for many years our Public Affairs Officer (or Technical Liaison Officer, titles keep changing) had left TOPOCOM unfortunately, or such things would not have happened. The TOPOCOMMENT had headlined the news with "It Will Never Be The Same." So true! He and the officers before him had shepherded our clearances through the labyrinth with efficiency, pride in our achievements, pertinent information, and fatherly advice and assistance. Unfortunately, that position had not been filled. For a while, Mr. R. G. Ingersoll Waite (Command Information Officer and Editor of the TOPOCOMMENT) and Mrs. Virginia Lee, Assistant, who had both worked along with Kephart on the clearance chores, carried on as efficiently and pleasantly as ever, supported for a while by Mrs. Kay Sacks as secretary. But then they were restricted to the newspaper job while our manuscripts in search for clearance were all of a sudden led through another subunit, the Publications Office, which until then only physically produced government publications. Do not confuse this subunit with Mr. Kephart's overall Public Affairs Office, which handled all public relations including clearances, transmittals, publicity, conference arrangements, etc. The takeover of some of these public relations functions by a production unit was aided by the similarity of the office names, which muddled the awareness of the difference. I was suddenly confronted with that difference when I was waiting for the clearances of two abstracts, in response to an invitation from Australia. I had submitted them through Mr. Waite as usual and I had already been notified by telephone from MDA Headquarters that they were approved and in the mail back to us. They had to meet a due date and I kept wondering how long mail took from the first floor to the fifth floor. Mrs. Sacks confirmed their arrival but said they were channeled now through the Publications Office and yet another office and held up there. I went to that office to ask for a xerox of the already approved clearances since I had to meet that due date. The reception I received was as unexpected as unusual. It turned out that this office held my approved clearances back as a reprisal for my Panama trip report. I had heard already through the grapevine that attempts had been made to either rewrite the passage about the fate of my papers in Panama and my appeal and suggestion for future improvements, or to keep the whole report from reaching the Executive Office.

Maybe the Publications Office should look very closely into every possible aspect to see whether it may not be in the interest of this agency that you publish so much and go to so many meetings to present papers," I heard this man say to me and I hardly believed my ears. I had never had any dealings with him before, barely knew his name, and thus he could not possibly have any personal axe to grind; he must have been asked to front for the Publications Office and to grind their axe. He obviously tried to intimidate me, but such an absurd threat was a poor way to try that. He was in no position to keep me from publishing or presenting papers if my superiors (and he was definitely not one of them) wanted me to do that. But he could delay transmitting my clearances beyond due dates as he tried now, and he could do some mischief in a million ways if not stopped now. What should I do? One of the ground rules for survival is: never underestimate a bragging adversary, even if he makes himself sound like a fool. I went to my division chief and my department chief, reported the ominous threat for the future, and asked them to get my abstracts released. Col. Jonah took the telephone, and the following day this would-be adversary personally brought me the abstracts. The clearances showed a date of more than two weeks earlier. I never had any business with this man again; he had other positions later on and probably does not even remember this incident. I do, because the Publications Office repeatedly reminded me of it by trying to do just that kind of mischief in the next several years. Eventually, some officials did see the light, and this office was taken out of the clearance circuit again at long last.

3. PEOPLE OR PUPPETS, WHICH WAY INTO THE FUTURE?

Since several years the government, including the Defense Department, had shown active interest in developing executive talent by encouraging management courses. As mentioned earlier, I had been exposed to a bunch of such courses in 1971/72, which resulted in a rejected suggestion to apply what was taught, to our agency. The in-house courses were useful through their close bearing on in-house situations. Our Personnel Office provided speakers on regulations concerning personnel actions, criteria of Civil Service standards on grade evaluations, procedures for promotions, incentive awards, safety controls, etc. While we knew all this from experience, it was an opportunity to get clarifications and to discuss difficulties. An amusing homework consisted of interviewing your supervisor (who had presumably taken these courses already) about the role of a supervisor and the division or delegation of authority from one level to the next (exposing the conflict in expecting full responsibility without granting any delegation of authority nor even a sufficient flow of information). There was a film about a university research program where an administrator insists on taking over and pushing for competition to the detriment of quality, and gets a medal for himself. The reactionary element in our agency advertised itself in a speech on discipline and the penalty options and procedures. In a paternalistic talk-down the speaker compared the supervisory relationship naively to that of parent to child and seemed to regret that you can't spank them anymore nowadays; and firing employees is difficult because of all those appeals and grievances channels. He also was against the coffee-break allowance, because it costs the company too much, as he proved by tabulating the minutes spent away from one's desk for coffee and other unproductive reasons. On the other end of the spectrum of speakers was the remarkable performance of Father Joseph C. Martin (Department of Health and Mental Hygiene, Baltimore, Md.) who spoke in our anti-alcoholism program with matchless wit, clarity, and compassion. The Top Management Seminar of AMETA (Army Management Engineering Training Agency), usually given in Rock Island, Illinois, was conveniently given in Fort McNair, Washington, D.C., when I attended in 1971. It offered a procession of illustrious speakers on current management concerns, valuable discussion periods, and also comprehensive reading material on the history and philosophy of management concepts. It was an interesting experience in several respects, positive and negative. I was mainly interested in ideas about employee motivation, specifically in a research environment. It is always pleasurable to find speakers or reading material supporting your own views, and you only wonder why such emphasis is needed for the obvious. A manager must earn respect as a leader, it does not come automatically with the position," (of course). "Management is not a matter of occupation, but a particular form of social activity; not the perquisite of any single class, but an inherent ingredient in the lives of every one of us. All people are in varying degrees executives in their own right," (of course). Douglas McGregor: Theory X (workers are inherently indolent, not very bright, dislike responsibility, need to be coerced) is being modified or superseded by Theory Y (workers are intelligent mature adults, look for development of their potential, welcome responsibility and creativity) or by a 'managerial grid" finding the optimum of sensible compromises (of course). And on abuses: "Chief ways in which power-oriented supervisors control and thwart strong subordinates is to deny them the full knowledge they need to take action,' (of course; have I not seen that?). Motivation depends on value systems that change, and management approach must be responsive.

When survival needs are satisfied, social and psychological needs are more potent as motivators, (of course).

But then you also listen or read about what you would have considered reactionary views that had been left behind (you thought), and it gives you a chill to recognize one or the other feature as being introduced in your agency as the newest thing to de. Had I not heard there the praise of "scientific management" and the need to introduce it at last? Well, F.W. Taylor's ideas of half a century ago were then certainly an important milestone with respect to standardization of tools and factory procedures, working conditions, and maybe timing a narrowly skilled work force on the assembly line. Several years back, some such timing program had been tried on our research division, with a man walking the halls and clocking our coming and going during the day—until we rebelled against it as a nuisance and insult. As long as today's efforts consisted in clear formulations of assignments, due dates, and coordination of efforts and flow of information, we were all for it. But Taylor, a mechanical engineer, saw workers as machines to be clocked and manipulated. Championing his views of a complete division between work force and management, with a superiority complex of the latter, sounded a little ludicrous in our research environment where the manipulatees could easily be more knowledgeable and efficient than the manipulators. The ideal of our management automation group was to have a printout from which the front office could tell any day who was doing what in any office, and what he should rather do in which order—like moving puppets on strings, or drilling the Lippizaner horses to dance in the Viennese riding school. "Value Engineering" was another phrase I could now identify. It was being tried in the office, but since nobody could explain what it was or was interested enough to find out, nothing much besides irritating people was done about it. Now I read in the Executive Skills Handbook that it was one of the newest management tools for "efficient identification of unnecessary cost. It is a proven system which delves to the root of function...to provide the simplest, easiest and least costly ways in which the essential function can be performed." (Was that why we had to spend endless costly hours to write and rewrite function statements for branches and divisions without knowing why one formulation was rejected for another?)

One could also learn from the Handbook that" communicating" was essential to effective management, that "directing'" was downward communication, "controlling' depended on upward flow of information, "coordinating" required lateral communication, and informal communication was also a significant factor. Tons and tons of words were collected there in empty sentences to teach us this level of wisdom.

In 1973 there would be again a bunch of management courses. I was wondering whether any more would be worth my time or would be more of the same, with the same inapplicability to our present state of health. But a protective secretary's wisdom counseled to play the government game: You've got to have these courses in your record, she said, or it might count against you on some future occasion; courses, especially management courses, are the in-thing now; you've got to know the management lingo to know what they are talking about. She was right, of course, and she signed me up where she could. Joy Trickett was very knowledgeable in government ways; she had looked out for me on many occasions throughout the years, always protective, helpful, informative, and pleasant. She was "Miss Division Chief" for me, because I could do without a division or department chief but not without her. She knew or could find regulations, she knew and kept track of what was to be done when, she knew how to find out where and why some administrative wheels were stuck and what to do about it. She also had a sound judgment and, of course, she knew much more of what was going on in critical times than I did. Joy signed me up for a new Management Course for Supervisory Scientists and Engineers, which sounded like I should take it; but although it was scheduled twice, someone else apparently decided against my participation. But I did take a course on "Motivation and Productivity" at our own Training Center. Its able chief, Gil Monck, based the discussions on well-selected excerpts from the prolific management literature. It was an excellent way to find out what these management gurus had on their mind, without having to spend excessive time to look up their whole books or articles. Most of the topics were already familiar to me as I had expected, but there was one excellent article that alone made the whole course worthwhile, since otherwise I would never have come across it or chosen to read it by the title. It was witty, lively, very intelligent, to the point, and fun to read. It was called "Group Relationships and Participative Management" by James N. Mosel, Professor of Psychology at George Washington University {*Perspectives in Defense Management Magazine*, March 1968). The author won my heart and attention right away by a number of very sensible statements: Management was not a science (as my unbelieving ears had heard some others claim) but a doctrine, since its sentences are imperative, not declarative. Taylor's pyramidal organization with its strict dichotomy into line and staff was historically taken from the military example, but is not the only possible form. There could be circles, as in working committees, or other forms. In Thailand, for example, there is a very different, yet very effective organization where jobs are not classified in terms of job duties but of the characteristics of the person performing them. There is no line-staff distinction. Thais say that otherwise their line functionaries would not take staff functions seriously and thus not heed the staff recommendations. (Sensible people some of us researchers feel so too.) Historically, the maintenance of the classical "scientific management" needed an increasingly authoritarian chain of command with carrot and stick enforcement techniques, but that did not work when applied to the rapidly rising scientific research in and after World War II. It turned out that the nature of scientific research or any creative undertaking was incompatible with the classical doctrine of scientific management.

An entirely different approach was needed: Instead of the classical way of predetermining a job and carefully programming it to ensure minimum time, cost, and effort, the incumbent would be allowed an area of freedom to determine on his own the nature of his task and thus characterize the job. (Was that not what Dr. O'Keefe and Mrs. Rust had been talking about some fifteen years earlier, which had led to my "researcher" title? So they had been implementing modern management concepts then, forgotten in the current developments at our agency, which went into the opposite direction.) The article contains an exhilarating story of a collision between a conventional job analyst and a way-out pure mathematician. Briefly (you can look up the story and enjoy its full text on pp.86, 87 of the reference): the job analyst wants to write a conventional description of the duties and gets the formulation of an abstract mathematical theorem to be proved, of which he does not even understand the words. Nonplussed, he wants to see the products—which, he is told, don't really exist since they are mathematical ideas. At a dead end, he tries another approach by observing what the mathematician does, and he sees him drink coffee, look at books, and write on the blackboard. Next, he asks the boss what this man is doing and is told that no one knows, else he would not have been hired; he was the only man in the country who understood this sort of thing. Next, the analyst asks how the boss will know whether the job was being done well, and hears with disbelief that the boss does not have the slightest idea but waits till the employee tells him that the job has been done. right.

This is an extreme case, of course, but the author makes the point that different types of jobs require different types of management, with classical management of routine jobs at one extreme, and participative management of creative jobs at the other extreme, and a whole continuum between. The managerial changes at my office fell into a clearer pattern for me, explaining by a larger perspective the all around rise in irritability and cynicism.

The course on "Personnel Management for Executives" (PME) given in Annapolis by the Army Regional Training Center, was of a different design, with an obligatory live-in arrangement. It addressed itself to the human make-up of a manager, as indicated in the two mottos on the handout folder: "A man's personal philosophy, his way of looking at the world and the men and women around him determine his success as a manager of things and people more than any other single factor...." (Dean Stanley F. Teele, Harvard Business School), and "...(The manager) directs people or misdirects them. He brings out what is in them or he stifles them....He may do them well, or he may do them wretchedly. But he always does them." (Peter Drucker). There were discussions of case histories, mostly taken from the Harvard Business Review or from The Administrator; management films; and a number of high-quality guest speakers, among them Father Martin, whose excellent anti-alcoholism talk I did not mind at all hearing again. Of the other speakers, I was particularly impressed by Sister Margaret Gorman (Chairman of the Psychology Department at the College of the Sacred Heart, Newton, Mass.; and also Psychological Consultant to the Army and Air Force) who talked about "Understanding Youth." She pointed out that people attach different meanings to words according to their background, which makes messages ambiguous. Communication goes through a personal filter, a reference frame of past experiences, which prevents a listener from hearing what you said and makes him hear instead what fits into his frame. The reference frame of a young person today, e.g., does not include personal memories of the depression or World War II or pre-TV values as does that of an older person. Understanding and overcoming ambiguities requires listening to the other person's viewpoint in his frame of reference. Another excellent talk was that of anthropologist Dr. Ethel J. Alpenfels (New York University) on Cultural Influence on Behavior, with many examples of cultural differences which govern gestures, linguistic phrases, manners, valuations of time and space, etc.

Such lectures of general interest and applicability, given in management courses, seemed to acknowledge that managers are people like other people, with the same perplexities in human relationships. Why is it assumed that these life-long perplexities can be resolved for managers in brief lectures?

The discussions of managerial case histories were also aimed at a better understanding of psychological responses, but since they were taken mostly from private industry, they were not easily relatable to the different government situations. Suggestions by several course members that *The Administrator* and the *Harvard Business Review* be replaced in government courses by a more meaningful collection of government case histories were acknowledged by the course leaders as useful but hard to follow, since such material was difficult to find and would probably meet with resistance to produce. In my Trip Report after this one and a half week's TDY sojourn in Annapolis, I referred to Col. Jonah's plans for follow-up sessions within our department to make these management courses practically applicable, and suggested to use these sessions as a basis for producing such a collection of more relevant course material for future PME courses. Nothing came of these plans, since Col. Jonah was suddenly assigned elsewhere, as suddenly as he had come just one year, before. He was the last of our relevant leaders.

On 1 August 1973, I had a phone call from Commander Eleanor Hunn, Military Personnel at DMA Headquarters, asking whether I had applied for the Federal Executive Development Program to be conducted in Charlottesville, Va.—No, what's that?—Application forms supposedly had been sent to all CS-15s, did I not receive them?—No, I didn't.—Commander Hunn was involved, among several other things, in championing women's rights in government and suspected a non-accidental oversight. She would see to it that I got the forms, the due date was 31 August. It was very important, she said, that I as the only female GS-15 at this agency establish a female presence among the applicants, not only in my own interest but also in the interest of other women later on. On 27 August, the application forms had not reached me yet, although extra forms had been sent from DMA to TC for distribution. Commander Hunn asked me to check with the Personnel Office; only then did I get the forms, and filled out the lengthy Executive Inventory Record in a hurry. Of the government-wide 3200 applicants only 25 participants were chosen. I was not among them, but I felt remotely connected with the seminar as my viewgraphs and briefings on the South American Datum project were used later on by a participant from DMA Headquarters in a homework assignment. I gathered that the idea of these periodic six-weeks live-in seminars in beautiful Charlottesville (where we had lived or a few years in an earlier lifetime) was to give an opportunity to explore managerial topics in greater depth and perspective, with freedom to pursue one's own choices with the benefit of stimulating discussions. Whether it succeeded to make these managers wiser people remains an open question. Whether it had any impact on government ways is also a question.

The various enlightened training efforts did not seem to rub off on the practical performance of middle management (at least not in my agency) which curiously went into the opposite direction towards dehumanization, in two probably interrelated ways: (1) Attempts to manipulate not just people's job-related actions but their minds, in a naive expectation that this can be simply accomplished through lectures and courses with compulsory attendance (like force-feeding). These were quickly branded as brainwashing' and thus, irrespective of the merit of the message itself, made ineffectual in their low level crudity, which precluded any honest discussion of obvious fallacies. I remember estimating and adding up the high daily salaries of the captive audience, multiplied by the days of the course, times the number of the course repetitions to cover all managers from the Executive Office and the department chiefs down, plus the commercial fee paid to the youngster who lectured us from the wisdom of his relatively short life and the clichés of the times "to acquire an understanding of the concept of human resource management." He set our benighted minds straight that education and experience should not be considered when hiring a new employee, but he did let slip in that he was a PH.D. candidate. "Not that I think this has any value," he explained, but it does open doors." (2) Attempts to strengthen collective management powers by expecting conformity and mutual backing as an antidote to individual job-insecurity. I remember two distinct, particularly striking cases where an apparently normal person went into a meeting, say Monday morning between 9:00 and 10:00 a.m., and came out transformed into a conformist, suspending personal inquiry and judgment of right or wrong, true or false. Each must have been given a quickie lesson in the virtues of managerial solidarity. These attempts, however, boomeranged into loss of leadership and relevancy. It seemed as if managers left their personhood at home and put on office uniforms as faceless cogs in machinery, incapable of making and defending responsible individual decisions but instead trying to outguess their superior's wishes (unsure in turn) or, even safer, seek collective decisions to keep out of responsibility. If you don't believe it, just listen to some comments I have heard from high-grade managers: "The safest path to take in government is to do nothing." "Just be careful to stay in the middle, so you won't fall into a ditch on either side." "I work for this man, so I click my heels and say 'Yes sir.'"

The trend towards dehumanization and futility must have had an increasing impact also on middle-grade employees whose manipulated personhood reacted with "career stress." The *Washington Post* of 13 May 1979, in a long article on the federal government's billion-dollar subsidy to the commercial education industry (in a time of severely curtailed government spending for projects and salaries) also tells of popular, week-long, government-paid courses in a Williamsburg hotel to teach federal employees how to breathe and relax each muscle to enable them to cope with mid-career stress. So it is not the cause of excessive office-stress that needs to be alleviated, but the reaction to it. Is there not something wrong somewhere?

13

A NEW FRONTIER
(mid 1973–mid 1975)

1. DEFLECTIONS AT SEA

It had started with a request for help from the Datum Branch in early 1972. Jim Walker, its Chief, gave frequent support to the geodetic needs of our test range on the Kwajalein Atoll, in the Marshall Islands, and now, "they" wanted to improve their first-order control network and densify the deflection data. The deflections of the vertical, as you may or may not remember from reading the earlier chapters (where I told how I myself gradually found out about these little creatures), are important little indicators of what the variations in the gravity force of the Earth from place to place might do to, e.g., a missile trajectory (we worked for the Defense Department, you know). They indicate by how much, at a specific station, the vertical direction to the zenith in the sky is deflected by the irregularities in the Earth's crust mainly in the neighborhood of that station. If the entire area of and around the test range were covered with deflection data, it would permit an important refinement to trajectory testing.

The amount of the deflection of the vertical at a specific point can be established by the difference between the astronomic and geodetic position (astrogeodetic deflection); or it can be computed from a dense gravity survey over a whole region (gravimetric deflection). There were sixteen astrogeodetic stations on the atoll, but the expense of establishing a dense net of such stations would be prohibitive. Jim planned to compute gravimetric deflections instead, and he had acquired a map of a gravity survey collected on ships going by in all directions. He came to us for our machine program for computing such deflections, and he wanted Phil Wyatt who had written that program to assist Therese Jaspers, Datum Branch, in its use. It soon turned out that the gravity survey was not sufficient for the purpose. While more gravity measurements had been made here than anywhere else in the Pacific Ocean, they were very dense along the random ship lanes, but nothing between, and one needed a good distribution for the whole region.

What else could one do? I pointed out that the direction of the large observed astrogeodetic deflections around the atoll could roughly be predicted by just looking at the shape of this landmass in the middle of the water. Suppose you put a plumb line (which indicates the vertical direction) on one side of a big mountain, then the bob at its lower end will be attracted by the mass of the mountain, and its upper end (the zenith direction) will point away from the mountain (the vertical will be deflected outward). Imagine doing this all around the mountain, and you will find the vertical deflected outward all around. And that is what you saw around the atoll. The density of the land mass is about three times that of the water, pulling the bob of the plumb line inward and pushing the zenith outward. I felt that at least a major part of the deflections was caused by the land-water distribution, and I was curious to see whether that could be proved, and if so, whether one could not turn the tables and compute deflections from the land-water distribution. I offered to do this if Jim would sign over the job to me as a request from the Kwajalein test range, with an appropriate job number. This he did, and so began a new fascinating study with several ramifications.

As a starter, I needed maps with bathymetric (depth of the ocean) information for a region of about 300 miles from Kwajalein out in all directions. Then I plotted on these maps the sixteen Kwajalein stations whose deflection values I wanted to duplicate from the bathymetric information to prove the validity of my proposal. Well, in real life things never go as smooth as you read in these beautiful textbook examples, and it is the unexpected trivial snags that take up your time and hold you back. In this case, the stations on the island when plotted by their coordinates, annoyed me by falling into the water. They shouldn't do that, should they? It took some time of checking, reading the station descriptions in the field books, talking to our surveyors who had been there, checking the different coordinate systems on the different maps, until we could reconcile the apparent contradictions into a consistent system. Then we had to transform the bathymetric information on the maps into something the electronic computer could understand, that is, into a suitable scheme of numbers. To that end the whole region was divided into rectangular columns of 5 km depth from the ocean surface down (the maximum ocean depth of the region), and the actual depth of water in that column was recorded; the remainder to the 5 km was then assumed to be rock of a certain density. Columns under the island had, of course, no water part, and those of the lagoon only a few feet. Next, we needed formulas to compute the attraction of the water and rock parts of a specific column anywhere on the plumb line at a specific astrogeodetic station, and express it as a deflection of the vertical there. The computer would repeat this computation for all other columns and sum the effect for that station; this sum would represent the deflection at this station due to the bathymetric data that had been used. The procedure would be repeated for all 16 stations. Certain assumptions must be made specifically, when you work with numbers and not just tell about the general procedure. For instance, you don't know the actual density of the rocks under the ocean and it may be different here and there. So you start with a reasonable assumption taken from the literature and vary it for this area to see what the effect on the final result would be. And there are several other assumptions that you have to consider and vary and test, such as the possibility that there is some isostatic compensation for the atoll underneath the 5 km depth. Since you can't go look, you will try some of the hypotheses on isostasy in the literature and see what happens in Kwajalein. The proof of the pudding is a comparison of the computed deflection with the observed deflection. Ideally, they should be the same but seldom are, yet certain assumptions will make the difference smaller.

Things seemed to go pretty well and I felt confident to be on the right track, even though there was much to test before decisions could be made. When Col. Jonah appeared on the scene and reviewed all our projects, he was very skeptical. He thought the maps I used were very old and not good enough. Since there were no others available, I pointed to my surprisingly good preliminary results. If one can get such good results with these maps, how much better results will one get with future better maps; it was even more reason to work out the procedure for this new bathymetric method, since deflections at sea were now in the foreground of requirements. To this Col. Jonah agreed but he wondered whether the Navy would not object to my interference in their watery sphere of interest. He wanted me to go over and ask them.

When I went to see Tom Davis at the Naval Oceanographic Office and told him what I was doing, he was delighted. His little research group, then five people, was also working, among other things, on deflections at sea because it was a high priority requirement. But their approach was different. They tried to interpolate between the lines of measured gravity values collected on ships, by devising a correlation function between gravity and ocean depth. This way they hoped to develop a dense gravity survey for a region, from which deflections could be computed (as Jim had tried for Kwajalein but with insufficient data). If I wanted to try my method, it was all to the good, since the product was needed. They also mentioned that they were using my method of computing gravimetric deflections and considered it the best among competitors. That was nice to hear. Col. Jonah was satisfied too and I was greatly encouraged. Not only was my impression correct that deflections at sea were an important requirement now, but also that my bathymetric idea was sound; the Navy too tried to utilize bathymetry, but their approach was a two-step plan (first get a dense gravity survey from the bathymetry, then get from that the deflections), while mine was a direct, one-step plan (use the bathymetry directly for deflections, skip the gravity with its specific difficulties altogether). If it worked, it would obviously be a more economic plan. 'Deflections at sea fitted also into the plans for exploiting altimetry data which were expected from NASA's future GEOS (Geodetic Earth Orbiting Satellite). This satellite answered to the brief name of GEOS-C (as the third of a series), which also stood for a second full name: Geodynamic Experimental Oceanographic Satellite. It was to be launched in July 1974, with a radar altimeter on board, to measure (in essence) its distance from the nearest point on the Earth while it circled around it. Specifically, it would be used to explore the ocean surface, which was practically a terra incognita for land-based classical geodesy. The millennia-old concept of the geoid as the shape of the Earth had always been visualized as being principally equivalent to the ocean surface (about 5/7 of the total Earth's surface) with a continuation through the continents; there one could neglect or compensate for the mountains above the zero elevation line (the mean sea level) as insignificant in relation to the so much larger size of the Earth. Yet, the oceans presented obvious and formidable difficulties for geodetic techniques. But now there seemed to be a concerted attempt to conquer the oceans geodetically, and it raised all the excitements of a new frontier. There was now satellite geodesy, which determined long distances across waters and established positions of islands and ships with increasing accuracies. Geodetic markers were being placed on the ocean floor, with a vision towards a marine control net after the manner of land control nets. The Navy was collecting bathymetric data with increasingly sophisticated methods. And the satellite altimeter would eventually give marine geoidal undulations and deflections of the vertical, besides monitoring oceanographic features such as wave heights and ocean currents, all impossible in ancient and not quite so ancient times. So there were new things under the sun, after all.

Already in 1964/5 I had heard Ben Yaplee and Al Shapiro of NRL discuss the idea of utilizing their experience in getting radar distances to the Moon for getting radar distances from a moving satellite to the Earth underneath. Could we get geoid information out of that?—they had asked. Also others mulled over the merits of placing an altimeter on a satellite. Meetings working groups, and reports explored the various instrument requirements, data collection, reduction and interpretation, accuracy expectations, and significance of results. Eventually, in 1970, NASA decided to put an altimeter as a technological experiment into the Skylab project. The experience of its performance should benefit the geodetic application of the altimeter of the GEOS-C project, which was established in 1971 with a launch target date of 1974.

The huge project required also extensive planning for shared responsibilities between participating agencies, particularly an understanding between NASA and the Defense Department about specific mutual support in funding, equipment, manpower, and purpose. Any other qualified researchers were also invited to submit research proposals for utilizing GEOS-C data, specifying the type of data and support they would require. The great and enthusiastic response made it imperative, however, to limit this offer and set up a system of screening, evaluating, and ranking these proposals. NASA set up 14 subject categories and six evaluation panels. In February 1973, Mr. Don Rice (Coast and Geodetic Survey) called me by telephone and asked me to serve on his panel supposed to screen proposals connected with geoid determinations. In the following weeks I looked at twelve proposals and tried to rank them with respect to scientific merit, funding needs, and probability of successful completion in reasonable time. The panel members discussed their findings in panel meetings, and Don Rice as chairman summarized them in a report to the NASA Project Managers; these had to make the final decisions.

Research proposals originating in the Department of Defense were not subject to this screening effort, but were evaluated within the DoD with respect to its specific missions and responsibilities were divided between the several DoD components. The DMA Topographic Center, besides giving support in specific ways, proposed four research projects of exploiting GEOS—C data: improvement of orbit determination (Randy Smith and Charlie Schwarz), improvement of the gravity field (Charlie Schwarz, later Archie Carlson), supplementing astrogeodetic geoid information and evaluating an altimetry derived geoid with respect to a world datum (Irene Fischer). I had Kwajalein and the bathymetric approach in mind, of course.

My Kwajalein project, even if it should drag out beyond the originally anticipated target date through unexpected difficulties, was thus securely established as supporting the department's missions in three ways: as a request from our test range in Kwajalein, as a response to a navy requirement, and as two tasks in the GEOS-C exploitation plan. Accordingly, I asked for and received something better than a job number for an outside request with a relatively short due date. Robert Guker (Advanced Technology Division of the Directorate of Plans,

Requirements, and Technology; in short, the T in PRT) secured for me an "Engineering Applications Project Assignment" (EAPA), that is, the funding for a long-range, in-depth study "in direct support of the TPC Center's operational programs". What more could I want? I had now the permission to explore this bathymetric idea in its ramifications, and to produce eventually charts of detailed deflections of the vertical and of geoid undulations as "ground truth' information to test the future GEOS-C data. That satellite was not even launched yet; and after launch there would be a few months of testing and calibrating the new technique against "ground truth" in a designated calibration area in the Atlantic; and after that the data flow would be directed through the designated agencies for processing; and after that one could hope to get requested data for specific areas. So there was plenty of time to get my Kwajalein charts up to notch. For Col. Jonah's show business of advertising our projects, and for my own future briefing needs, I designed three colored illustrations for briefing boards, viewgraphs and slides to explain (1) why deflections at sea were needed, (2) the Navy's two-phase approach to get them, and (3) my direct one-phase approach. I took the designs to Mr. Vincenti in the Pictorial Arts Section for advice and professional handling. While I always had excellent service from the visual aid draftsmen for my numerous technical illustrations, I was not prepared for Mr. Vincenti's sketches to explain such difficult concepts. They were strikingly beautiful masterpieces, and were then carried out to perfection in glowing colors, demanding attention in any exhibit. Needless to say, I wrote a delighted official Letter of Commendation through my department chief to his department chief.

Work on the bathymetric deflections had progressed in the meantime to good results beyond expectations. The sixteen observed deflections, or rather their meridional and prime vertical components, that is thirty-two values, ranged in magnitude from about-25 seconds of arc to +25 seconds of arc, which is very much as far as deflections go. In the first step towards duplicating these values by computation from the maps' depth information, the observed values were reached within 2 1/2 seconds of are on the average. The next step consisted of the refinement of considering variations in the density of the rocks under the ocean bottom. Various reasonable assumptions about such density models with and without isostatic compensation were studied to see whether they would further reduce the residuals (the difference between observed and calculated values). The observed values were thus used as guides in selecting useful density models. Our selection produced residuals of less than 1 1/2 seconds of arc. Now a third step was needed: Since all 32 values had been used in refining the model to produce such small residuals, one could not be sure that the good results would apply also to other places. But there were now seven new deflection stations in Kwajalein, that is 14 newly observed values, which could be used for testing. With the same density model, the average residual at these new stations was about one second of arc! This compared pretty well with the Navy's stipulated accuracy requirement of four seconds of arc. Furthermore, new technological attempts to provide deflections at sea at that time were quoted in the literature with an internal consistency estimate of 3 to 15 seconds of arc, and with discrepancies in results by different techniques of up to 70 seconds of arc.

For a general application of my computational bathymetric method in the oceans, however, one would have to find some alternatives to the guiding function of the observed astrogeodetic deflections on Kwajalein, since there are usually no such things in the middle of the oceans. Maybe, GEOS-C data could be utilized; this would be a future research topic. The use of observed values as guides for constructing suboceanic density models was another significant difference from attempts reported in the literature to solve the problem by searching for a mathematical correlation function between gravimetric and bathymetric data. Since gravity measurements are affected by suboceanic mass variations just as deflections are, the success of a general correlation would depend on the magnitude of the neglected suboceanic effect. Just compare two oceanic areas: one with rugged ocean bottom topography and relatively little or no variations underneath (seamounts, Kwajalein), and another with smooth ocean bottom but oil fields underneath (as in the North Sea). The correlation would be strong in the first case, but fail in the second case.

When Dr. Ron Mather, Secretary of IAG Section V, invited me to speak at the Symposium on the Earth's Gravitational field and Secular Variation in Position, November 1973, in Sydney, Australia (sponsored by the IAG and the Australian Academy of Science), my attendance was approved by my agency and I made two offerings: the new bathymetric feasibility study at Kwajalein, and a mid-period report on the activities of my IAG Special Study Group V-29. Its membership had risen to over 30 and I had kept in touch with several. Now I sent a circular to all members for contributions to that report. I received them, almost by return mail, in the form of recent reprints, new manuscripts, paragraphs describing current activities, or just informative, cordial letters.

While combining this information into a coherent report, formulating my other announced symposium paper "Deflections at Sea" (J.G.R.,vol.79, no.14, 1974) and designing its several illustrations and informative tables, I also responded to the AGU appointment as the geodesy member of the AGU Geophysical Monograph Board (which did not demand much time), to an invitation to speak even before the Australian trip at the GEOP (Geodesy/Solid Earth and Ocean Physics) Research Conference, August 1973, in Boulder, Colorado,(which took some more time and thought, on relatively short notice) and to several invitations for 1974, with titles and abstracts to be submitted now or soon. More about these later on. Mr. Byars was going to the VI Brazilian Congress on Cartography, Rio de Janeiro, in mid-July 1973, and wanted to distribute there my paper on "The Basic Framework of the South American Datum of 1969"; so a number of copies had to be readied with the appropriate cover sheets for presentation to that congress. He brought back the information that Brazil considered using the SAD 69 for mapping, with gradual conversion over the next few years. Good!

Production of my two handouts for Australia were in the hands of Mr. Joe Love and Mrs. J. Megaw of the Technical Publications Office who fortunately distanced themselves from the antics of that office before and after my Panama trip a few months earlier. These two people went out of their way to be cooperative and helpful in producing these papers on time. On the day before Thanksgiving, my husband and I were off to "down under", to Australia. We had intended a Thanksgiving rest on the Fiji Islands after the tiring flight over so many longitudes, but that turned out slightly different from plans. Our luggage was not there and the heavy winter clothing on our backs were all we had for this tropical heat. No swimsuits to recuperate in the hotel's lovely garden pool. But we managed some sightseeing with an Indian cab driver. We heard about the tensions between the Fiji and East Indian inhabitants, watched the communal replacement of a grass roof in a Fiji village along with the weaving of bamboo strips for new walls, and took a trip into the empty interior to a lone Indian farm, which had some intriguing human figure statues made from compressed cypress needles standing in front of special huts.

When we arrived in Sydney, the luggage had not caught up with us yet. A cousin who fortunately lived there and met out plane, offered me her wardrobe, but it did not come to that. The airline had apparently confused the destination airport Nandi on Fiji (NAN) with Kansas City (KAN), had taken the luggage off at this one transcontinental stop, and upon inquiry had sent it on to Nandi where it arrived after we had left; it was now on the way to Sydney and would be sent to the hotel; so it was. Worse yet was the scare that the immigration officer did not want to let me enter, threatened to send me all the way back (imagine!) because my official passport did not have a visa. My husband's private passport did have a visa. I tried to plead with the officer that it was not my fault, that the travel office at my agency had been told that on U.S. official passports a visa for Australia was not required, that I was an invited guest speaker at that symposium, that sending me back would punish the wrong culprit and start embarrassments all around, etc. Eventually, he softened and let me enter, "But tell your travel office to get correct information; next time we'll send them back."

Participation at the symposium was truly international, with about half coming from twenty countries all over the world. Topics ranged from astrogeodetic methods and combinations of satellite and terrestrial data to absolute gravity measurements and gravity variations in time, sea surface topography, plate theory, four-dimensional (time-dependent) geodesy, and to the need for an unprecedented measuring capability on the 10 centimeter level for monitoring time-dependent changes of geophysical and oceanographic features. Among the many distinguished scientists present was also Dr. William Markowitz, who invited me to write a comprehensive review article for his new international journal, *Geophysical Surveys*, and we agreed on the topic of the Figure of the Earth. That sounded like fun, and I mentioned it in my Trip Report as an "Action Item and Commitment" in order to secure early office permission. In that Trip Report I also wrote that "there were again appreciations of the quality of my colored slides and viewgraphs (the latter at the GEOP Conference in Boulder, last August), which I am happy to ass on to the Visual Communications Branch at TC, who have always been exceptionally responsive and efficient." I could not resist to continue: "Mention should also be made of my relief to find the 150 copies of each paper actually there at the Conference, after the embarrassing experience in Panama last spring. Thanks are due specifically to Mr. Joe Love and Mrs. J. Megaw for their responsiveness. I might add that even carrying some copies in your luggage (other than the slides and notes in your purse) is not a sure thing in this uncertain world, as my suitcases were lost on the way by the United Airlines and caught up with us only after several days. Speculating which is worse, to have no papers at a Conference presentation except the one you carry in your head, or to have no clothes except those you travel in, I counted my blessings when the suitcases did show up eventually. Somebody else at the Conference was not so lucky his suitcase had not been found yet after three weeks." The Trip Report came back through channels with the CC's pleased comments: "A good MOA" (Memorandum of Accomplishments) and his specific approval n the margin for the article commitment and the commendations. Happily, I started planning for that article and wrote official Letters of Commendation for Joe Love and J. Megaw in the Publications Office, and for Kurt Streit, N.W. Spaid, and R. Souls in the Visual Communications Branch. After the symposium in Sydney we had a few days annual leave to get at least a glimpse of beautifully planned Canberra (and a very pleasant visit with Tony Bomford and family), of Melbourne and the Healesville Sanctuary (emus, koala bears, kangaroos, etc. roaming freely about), Rotorua in New Zealand (Maori culture, geysers, and hot springs for health bathing as well as for cooking your turkey etc. in the street), and colorful Tahiti.

2. THE CHALLENGE OF THE OCEANS

As geodesy expanded into the marine environment, contact with oceanographers seemed desirable. The GEOP—IV Conference in Boulder, Colorado, August 1973, (the fourth-in the series of Geodesy/Solid Earth and Ocean Physics Research Conferences) was devoted to the theme "The Geoid and Ocean Surface" and was intended to promote an informal discussion between geodesists and oceanographers on topics of common and overlapping interests, in order to explain methods, concepts, and goals of each group to the other. Dr. R. Rapp's keynote address gave a comprehensive overview of gravimetric geoid determinations and pointed to future combinations with oceanographic and satellite altimetry data. Other geodesists talked about methods and current status of leveling, satellite geodesy, and analysis techniques. I had been invited to explain the mysteries of the astrogeodetic method, and I showed a string of 16 beautiful viewgraphs, which told the story very graphically with a minimum of running commentary. Several people suggested that I should make this talk available as a publication for the benefit of a wider audience, which I did later on.

The other way around, I found it very illuminating to listen to the tutorials of the oceanographers, in particular to Professor R. B. Montgomery (Johns Hopkins University), who pointed to the type of oceanographic features such as shifting ocean currents, wave heights, changes in sea level, etc., where satellite altimetry might possibly be useful; but only if the measuring capability were at the 10 cm accuracy level. Such high accuracy requirement seemed to be a sine qua non condition, in view of the fact that e.g. seasonal variations in ocean water densities caused sea level variations of 10 to 15 cm, and so did atmospheric pressure changes. Well, the geodesists asked, if satellite geodesy succeeded in establishing such a precise geoidal surface over the oceans, could the oceanographers use it as a reference for monitoring these oceanographic features, in terms of deviations from it? Professor Montgomery was not so sure; how would that geoid be defined and how would one relate measurements to it?

The classical definition of the geoid as "that level surface which, on the average, coincides with mean sea level in the open undisturbed ocean" did not make any sense to him whatsoever, since there was no such thing as an "undisturbed ocean" in real life. Come to think of it closely, that definition did not make too much sense to us geodesists either, but up to now we had not been challenged to give it a practical interpretation with a better than two meter precision. Without that precision, it had been a useful, graphic elucidation of the idea that the geoid was not subject to the effects of wind and weather and tides, but solely to the gravity effects of the rotating Earth. But now, satellite altimetry would have to be more precise in connecting this concept with actual precise measurements. The altimeter is located in the satellite, whose position in space is known with respect to the center of the Earth through orbit theory. It essentially measures its distance to the nearest surface point on the instantaneous ocean surface, which is not the same as the geoid. In order to reach the geoid, one must make corrections to the measured distance for the variable oceanographic features mentioned above. These features were the *piece de resistance* for the oceanographers, but just "noise" obfuscating the real thing for the geodesists. Well, one man's junk is another man's treasure—which is as good a reason as any for cooperation. Professor Montgomery insisted that establishing the geoid and heights of sea level in relation to it was an intrinsically oceanographic task for which that pale abstract geodetic geoid was of dubious value; but he conceded amiably that he would look at the animal when we had one.

Less amiable, rather bewildering and emotionally charged was the mystifying controversy over the slope of mean sea level; was it sloping down or up towards north? Emery Balazs (NOAA) had reported on the recent leveling adjustment along the U.S. coasts, starting at Portland, Maine, and Neah Bay, Washington, respectively, which showed that the local mean sea levels at the tide gauges along the way toward south on both coasts tended to be more and more below the starting level at Portland or Neah Bay respectively. The whole change was on the order of a meter. It was a little less than what previous leveling adjustments had shown, but the trend was still the same. To me this report represented a good demonstration of the well-known fact that "mean sea level" in spite of its name was not necessarily level," since it depended on the local situation around a tide gauge, and the procedures of averaging hourly or daily readings over time periods of various length. Beyond that, I was only mildly interested in this millimeter business of leveling, and thus totally unprepared for the almost violent reaction from the oceanographers. Tony W. Sturges (Dept of Oceanography, Florida State University) objected that this was all wrong, clearly must be wrong, because oceanographic methods of leveling showed beyond doubt that the sea surfaces sloped down towards north; and he showed graphs of meridian al ocean sections in the Pacific and Atlantic near the coasts, all of which bulged in the equatorial regions and sloped down towards both poles almost symmetrically.

Had I heard right? How do they level in the water measuring what against what? How do they compare leveling in the ocean with leveling on land, that is, at a different place miles away? If they do, what is the common reference basis? Why were these funny-looking graphs practically symmetrical (give or take some detail) in the northern and southern hemisphere, reminding me vaguely of something basic in the geodesy textbooks? Why were they supposed to show the "correct slope" and thus proved the geodetic leveling on land to be "obviously" incorrect? Why was there no quarrel for the east-west directions and why were there no funny bulges in that direction? It was all very new and confusing to me but I sensed that the controversy was not so new to the others. It had been mentioned briefly in Rapp's address, in a matter-of-fact tone of voice. Maybe, I needed to witness such emotional fireworks to appreciate the problem.

While still wondering what was going on, I was suddenly drawn into the dispute, when some people in the heat of the fight appealed to me whether I could not resolve the conflict by astrogeodetic methods; maybe these could clearly determine the geoid and thereby establish the slope of the water surface against it. Well, I reasoned, these methods were based on clear concepts of measuring the height of a specific point above the geoid (or above an equally useful specific level reference surface-such as those through Portland or Neah Bay) and of the height of the geoid above a well defined ellipsoid; did they have an ellipsoid to describe the shape of the geoid? No, they didn't; they didn't need to, because their approach was different. What was it, specifically? Professor Montgomery tried to explain again the method of steric leveling (calculated from sea water densities); he used reference surfaces at about 2000 m ocean depth, which were "nearly level," although in some regions there were still strong currents at that depth, and therefore some researchers preferred another depth. I could not quite follow his patient explanations because I was plagued with more questions in my mind, such as, what do they mean by "nearly level" when we are dealing with a centimeter range in deviations from "level". How do they know that 2000 m down it is (what is?) "more nearly level" than at 4000 m or 100 m? And what do they do near the coast where there are no 2000 m ocean depth and yet they claim to know the relation to the leveling on land? It was all very confusing and I would have to learn quite a bit more about a lot of things. A surprise presentation was a report on altimeter data from Skylab, launched in May 1973, which had been successful beyond expectations. This altimeter had been meant as a mere instrument experiment, not yet for geodetic application. But it had been possible to extract from these preliminary data a geoidal profile in the ocean along the satellite path, which showed a remarkable agreement with an existing gravimetric profile computed by Dr. Talwani (Lamont Geological Observatory). Of course, the centimeter squabbles with the oceanographers about the difference between mean sea level and the geoid were not considered yet, but the demonstrated feasibility of getting something meaningful from the first try raised high confidence in the performance of the future GEOS-C altimeter.

My calendar for 1974 contained a flurry of deadlines for abstracts, papers, and conference presentations besides my major concern of moving ahead with the Kwajalein work. The topics for these numerous products were aspects of my several simultaneous projects, geared to the specific invitations and audiences. They can be subdivided into four groups:

A. Papers on Kwajalein were pertinent for several different audiences. (1) At the GEOS-C meeting in April 1974, I gave a formal talk on an "Astrogeodetic Geoid in Ocean Areas" and stressed the use of my upcoming deflection—and geoid charts for Kwajalein as ground truth checks for the expected GEGS-C derived deflections and geoid profiles there, and also for densification between the wider spaced satellite traces. The GEOS-C launch date had fallen back from July 1974 to spring 1975, and I was asked to formulate a further task within the "A Plan for Exploitation of GEOS-C Data." I proposed "Altimeter Evaluation by Comparison with Sea Bottom Topography and with geoid profiles Computed from Bathymetric Data" as a response to the doubts expressed by some people whether the preliminary Skylab data really could show seamounts that clearly. If I had a geoid chart of Kwajalein and a GEOS-C profile across Kwajalein, one could compare it with the height of the atoll above the ocean floor and its bathymetrically computed geoid undulations. (2) At the International Symposium on Applications of Marine Geodesy, Batelle Laboratories, Columbus, Ohio, in June 1974, I discussed the technical details of the bathymetric method ("Deflections of the Vertical from Bathymetric Data", *Proceedings*, Marine Techn. Soc. 1974). (3) At the Fall Convention of the American Congress on Surveying and Mapping (ACSM), September 1974, I had hoped to present the finished deflection charts, but they had not been completed at the time of the clearance due date several months earlier, and so I could only explain the compilation of these charts and their estimated accuracy ("A Detailed Deflection Chart for an Oceanic Region from Bathymetry," *Proceedings*, ACSM, 1974. (4) At the FIG Conference (International Federation of Surveyors), concurrent with ACSM in Washington, D.C., I distributed a bathymetric paper for the Commission IV, Hydrographic Section. (5) The Kwajalein Test Range received, in November 1974, several sets of the completed large-scale charts and an accompanying detailed report. Since this did not need to go through clearance, it could not be given out generally. Therefore, (6) In November 1974, I also sent a final paper on the whole project through clearance for publication in the *International Hydrographic Review* and for presentation to the upcoming 1975 IUGG General Assembly.

B. The role of the astrogeodetic geoid in geodesy was pertinent for different audiences with different purposes. (1) My talk in Boulder, Colorado, was turned into an article "Is the Astrogeodetic Approach Obsolete?" (*Surveying and Mapping*, June 1974); and translated into Spanish by P. Campus V. for the *Boletin Informativo*, Inst. Geografico Militar, Santiago de Chile, 1974). (2) "The Role of the Geoid in Datum Definitions" was my invited contribution to the discussions at the International Symposium on Redefinition of North American Geodetic Networks, New Brunswick, Canada, in May 1974 (*The Canadian Surveyor*, Dec. 1974). There I analyzed the concepts and procedures of choosing a new geodetic datum for its traditional and primary function of recording and adjusting practical field work of horizontal geodetic control; and I pointed out the three-dimensional implications of a chosen conventional datum definition which were not explicitly stated or realized, and turned up historically as a surprise long after the choice had been made. (3) "The Determination of the Geoid for National, Continental, and Global Purposes" was an invited presentation at the First Pan-American and Third National Congress of Photogrammetry, Photointerpretation, and Geodesy, in Mexico City, July 1974 (*Memorias* of the Congress, Mexico 1974). It addressed itself to the different purposes in establishing a geodetic datum for areas of local, continental, or global extent; and to the role of the geoid in predicting the effect and thus guiding the choice of a new datum. The example, par excellence, for this audience was, of course, the new South American Datum. (4) At the FIG Commission V, September 1974, I presented "A Continental Datum for Mapping and Engineering in South America" to acquaint this audience with the fact of the new SAD 69 and its significance (*Surveying and Mapping*, Dec. 1974). (5) For the benefit of Chilean audiences, the *Boletin Informativo of the Instituto Geografico Militar* published a Spanish translation of my SAD 69 paper: "La Estructura Basica del Datum Sudamericano de 1969", in 1974.

C. The animated controversy between oceanographers and geodesists that I had encountered at the Boulder GEOP Conference had kept bugging me in its mysteriousness ever since. Some correspondence had been circulated in the wake of the conference discussing things further because of their importance and urgency; I had asked Professor Montgomery for a copy of his conference talk so I could study it more closely. "I wonder", I wrote to him and Sturges, "whether there is not a difference in concepts and reference systems...If that were so, then one would want to try and eliminate or specify underlying tacit assumptions and concentrate on the message of actual observations..." My first try in that direction was formulated in an article "Does Mean Sea Level Slope Up or Down Towards North?" (*Bull. Géodésique* No.115, March 1975). Its impact made me look more and more closely at oceanographic literature in the following two years.

Last but not least, there was my commitment to write a comprehensive review article "The Figure of the Earth—Changes in Concept" for Dr. Markowitz, editor of the *Geophysical Surveys* (vol.2,no.1,1975, 3-54). It was pure fun to trace this great story from the ancient mythical wonder about the world (what with the elephants standing on a turtle swimming in milk, or the poisonous waters isolating a forbidden place which holds the secret of immortality) through the millennia to the present wonders of satellite wizardry and the mysteries of oceanographic leveling, but it also meant much work; and I soon transferred this effort to home, for lack of time in the office. Dr. Markowitz proved to be a model editor: enthusiastic, encouraging, with valuable advice, and a pleasure to work with.

The many intense activities along diverse lines and the many timely products rewarded me with a sense of vigor and joy in both the action and the achievement, but they were also appreciated by the agency, and recognized again in the form of my nomination by DMA for the Federal Woman's Award, now for the third time. Apparently, someone in the Technical Publications Office did not approve of my many activities and their ready recognition, and tried sniping from the sidelines in conformance with that earlier threat. Col. Jonah who at that time had rescued my two abstracts for Australia from reprisal with a mere telephone call, had unfortunately been reassigned somewhere else in July 1973, and the caretakers could not or would not cope with that little office. It was their and not my time, however, that was lost in repeated fruitless efforts to speed up my and others' several clearance requests. Just to watch this spectacle gave an interesting mini-demonstration of the forces at work, which apparently had nothing to do with getting a product out or even showing work efficiency. To give concrete examples, I only mention a few referring to my own papers: requests for copies of my Technical Reports went unanswered for weeks; a symposium paper missed the organizers' due date by a month and the abstract had to be telephoned to Canada: several manuscripts were lying around in the Publications Office for five to ten weeks instead of the two to three days transmission time for logging in and logging out, that had been sufficient under Mr. Kephart. Each time, the repeat-edly inquiring department chief was told that the paper had been passed on or even cleared, which then turned out to be a (deliberate? or uninformed?) "misun-derstanding'; in one case, the clearance documentation for another symposium abstract had not been received in the department seven weeks after the organizers' due date, when repeated inquiries by the department chief and the division chief finally established that the abstract had already been sent to the organizers (when?) without notifying us, and that according to a new policy (by whom?) neither notification nor evidence of clearance and handling would be given to the department or the author in the future. This sounded like mockery (it probably was) and, I guess it broke the camel's back. I was asked to write a memorandum to defend the department's interest. With the proper Federalese buzzwords of the need for "(a) streamlined, (b) purpose-directed, and (c) cost-effective procedures to save funds, time, and manpower—a extremely scarce at this time" I suggested the obvious: the previous, well-working regulations should be re-established "which effectively separated the function of the Public Affairs Office (to get out-side clearance as quickly as possible) from the functions of the Technical Publica-tions Office (to produce specific TC publications other than scientific papers destined for publication in journals)." After some commas and words had been changed around in this memo, the department lost its courage to pass it on to higher echelons, quietly sewed up again the camel's back and submitted to the predictably increasing humiliation of having to go begging again and again to the door of that power-crazy Publications Office. It was an inspiring sight to behold. "One never knows for sure" I was told, "who may be backing that office."

Needless to say that the mockery continued as if designed to show up and enjoy the confusion and appalling helplessness and insecurity of the much higher ranking managers at the other end, which these had just demonstrated. If that was the motivation for the mockery by that little office, one had to hand it to them for efficiency indeed. Would that they had shown such efficiency in getting out our papers on time! But the fact that by the same token, the Publications Office demonstrated their own work ineptitude, wastefulness, and disregard of tarnishing the agency's image on-the international geodetic scene, did not seem to matter to them. Here was a mini-lesson that actions may be inspired by a per-sonal power play and not necessarily by a common goal of work efficiency in the agency's interest, which could be enforced by greater hiring and firing power of management, as naively expected in the current Civil Service reform. Who would have dared or even wanted to enforce it in this case with such big weapons, for example? That would have been comparable to using nuclear weapons in a local conflict. The cases where big weapons are applicable are rare and extremes and could be handled under the previous system too. Waste and inefficiency come more often from the accumulation of sand and pebbles in the bureaucratic gears, interconnected by insecurity and mediocrity. In our little example, even a modest and more commensurate request to change or limit functions in order to achieve some streamlining, did not go through for fear of retaliation.

3. OCEAN MILES...

"On the third day You commanded the waters to be gathered together in the seventh part of the earth, but six parts You dried up..." (2 Ezra 6:42). Thus one knew in Columbus' times that there surely must be a lot more land than water. Working on the review article "The Figure of the Earth...," I enjoyed using such insights into the geodetic knowledge and its sources at the time of Columbus. "Posidonius...reducing Eratosthenes' early correct estimate by over one-fourth;...Ptolemy's acceptance of a smaller figure of the earth's dimensions...afforded Columbus encouragement to attempt a westward sailing to the Indies." (Encyclopedia Britannica 1973)

I wrote to the Britannica for enlightenment, since my geodesy books had not told me of such a big and fateful reduction of Eratosthenes' famous determination of the Earth's circumference (250,000 stadia). In fact, Posidonius' result is mentioned only in fine print, if at all, with an almost equivalent number (240, 000 stadia). The answer from the Britannica was kind, extensive, and insufficient: The quoted paragraph, they wrote, had been changed in the forthcoming new edition to read: "One fundamental error that had far-reaching effects was attributed to Ptolemy, an underestimation of the size of the earth. He showed Europe and Asia as extending over half the globe,...So lasting was Ptolemy's influence that 13 centuries later

Columbus underestimated the distances to Cathay and India partly from recapitulation of this error." So the reduction by Posidonius had been dropped in the new version and Ptolemy had been saddled instead with a "fundamental error that had far-reaching effects—an underestimation of the size of the earth." The letter proceeded to recount in detail the well-known uncertainties about the several stadia in antiquity, that is, which of these were probably used where and by whom, and how they might be converted to our modern length units. More important to me, it also gave a partial background for the previous version, namely, that Posidonius had been quoted in the ancient literature with two results: the familiar 240,000 stadia by Cleomedes, and 180,000 stadia by Strabo;

the second smaller number had also been used by Ptolemy, but scholars were still divided in their opinion whether or not these two numbers were actually different results or only expressed in different stadia (stades).

The Manual of Greek Mathematics (Thomas L. Heath) on my bookshelf at home did briefly mention the two numbers in connection with Posidonius, saying that they depended on whether the distance from Rhodes to Alexandria was taken to be 5,000 or 3,750 stades, but it dismissed the whole thing because of the gross inaccuracy of two degrees in the accompanying astronomical distance. And I had never paid much attention to that paragraph, because it was so inconsequential. But it was not inconsequential anymore. I was writing a comprehensive article on the whole panorama of the quest for the size of the Earth, and if this one number had such a far-reaching effect on Columbus, I ought to know and write about it. Of course, it is reported gleefully in history schoolbooks that America was discovered by mistake, Columbus thought he was reaching China by underestimating the miles in the supposedly open ocean between Spain and China on a much too small Earth. But it is one thing to tell about Columbus' ignorance that turned into success, and a totally different thing to have him rely on an authoritative ancient revision of Eratosthenes, which was not mentioned in geodesy textbooks. The one crucial number, 180,000 stades, was blown up to great importance in the general world of literature, history, and encyclopedias, and. was still today a matter of emotional controversy in historical-linguistic circles (as I found out), while it was practically non-existent in the no-nonsense geodetic-technical literature (to which I felt partial). It definitely sounded fishy, and I just had to find out: where did that number come from? how was it derived? why would Posidonius give two results? why would he revise Eratosthenes? The mystery started me off on voracious reading of everything connected with it in ancient and modern literature, crowding my evenings and weekends with chasing leads and references in a near-frenzy, in view of the short self-imposed due dates.

Unfortunately, no statement by Posidonius himself on what he did or did not do, has survived; only what others reported. I soon found the place where Strabo mentioned the 180,000 stadia, alas, with no derivation; and a second statement that this makes the ocean distance from Spain to India less than 70,000 stadia. I hunted for a more complete explanation through all the 17 books of Strabo's *Geography*, to no avail. I looked up Pliny, Hipparchus, Marinus of Tyre, Ptolemy, and others, and I read modern histories of Greek science, discussed and criticized by D. R. Dicks, H. Berger, H. Prell, H. v. Mzik, and others, and consulted the standard Pauly-Wissowa encyclopedia, *Der klassischen Altertumswissenschaften*. Soon the quotes repeated themselves because the finite number of existing ancient reports have been hashed over for centuries; but the evaluations may differ and critics are criticized by later critics. After a while, you begin to feel as an active partner in an exciting discussion group, never mind that the other participants are long dead or at least not present in your living room. You argue with them just the same, agreeing with some, disputing others, pondering arguments, sensing fallacies. Nobody really seems to know; the plot thickens—"who's done it?"

While looking for cues, you can still enjoy the company. There was Strabo (64 B.C./21 A.D.), a most enjoyable, lively, and informative writer, represented in H. L. Jones' splendidly vivid translation. Strabo described very clearly the geographic knowledge of his day, and how and by whom it was accumulated. It was delightful to learn some ancient geodetic-geographic terms such as "the cinnamon producing countries" for the southern boundary between the inhabitable world against the uninhabitable torrid zone; or astronomical terms such as Homer's "stars rising from their bath in the ocean, and the Bear who alone has no part in that bath" (for the circumpolar Dipper). It gives you a feeling of the universal scientific quest underneath a picturesque language with different associations. But Strabo also shares his personal views with us, and polemics creeps vividly into his judgment. So he tells that Pythias of Massilia claims to have observed on Thule, the most northerly of the British Islands, that the sun does not go down, but Strabo does not believe him: he doubts that there is at all an inhabitable island that far north, since the northern limit of the inhabitable world is known to be Ierne (Ireland), "where complete savages lead a miserable existence due to the cold. Pythias misleads people everywhere else," he warns, so he must be in error here too." Strabo carries on a polemic also against Eratosthenes' three-volume *Geography*, while at the same time using Eratosthenes' data and judgments as the accepted storehouse of common knowledge. He reproaches him for spending too much time on philosophic speculations about the Earth as a whole and the relationship of land to the surrounding ocean instead of concentrating on the inhabited world as he, Strabo, does.

The reason for this reproach becomes amusingly clear: Strabo freely admits that he does not understand much of these mathematical pursuits. And we can believe that since he has no idea of how Eratosthenes determined the size of the Earth, probably was not even interested. He naively assumes that the geometricians can measure the north-south width of the inhabitable portion of the Earth by adding up the report of travelers, then estimate its ratio to the distance south to the equator and north to the pole, add up and multiply by four. It is not surprising, therefore, that he does not know or care about any derivation of the

180,000 stades estimate by Posidonius, does not realize its revolutionary significance (if correct), and does not even use it anywhere to replace Eratosthenes' numbers. He only mentions briefly that "Posidonius suspects that the length of the inhabited world, about 70,000 stadia, is half of the entire circle…so that if you sail from the west (of Spain) in a straight course, you will reach India within 70,000 stadia." Eratosthenes calculated about the same 70,000 stadia of inhabited world to be only one third of its circle, the remaining ocean distance being twice as much. So there was a real difference, not just a change of units as some scholars suggested. Strabo enjoys telling us, albeit without any details, that some later writers disagreed with Eratosthenes' size of the Earth, but he admits that Hipparchus used it just the same saying that "it deviates but slightly from the truth and is good enough for the purpose of making a map."

On the other hand, Strabo understands very well to describe practical aspects, explains the use of the terms longitude and latitude (length and width, since the inhabitable world around the Mediterranean area was known to be more than twice as long east-west than wide north-south), talks about the problem of mapping parts of the spherical earth on a flat sheet of paper, and how to evaluate travel reports. He warns, e.g., that "it is necessary for us to hear accounts of this country (India) with indulgence, for not only it is farthest away from us, but not many of our people have seen it; and even those who have seen it, have seen only parts of it, and the greater part of what they say is from hearsay. And even what they saw they learned on a hasty passage with an army through the country. Wherefore they do not give out the same accounts of the same things, even though they have written these accounts as though their statements had been carefully confirmed. And some of them were both on the same expedition together…; yet they all frequently contradict one another. But if they differ thus about what was seen, what must we think of what they report from hearsay?"

Strabo's description of Egypt contains a detailed account of its cadastral system and the necessity of periodic re-surveying for tax purposes in the wake of the yearly inundations of the Nile. He points out that this is why one says that the science of land surveying originated here. But he is not aware that by the same token he spelled out one half of the reason why Eratosthenes' earth determination was so superior to all others in antiquity: a better distance measurement between Aswan and Alexandria than was available anywhere else in the world. Neither does he know about the second half: the best calculation of the accompanying astronomical distance by means of the shadows on a sundial. Strabo describes the nilometers and their function, and also the well at Syene marking the summer tropic. "Coming from Greece toward south, it is at Syene that the sun first gets over our heads, and causes the gnomons to cast no shadows at noon." He is not aware that this phenomenon was used in Eratosthenes' calculations.

Hugo Berger (*Geschichte der wissenschaftlichen Erdkunde der Griechen*) in 1903 is no less temperamental than Strabo 19 centuries earlier. Berger does not like Strabo one bit. He is irked by Strabo's mixing accurate and approximate numbers, which makes his apparent precision in distance statements useless and even misleading. Berger (and Letronne) think that Posidonius never made a new determination of the size of the earth, because it is unconceivable, they say, that he would have abandoned Eratosthenes' superior method and revived older inferior ones instead. Furthermore, using the smaller distance between Rhodes and Alexandria, determined by Eratosthenes, would have made him guilty of circular reasoning. What happened was, they think, that Posidonius tried to stem the Roman anti-mathematical attitude by explaining to them the Greek ideas with an oversimplified example or two, because Eratosthenes' sophisticated methods were too complicated for a Roman to grasp. "Had Posidonius meant to show hereby a new earth determination of his own, he would have been branded an idiot for all times. Strabo who did not understand a word of all this, just like any Roman, took these numbers about a so-called smallest earth measurement from Posidonius' writings out of context." Having thus cleared Posidonius of the crime, Berger proceeds to present another "murderer" or rather two: He says that Marinus of llyre picked up this meaningless small number of 180,000 as the most recent one without understanding its provenance; and Ptolemy, the mathematician, inherits it blindly without any criticism whatsoever in an "appalling negligence", thus jeopardizing the reputation of Posidonius' sanity. It is only fair to let Marinus and Ptolemy defend themselves. Marinus of Tyre (1st to 2nd century A.D.) can speak to us only through the critical account of Ptolemy (2nd century A.D.), and there is plenty of criticism. But it refers to evaluations of reports from distant countries outside the well-known much traveled areas; and not to a sensational reduction of Eratosthenes' earth dimensions. We are treated to illuminating examples of processing raw data into distances and directions for the global map which both of them and others in antiquity labored to draw or improve upon. The raw data were travel times on foot or by ship, observed celestial phenomena, types of crops or animals seen, and hearsay stories. The processing consisted of converting the travel times into distances, reducing these from the actual path to straight-line distances (e.g. reducing the sailing along the shores of a bay by one third to get the pertinent distance across the bay), reducing them further for estimates of possible bragging and rest periods, checking the probability that certain stars could have been seen from certain places, and the compatibility of observed flora and fauna with the climate and other information. We read, e.g., of the identification of far away exotic Africa: "where the rhinoceros mate" and that reminded me of a German expression for a far away place: "wo die Füchse sich Gute Nacht sagen" (where the fox wish each other good night). Maybe there is a direct connection.

For the last processing step of adding the new place to the map, its position was expressed in latitude and longitude degrees following Hipparchus' ambitious ideas of a unified world system. It is here where the controversy of reducing Eratosthenes' result might have come up. But it did not. Ptolemy says that one did not really need any particular length unit in order to draw a map in correct proportions as long as one is consistent. "But if one wants to use the customary, well-known units, then one must apply the same units also to the earth's circumference by comparing a measured distance with the corresponding celestial great circle arc." He and Marinus used a conversion factor of 500 stadia to one latitude degree, "which conforms to known measurements." This results in 180,000 stadia for the circumference. (For Eratosthenes' 252,000 stadia circumference, the conversion factor was 700 stadia to a degree, although Eratosthenes himself did not yet use degrees but hexacontades, that is, 60th parts of a circle). It clearly had nothing to do with Posidonius. Strabo's and Ptolemy's 180 000 were only coincidentally the same numbers, but meant different magnitudes, one incorrect and the other correct, due to different units. The two suspects, Marinus and Ptolemy, who had been accused so acrimoniously, were innocent. Dr. R. Dicks (*The Geographical Fragments of Hipparchus*, 1960) was another very sharp-tongued critic of an ancient writer. He had only contempt for Pliny, the Elder, (1st century A.D.), and he cites several colleagues of the same sentiment. Pliny's statement that "Hipparchus…added a little less than 26,000 stadia" (to Eratosthenes' figure for the Earth's circumference) had given the scholars no end of frustration. Dicks said that it contradicted

Strabo and that most modern commentators agreed with him, Dicks, that "Pliny's remark was merely an egregious blunder." He approvingly quotes H. Berger that the 26,000 may have come from combining different numerical results of Eratosthenes', "which were misunderstood and confused by Pliny"; and for reinforcement he adds D. J. Campbell's judgment that Pliny suddenly changed the subject from various distances to the earth's size, implying that the poor guy did not know what he was talking about. I looked up Pliny's *Natural History* and read the pertinent passages with pleasure, and decided that the confusion was definitely on the other foot, that is, on the commentators' side. What I read, made perfectly good sense to me, no confusion, no undue change of subject, and instead of a contradiction to Strabo I found that this statement gave a welcome and complementary explanation to Strabo's remark, namely, that others disagreed with Eratosthenes' result, but Hipparchus said, "it deviated but slightly from the truth." How would Hipparchus have known? Why, by checking and refining it himself, of course, as we would do also today with any important result. Moreover, Pliny's sentence about the 26,000 stadia was preceded by a very useful statement about Eratosthenes. He said (in H. Rackham's translation): "These are the facts that I consider worth recording in regard to the earth's length and breadth. The total circumference was given by Eratosthenes (an expert in any refinement of learning, but on this point assuredly an outstanding authority, I notice that he is universally accepted) as 252,000 stadia,...an audacious venture, but achieved by such subtle reasoning that one is ashamed to be skeptical. Hipparchus, who in his refutation of Eratosthenes and also in all the rest of his researches is remarkable, adds a little less than 26,000 stades." Since Pliny lived after Posidonius, and either at the same time or probably before Marinus and Ptolemy, his account implicitly refutes the theory of a sensational reduction in Eratosthenes' number. He would have known about it and commented on it. But there is not a word. Pliny also entertains us with an amusing little story about a celebrated geometrician, whose heirs said that they had found a letter in the tomb a few days after the funeral, signed with his name and addressed to the living. It said that he had descended from his tomb to the center of the earth a distance of 42,000 stadia. So the circumference must be six times as much, that is, 252,000 stadia. I suppose, a significant revision of that number would have entered such a joke too.

All this excitement about the 26,000 left its mark on me too. I kept speculating what Hipparchus may have done. What would we do in such a case today? Well, we would go over every statement with a fine comb, consider all uncertainties of measurements, maybe insert updated information, and weigh all alternatives. Hipparchus probably did that too. Let's see whether one can double-guess him; and I think I did just that. Hipparchus is known to have used a more precise latitude for the summer tropic than was generally assumed in Eratosthenes' time, and thus must have calculated that Eratosthenes' terrestrial distance was a little too short and, therefore, the circumference a little too small. Considering also the uncertainty, quoted by Eratosthenes himself, of the exact location of the well with respect to the summer tropic (since the crucial lack of a noon shadow on solstice day applied to an area of about 300 stades), he could have recalculated the result, using the alternatives of the upper and lower limit of that uncertainty range. When I did that, I arrived at a potential correction of 15,900 stadia. That did not look like 26,000 but I was apparently on the right track. What could I have overlooked? Staring at Pliny's Latin quote: "Hipparchus…adicit stadiorum paulo minus XXVI M" it suddenly hit me: my 15,900 would be "stadiorum paulo minus XVI", a little less than 16,000. It would not have been the first time that a copier doubled an X by mistake!

There were several more such peculiar comments about ancient numbers by some historians, which made me look again and play a little with these numbers in order to understand the ancient explanations better through plausible recomputation. Berger and Dicks, for example, discounted ancient quotes that Eratosthenes remeasured the arc between the tropics (twice the obliquity of the ecliptic) as 11/83. They judged such a number to be "alien" to Eratosthenes' relatively simple tool kit, and therefore it must be Ptolemy's number whose clear quote they interpreted as ambiguous. That judgment hit me as absurd, since Ptolemy quotes his estimate in degrees as was already usual at his time, and only then he compares it to that fraction. It would be rather strange and senseless to transform a number expressed in the advanced system of 360 degrees for a circle, into such a crazy-looking fraction unless for comparison with an older estimate in that form. But this pits only a mathematician's conjecture against that of historians. I could strengthen mine considerably, however, by showing how Eratosthenes' well-known sundial method and his hexacontade units might have led to 11/83; and even more, how this same procedure, a little cruder yet, may have resulted in the pre-Eratosthenes' generally accepted value of the obliquity as 1/15 (four hexacontades out of 60 of a full circle, and the approximate latitude of Syene). This value, described as the side of an inscribed 15-gon, was another mystery for some historians who did not know how numbers work, and could not follow a quoted explanation (even if they quoted it themselves) like the one (by Achilles Tatius) about how the length of the longest day at varying latitudes was determined. The difference in approach to understanding what the ancient scholars said they did and why, appeared even more incongruously in Berger's summary evaluation that they failed in their attempt to determine the size of the Earth because of their hopelessly inadequate means. To me as a geodesist actually working on essentially the same question of the size of the Earth, this evaluation seemed grossly anachronistic and a basic misunderstanding of purpose and design in this type of work. Historians would do well to familiarize themselves somewhat with a subject matter before making such absurd judgments about it. My irritation was sublimated in an (extra-curricular) article: "Another Look at Eratosthenes' and Posidonius' Determinations of the Earth's Circumference" (*Quar-*

terly Journal, Royal Astron. Soc., June 1975). It could not reach Berger anymore, of course.

The mystery story of the 25% reduction in the size of the Earth, that had prompted this detailed excursion into antiquity, had not found its solution there. All cues had led to a dead end. But the corpse was still there: Columbus' underestimate of the ocean width was still unexplained, and America was still discovered through that mistake. As in all good mystery stories the solution came from an unexpected side that was there all the time. True, Columbus had collected ancient quotes such as Aristotle's surmise that one could sail from Spain to India "in a few days", and Esdra's wisdom that the waters—took up only one seventh part of the earth, and liarinus' overestimate of the uncertain length of Asia in longitude degrees so that few longitude degrees were left for the ocean width, and several other references, all needed and welcome to prove that the westward sea passage to India was so much shorter than the eastward land route, in a plea for funding his plans for a sailing venture. But that was not enough. He needed persuasive numbers to find and convince patrons. There was an impressive juggling of numbers in an attempt to allocate an overlong land part and a very narrow ocean width on the circle around the (unchanged) Earth. It involved dazzling conversions back and forth between degrees and miles, and in between an easily unnoticed switch between miles and miles; that is, between the medieval Arabic mile (more than two kilometers) and the Roman or Italian mile (about one and a half kilometers). So here was the 25% reduction through switching miles. It had nothing to do with the dimensions of the Earth.

4.... AND OCEAN CENTIMETERS

During the same time period that I wondered about ancient and medieval miles across the ocean, I also wondered about modern centimeters in sea level, hotly disputed between geodesists and oceanographers. After the exciting GEOP Conference in Boulder, August 1973, which had introduced me to that dispute, there had been more informal communications between several of the participants; and I had studied Montgomery's papers more closely and also begun to read various oceanographic textbooks in an attempt to understand the concepts and assumptions underlying oceanographic leveling and sea level slopes. With my bias as a geometry teacher, I felt that before even beginning to quarrel about whether a slope (of mean sea level or anything else) goes up or down, one had to make sure that the slope was defined as taken with respect to the very same reference line (surface) in both cases. Since the oceanographic reference surface (an isobaric =

equipressure surface) was thousands of meters down in the ocean while the geodetic reference surface (the geoid or another nearby level surface) was identified on land, they were certainly not the same. The oceanographers claimed that their isobaric surfaces were also level surfaces and thus "dynamically parallel" to the geoid, so that a slope comparison was valid. Pressed for a proof, they conceded condescendingly: "maybe not strictly level, but very nearly level, only a little different from level." Probably, they were right from experience, and I was just dense from lack of (oceanographic) experience, but I could not see why even "a little different from level" in the measurement could not produce "a little difference" in the slopes. After all, the whole controversy was only a matter of centimeters.

But suppose we do go along with our friends for the moment (I reasoned with myself) and assume tentatively that these isobaric surfaces are indeed level; what about these curious equatorial bulges in the oceanographic profiles? They had struck me at first sight as something systematically spurious, contradicting the geodetic notion of the geoid as a close approximation of mean sea level; that means that the discrepancies between the two should be random and small, say, less than a meter. But these bulges were not random at all; they peaked at the equator at about two meters for one author, and more than three meters for another author who used a deeper reference surface; and they appeared in all north-south profiles, but not east-west, and thus must have some connection with a latitude function. They reminded me of illustrations in geodetic textbooks where it is explained that, for an ellipsoidal Earth, the family (system) of ellipsoidal "level" surfaces (technically: equipotential surfaces in the Earth's gravity field) at the various heights (to which the geoid belongs as the member at about sea level) is crowded closer together toward the poles and is spread more widely at the equator. This is another way of looking at the fact that the intensity of gravity changes systematically with latitude (increases toward the poles, pulling the level surfaces more strongly together). Along the equator or along a fixed latitude, however, theoretical gravity is constant and thus two surfaces keep a constant distance between them. Does this distinction between north-south and east-west behavior have anything to do with the same distinction between north-south bulges and no bulges east-west?

Let us look at the north-south bulges a little closer. Let us look at two ellipsoidal surfaces, one inside the other, concentric, but about a thousand meters lower; and consider a meridianal (north-south cut through the pair: an oval line enclosed by another larger and a little flatter oval line about a thousand meters distant, a little more at the equator and a little less at the poles. Think of the inner line as-a section through the deep-sea isobaric reference line, and the outer as a section through a level line near the ocean surface. This higher line is the one which the oceanographers apparently take as equivalent to the geoid and use as the zero-line for their profile graphs. Suppose you would plot the distance of the outer line from the inner one along the middle part, say from 45° south to the equator and to 45° north; you would get a line with an equatorial bulge, wouldn't you? And this is strikingly how the oceanographic profiles roughly looked, with the actual data points of the sea level observations draped more or less randomly around this abstract, not actually drawn, line. The size of the peak, however, was only a fraction of what it would be in the above theoretical text-book case. On the other hand, the expression "dynamically parallel" means that the theoretical considerations have been inherently taken into account in the oceanographic computation methods (equivalent to the so-called "orthometric correction") so that there would not be a peak. That is why they interpret the straight zero-line of their graphs as equivalent to the geoid, and show the data along the bulges as a striking phenomenon of the ocean surface, deviating by up to two or three meters from the geoid. This posed a dilemma in my speculations: If these bulges were true deviations from the geoid, then their systematic, lati-tude-connected character was a novel and surprising discovery, not anticipated by any of the great geodesists, and would require explanations and corroborations which were nowhere in sight as far as I knew. If, however, these bulges were spu-rious as had been my first impression, then one needed an explanation of how, through what error or imprecision or latent assumption or whatnot, they had appeared. And this is where, in my speculations, the above textbook discussion about the two level surfaces might come in. If the "orthometric correction" had been applied fully, then the bulge in the upper line with respect to the lower line would not be there and the two lines would be "dynamically parallel" as claimed. If the orthometric correction was not applied at all, there would be a much higher bulge. But if that correction was applied only partially, say by imprecision in the procedures or by any other reason, then a bulge would appear of some lesser height. It seemed worthwhile to pursue this thought, at least to find out what magnitudes were involved and whether a possible explanation could be found by looking at the computation procedures.

I studied the numerical hydrographic tables which the oceanographers rou-tinely use for such computations and found that they were composed by V. Bjerknes in 1910 on the empirical basis of a small water sample taken in 1902 at the International Exploration of northern European waters; and while the ocean-ographically much more important variables of salinity, temperature, and pressure are handled with a numerical precision of several decimals, the intensity of gravity appears with only one decimal as a good, over-all representative value, but a fixed value with no variation in latitude. Thus, I reasoned, it could well be that the lack of a latitude variation caused the bulge, but that the use of a good mean gravity value kept that bulge at a fraction of the full theoretical size.

It looked as if I might be on the right track. If so, then the size of the bulge would be proportional to the fractional neglect of the orthometric correction. To compensate for that neglect, one would have to fit a new zero-line (level line) to the data points, taking out the spurious bulge. To do that, I assumed the fraction of the neglect as an unknown to be determined in the "least squares" fitting procedure. When the data were replotted accordingly, they appeared in less than a meter distance (as expected in the concept of the geoid) and the replotted oceanographic profiles agreed in their general direction of the sea level slope with that of the geodetic profiles: up towards north.

There were still local discrepancies and many other questions left to worry about at a later time or by other people (the geodetic levelists, for instance, reviewed closely their own procedures for possible omissions or imprecision), but this much appeared reasonably clear: the difference in reference surfaces cannot be overlooked; it has a crucial systematic effect. The zero line of the oceanographic profiles is <u>not</u> equivalent to the geoid which serves as the zero-line for the geodetic profiles. If equivalent zero-lines are used for both types of profiles, the general directions of their slopes agree. The claim that the deep ocean reference surfaces are "sufficiently level" and some are "more nearly level" than others, has not been proved but is an expression of wishful thinking. I searched the oceanographic literature for a proof and found, at a later time, that there was none. Distinguished oceanographers through several decades had wrestled with that problem, but attempts by some had been disproved by others. Our contemporary oceanographers, however, did not tell us that. On the contrary, a geodesist colleague wrote me that he had been told by "leading hydrodynamicists" that the idea that the deep ocean reference surfaces were not level was "scientifically unsupportable." Without proof? Had they not read their own literature? While I still felt like Alice in oceanographic wonderland, I also thought that I had made at least a sensible beginning; and I circulated the manuscript "Does mean sea level slope up or down towards north?" among several distinguished participants of the Boulder conference, with the question whether they thought it was indeed sensible. All the geodesists except one (who had some reservations which I thought I could clarify) said it sounded sensible to them, but one oceanographer reacted almost violently in his dismissal and tried to keep me from publishing. Yet he had been among those in Boulder urging me to enter the discussion, apparently with the foregone conclusion that I should point out the geodetic leveling as the sole culprit. I sent the manuscript for publication in the *Bulletin Géodésique* to Dr. Levallois in Paris, who had worked on some sea level problems before. He wrote that the paper seemed to be very important and he accepted it for publication in the March 1975 issue.

To know the date well in advance was just fine because I could still insert it into the references for my long review article "The Figure of the Earth: Changes in Concept" which would also appear in early 1975. My Eratosthenes article was mentioned there too and would be published in June 1975. Beginning 1975 was also the time to start preparing for the quadrennial IUGG Assembly in Grenoble, August 1975, and I sent a circular to my 30-member SSG V-29, asking for input into my obligatory chairman's report. This time it needed to cover only the last two years since the symposium in Sydney, Australia, where I had presented an Interim Report which would form Part I of the full Grenoble report. All the other IAG officials writing reports wanted input. There came requests from U. Uotila, President of Section V, R. Mather, Secretary of Section V, the writers of the U. S. National Report, and others. My Kwajalein study was represented by three publications of its different phases, but the final deflection and geoid charts had been completed later and were not yet published. Their delivery to the original customers of the Kwajalein Test Range and the accompanying final report had not required clearance. Nor did my presentation to the GEOS-C committee, April 1975, need clearance; but the use for the IUGG did. When I requested clearance with a view towards publication in the *International Hydrographic Review*, I ran into quite unexpected difficulties.

This time it was not so much the Technical Publications Office, which I have already painted pitch-black several times. That this office did not let me see, for more than a week, a list of questions received about the paper, and that they withheld from me for three weeks the finally approved charts, as almost routine sniping by then. This time, some one in DMA Headquarters attempted to refuse clearance although he admitted in an official letter that the material was unclassified and there was no security risk involved. So why? Because I had already published three papers on the subject and should wait with a fourth until I had the final product. The worthy, high Executive did not recognize the six charts as the final product when he held them in his hand; or did he, and tried to block them for a reason he could not admit? There had been previous peculiar attacks on this promising bathymetric study; they had seemed to me annoying, yet incidental games of the bureaucracy which had to be endured as the symptoms of a deplorable erosion of leadership and competence in some of our management. Now in hindsight, they all seemed to fit together into a pattern: a clumsy attempt to suppress a too successful study which, however, had official blessing and funds, not to mention the outside recognition of the papers.

Preparations were made, for example, to brief the new DIA Director, Admiral S. D. Cramer. There was to be only one briefer for the whole department, and a galaxy of briefing boards about the various projects. When I did not see the bathymetric study represented, I showed the organizer of the exhibition the stunning pictures made by Mr. Vincenti, the artist, and suggested that if you wanted to interest an Admiral you should show him something with water, specifically something of interest to the Navy and supporting the Fleet Ballistic Missile project. My hunch was right: After greeting the assembled branch chiefs, Admiral Cramer listened courteously to the drab routine briefing, then picked out with special interest only one project: my Kwajalein project. He asked me to come forward to tell him more about it, which I did with relish. Later I received a gracious letter from him: "…It makes me realize how fortunate I am to have outstanding personnel like yourself in my organization." The briefer, by contrast, sent me a sour message: he had not expected the Kwajalein pictures to be in the exhibition (why not?) and it made him nervous as a briefer to find surprises in the briefing aids. A competent briefer, indeed!

Less harmless because illuminating the spread of managerial inadequacy and furtiveness was the suppression of my "Map of the Month" exhibit, when the secrecy of the stovepipe system was used to hide whoever was behind it and why. The "Map of the Month" program had been set up to exhibit a new interesting mapping product in a conspicuous poster in the lobby for a month, with repeats in other buildings. My Kwajalein charts went under the category of "Special Purpose Maps." I had a board made up with small copies of the charts, the Vincenti pictures to explain the project, and brief explanations of the new method; there I pointed to the water-land distribution of the atoll region which caused large deflections of the vertical which in turn affected the launching and tracking of missiles. The person in charge of the "Map of the Month" program assured me that no further paper work through channels was needed since he had the authority for running this program, and he wanted to exhibit the board at the first vacancy. This was good, since in these times of managerial upheavals managers worried where and when they would exchange chairs and could not be bothered with real interest in work. The soon to depart department chief had explained his lack of involvement in our Geosciences Division (which then had no appointed chief either) by saying that his attention was needed more for troubles in other divisions, that we could take care of ourselves (true), and that we should stress visibility for our projects and international standing. Thus, the "Map of the Month" exhibit seemed just the thing to do. The board was up for one day; it looked beautiful, but then the fireworks started. One of the stovepipe members (I never considered myself as one although I was a manager too) tried to intimidate me with an interesting attempt at a lion's roar over the phone and on paper, complaining that he had not known of this exhibit (why had he not suggested one?) and that viewers would get the wrong idea that the water-land distribution caused deflections; therefore he had ordered the display taken down. Apparently, here was a kid feeling left out (unintentionally from my side) and enchantedly trying out his new clothes and a new voice, but at the same time broadcasting his geodetic ignorance: Mr. Geodesist (with a degree in geodesy) could not relate the density difference between ocean and atoll to the textbook phrase of "irregular mass distribution" causing deflections of the vertical. Why did he not inform himself before broadcasting such nonsense, or ask me, or read my Kwajalein papers, or just remember his geodesy courses? He tried to change Mr. Vicente's masterpiece, rewrite the legend and rearrange my explanations. He also changed the credit for this method as developed in my Geoid Branch to a credit for the department. (First they criticize your product, but when that does not work, they want to take the credit for the product.) To save face after his blunder and for the credibility of his revisions he promised to re-exhibit the poster. But he broke that promise. As I found out, a phone call had stopped the revision work at the Pictorial Arts Section without letting me know. Mr. Geodesist indicated that the resistance did not come from him but from someone in Headquarters whom he had to obey; I almost felt sorry for him and his predicament. He promised again to straighten things out and to put up the poster, broke the promise again, but refused to tell me who and what reason was behind such irrational orders. Since my final Kwajalein paper ran into peculiar clearance difficulties at the same time, I gradually began to put two and two together.

The list of questions concerning the clearance of the paper seemed at first harmless: Why did I spell metre instead of meter? (because this is now the international scientific spelling). What is the difference from the previous papers? (those reported on earlier phases and technical details of the method for audiences at symposia, while this paper brings the final charts, completed only recently). What is meant by "References below"? (References are listed on the last page of the text). But there was one puzzling question: was this method endorsed by DMATC as a practical and economic approach to the deflection problem? (This is a feasibility study for a new method. Results show that it would be practical and economical. The study was endorsed by Col. Jonah, by the Kwajalein Test Range, and by DMATC "Engineering Applications Project Assignments"). The thrust of that question became clearer when more questions were asked over the phone: What was "the TC position" on using that method now? Why publish it in the International Hydrographic Review? Questions might come from there and elsewhere why DMA was not using it, if it was so good. (Well, that would be a really good question; but so far it is only a feasibility study with only one special test area. A TC position stand is premature at the moment). The worries about an official position stand were then resolved by adding a footnote to the paper: "Opinions expressed in this article are those of the author and do not necessarily reflect a position of the Defense Mapping Agency." When I submitted the answers to all these questions in writing as requested, I added a plea: "In view of the FBM requirements for deflections at sea, DMA support (at least psychological encouragement rather than resistance) to perfecting this obviously promising alternative method for certain cases is respectfully requested. The international interest in dense bathymetric coverage (see *International Hydrographic Review*) will turn up studies like mine sooner or later. Why should not my Agency take and give credit for another timely 'first'. It does not take away anything from other efforts." I tried to strengthen that plea by calling for support on Dr. M. Macomber, who had a reputation for using his own judgment and not necessarily acting as a conformist. He was agreeable, acknowledged that there had been some misunderstanding, that he was satisfied with the explanations and the footnote, and would try to pry the paper loose from someone else's desk. Mr. Someone Else, however, said he still needed more time to study this five-page paper. How much time does a high-ranking geodesist need to read such a short paper accompanying my finished charts?

What was going on? None of our stovepipe "leaders" would come forward with a decent explanation, which fact in itself was a reflection on the quality of that "leadership." My colleagues around me speculated with me. One had given me the advice to call Macomber, but while that call gave us some illuminating insight, it still did not get the paper cleared or the poster displayed. So it must be some underhand politicking. Someone told me, so by the way, that Headquarters meetings these days consisted of petty bickering whether TC or AC should do one or the other part of a new big job. "You know, it is the old rivalry; each claiming they are better qualified, have more experience; AC saying that they are the official depository of all available gravity, TC saying they have done more work with land and coastal gravity." Was it possible that my demonstration of a new successful non-gravimetric method interfere with telling purse string holders and decision makers that only gravity, especially AC gravity, could do wonders in geodesy? A colleague suggested that something like that could well be the reason for suppressing publicity about my new method. It would also explain the nervous preoccupation with a "TC position stand" on using this method now. Whatever it was, I had to find a way to get that paper cleared in time for the IUGG. That was much more important than an in-house display of a poster, which by then had become a symbol of administrative insecurity and collective furtiveness.

As so often in my life, the Good Lord came to my rescue—amazing that the soot in our stovepipe was noticed up "On High." There came a quite unexpected phone call from Major Beers in Dr. Macomber's office at Headquarters. They had a request from George Mourad (convener of the Symposium on Marine Geodesy at the IUGG Assembly in Grenoble) for DMA support and contribution of a paper to the symposium program. DA would like to oblige, he said, and was looking for an appropriate paper. Would I have one or could I prepare one? Major Beers was all smiles, one could see that through the phone. I could not believe my ears: "Major Beers, you know that I have a paper: it is lying on your desk. If you can get it cleared, I'll be happy to present it at the symposium." The due date for submitting abstracts to the IAG in Paris was almost upon us, but Major Beers said not to worry, he would take care of everything, if I sent over an appropriate abstract as soon as possible. I sure did; and our new department chief sent it directly to Major Beers, who managed to get it cleared and forwarded to George Mourad as well as to Paris in a jiffy, with a clearance copy directly to me. So far so good. It felt good to know that there were also friendly souls in the stovepipe who used their common sense to circumvent obstructions and get things done. And then a second help came from the Good Lord in the form of a coincidence. The same day that Mr. Someone Else who had studied my five-page paper for four months by then, returned it refusing clearance, I also received a letter from Dr. Tengström, Sweden, chairman of the International Study Group Nr.16, asking about my recent work as input into his quadrennial report. With reference to this outside request, I immediately resubmitted the six charts alone for clearance, without the accompanying text; this made it look different and it would be hard to refuse clearance for unclassified charts. The text could easily be replaced later on by a speech to the symposium and a paper about already cleared charts. From watching the dates in my diary it appeared that the author of the refusal letter was not aware that a cleared abstract was already in Paris; or he would not likely have written such a foolish letter over his personal signature: "...no objection on the basis of security or technical content to its public release....concerned, however, with the number of papers...publication withheld until the studies were complete and...consolidated into a final report." What more final than this final report did he pretend to wait for? With the charts soon cleared ahead of time, and the IUGG papers (this one and the SSG V-29 Report) cleared through the Beers—Macomber route, the paper "Deflections and Geoidal Heights Across a Seamount" went happily to the *International Hydrographic Review* for publication and was immediately accepted for the next issue (Jan.1976). My previous three papers on the various phases of the subject had already been mentioned at their time in the Bibliography of the *Review*'s *International Hydrographic Bulletin.* The friendly acceptance letter "...The manuscript and figures appear impeccable..." contrasted nicely with the sooty letter from our stovepipe.

Even before the March 1975 issue of the *Bulletin Géodésique* with my article "Does mean sea level slope up or down towards north?" had reached me from across the ocean, an invitation came from Dr. G. W. Lennon, England, convener of the coastal section of the Marine and Coastal Geodesy Symposium at Grenoble. He had already seen that article, found it "refreshing and most interesting", and would like me to participate in the symposium with this topic. A quick reply was needed since time was late. Since no clearance was involved in using an already published article (thank Goodness!) I received permission to accept. It tied in nicely with a workshop on the sea level problem, which Henriksen was organizing for a discussion at the AGU meeting in June 1975. A this occasion, however, the geodesists could not convince the oceanographers that when talking about slopes one needed to identify a reference against which the slope was taken; and the oceanographers could not convince the geodesists that such discourse was irrelevant, superfluous, with no bearing on the problem because the oceanographic approach was different. There was no time left to advance the problem between geodesists. But the symposium in Grenoble would offer another chance with a wider audience to comb in and help. At this point, there were two depressing yet illuminating bits of information. George Mourad's bid for papers for his symposium had brought him a very full program, and someone from DMA Headquarters suggested that my paper should be withdrawn since I also had a paper at Lennon's part of the symposium. Good friend George, however, said he wanted both my papers there. I had not even been asked or told about a possible withdrawal, only heard of it later through friends. (A penny for your thoughts!) The second information came a brief month before I was to leave for Europe: my trip had suddenly been cancelled, an administrator would be going instead. No interest or significance was accorded to the fact that I had three papers there, the one on Kwajalein with the peculiar fight for clearance, one in an official role as study group chairman, and one as an invited participant in a debate about my own controversial paper on the sea level problem. *Dé jà vue*? South Africa, where an underminer had used incorrect information to persuade a new department chief to cancel my trip; I had gone on my own money in order to maintain my and my agency's international credibility for this long-standing invitation, but that had infuriated the underminer to further hostile acts. Then Moscow, where someone apparently wanted to prevent me from becoming an IAG Section president; any plans to pay for my own travel or to accept the IAG offer to pay for it, had been summarily forbidden. (Much later, I asked two high-placed officials what the real background of that maneuver had been. One promised to find out and tell me, but didn't; obviously, what he found was not o.k. The other just surmised: "Oh, probably just some clerk's misunderstanding." And now Grenoble! Not appearing at this debate on the sea level problem where I had played a prominent role and been attacked, was out of the question. Since the official reason for the travel cut was given as lack of money (which did not explain the unexpected substitution by an administrator into my travel slot), they could not very well forbid my paying my own way, and I asked for and received administrative leave.

In Grenoble, I tried to formulate the points of the problem and the alternatives of interpretation as succinctly as I saw them, and I also showed some charts of sea level topography made in 1941 by A. Defant, one of the great classic oceanographers. These charts and Defant's explanations seemed to show that my hunch was in the right direction: that the oceanographic reference surfaces were not necessarily level surfaces in the real ocean and that the problem of finding a level surface was still unsolved—contrary to what contemporary "leading hydrodynamicists" had assured my geodesy friend. There was great interest in the paper and a lively discussion, and friendly encouraging comments later from several sides. Admiral J. C. Tison, Director of the International Hydrographic Bureau, who was in the audience invited me to publish it in the IHR where the Kwajalein paper was in press by then. It appeared in the July 1976 issue in French and English: "On the Mystery of Mean Sea Level Slopes."

It would have been a hopeless idea to expect any of the official high-grade DMA representatives to deliver my paper in my absence and to stand in for me in the debate. I believe none of them was even interested enough to be present. To have a knowledgeable colleague of another agency deliver it (as had been the case in Moscow), would not have worked either in such a debate. Participation of scientists at technical conferences is endorsed by U. S. Government regulations which acknowledge the importance for both the scientist and the agency, but in practice the U.S. and the USSR seem to vie in restraining scientists from pursuing what they are hired to pursue, and impose an authoritarian attitude of "administrator knows best". The problem of "no-shows" at international meetings is lamented by one convener of an international symposium (C.T. Russell, EOS, July 29, 1980) worrying like other conveners about how many of the speakers on his program would show up as announced. In a table for this symposium, but probably typical in general, he shows that there were a significant number of "no-shows", partly withdrawals and partly presentations by colleagues, but only for the U.S. and the USSR. All other papers, mainly from Europe or Japan, were presented by the authors as scheduled.

In Grenoble as in other conferences, there were many more informative offerings and discussions besides one's own papers. There were, e.g., the respective SSG reports by G. Mourad (Marine Geodesy) and G.W. Lennon (Coastal Geodesy). Tengström's SSG to which I belonged, and my own SSG were described mainly through the specific contributions of the members along with an overall characterization of the topics. H. Mobritz' SSG-Report on Fundamental Geodetic Constants was especially important for discussing the currently most representative values for these constants which slightly differed, as expected, from the standard 1967 Reference System of Lucerne. Most important and stimulating as always at such assembles, was the opportunity to talk with many colleagues about points of common interest. None of these benefits would have reached me through an administrative substitute.

On the way home from Grenoble we made a hastily arranged brief detour to Karlsruhe, West Germany, for the ceremony of my receiving an honorary doctorate of engineering from the Faculty of Civil Engineering and Surveying at the University of Karlsruhe. The news of this honor had surprised me out of the blue shortly before our trip to Grenoble, and travel changes at the height of the travel season had been difficult. It was a memorable day at Karlsruhe, due to the exceptional hospitality extended there. The University of Karlsruhe is the oldest Technical Institute in Germany. For geodesists it has a special significance because W. Jordan, the original author of the "geodetic Bible" (the still important standard geodetic textbook: Jordan—Eggert—Kneissl, *Handbuch der Vermessungskunde*) which had so pleasantly introduced me to geodesy way back, had taught there a hundred years ago and had also started the still existing geodetic periodical, *Zeitschrift für Vermessungswesen*. Professor H. Draheim, Rector of the University, stressed that the University gives an honorary doctorate only for scientific merits and never one of the political type. I was told also that such honors are given out rather sparingly: the Geodetic Institute which had nominated me and Charles

Whitten (whose ceremony was a few weeks later), had done so only once before in its 100 years history, in 1950.

There was an official session of the Faculty for the ceremony with an address by the Dean, an address by the Rector Dr. Draheim about my work that had led to this honor, my response and my presentation of a paper for which I had chosen the "Mystery of the Sea Level Slopes." I had no difficulty expressing my thanks and appreciation in German but when it came to talking more formally about the sea level slopes and explaining the slides, I had difficulty with the language and asked for permission to continue in English, speaking slowly. Before the ceremony there had been a lovely "Jause" (afternoon coffee) at the beautiful home of the Rector, complete with several varieties of genuine rich Viennese "torte", with nostalgic significance for us. And in the evening, there was a delicious banquet and a tour through the Institute. Early next morning we had to catch our flight home from Stuttgart, after a beautiful ride through the countryside.

14

RACING AGAINST TIME
(mid 1975–end 1976)

1. BETWEEN ENCOURAGERS AND REPRESSORS

In July 1975, there was an occasion to invite our old-time boss, Mr. Floyd Hough, back to TC (the successor of the old-time famous AMS) for a day of festivities. TC had installed a series of six portraits in the lobby, a "Gallery of Distinguished Civilian Employees," and Mr. Hough had been chosen as one of these six. Mr. Frank Fleischmann who had been in Mr. Hough's front office back then, volunteered to drive all the way to Mt. Jackson, V., in the early morning to bring Mr. and Mrs. Hough here for the day, and drive them back at night. I just love to do something for this man," he said, "I'll never forget how he cared for everyone of his subordinates. There is something special about him..." Frank arranged a big luncheon in Mr. Hough's honor, and several of his one-time employees, now in other agencies, joined us for this heart-warming reunion. The following weekend, my husband and I were going to the mountains for a few days' vacation and we made a detour through Mt. Jackson to take the Hough along, because they had said they longed to see these places again, but were not able to drive anymore. They enjoyed the familiar scenery with us, including an interesting tour through the VEPCO plant in Mt. Storms, Va. After my return from Grenoble, Mr. Hough took great interest in my success there, like a proud father. Later in fall, he went to the hospital to have a cataract removed, then had to return there with a broken hip; and that was just too much. He died 6 January 1976. Another warm experience lapsed into memory lane.

So did the friendly "Good morning" voice that greeted us in the elevator, suddenly become a mere memory. Many of us missed Mr. Nathan High, the chief elevator operator, when he had to retire upon reaching the mandatory retirement age. He knew his passengers, knew which floor they had to get off, and often saved me from getting off the wrong floor. He even remembered my husband and friendly inquired about him, which made me feel good. He had a pleasant, positive attitude towards life and accompanied the elevator ride with little, cheerful philosophical remarks that helped to relax the preoccupied faces around him, and sent them along to face the office day with a smile.

And such a warm smile was appreciated more and more, the less it seemed compatible with the assorted chicaneries of daily office life. You had to learn to live with the fact that some people, unfortunately in a position of authority, would not respond on a technical or factual level; and be encouraged instead by the signs of support from the larger radius of the agency. You learned to distinguish between the repressors with their bureaucratic deputies and the encouragers. Most important, don't confuse the generally well-meaning agency with a few, not so well-meaning power-crazies. Somewhere I had found and enjoyed a useful little story of a young housewife—mother with three little unruly tots on her apron strings all day driving her crazy, so that each evening she found herself exhausted, frustrated, and with frazzled nerves; until one day she sat herself down to have a serious talk with herself: "Are you going to let these little ogres decide when you get upset? Certainly not! I'll get upset when I want to and not when they want!" I kept retelling myself this story many times, and together with Mr. High's serene "Don't let it get you down; the sun is shining outside", I managed to keep the perspective. After all, I was going to be here less than two more years before mandatory retirement, and there were several projects I wanted to finish before that—a race against time—; I could not afford to lose time, energy, and enjoyment of this work by getting upset about things I could not change. Remember: "I'll get upset when I'll have time and want to, not now."

On the positive side, for example, there were just then a number of appreciative comments about several of my recent papers, with many requests-for reprints, especially for the long review article, The Figure of the Earth—Changes in Concept". This particularly successful paper was the one that had had to overcome peculiar hurdles by repressors along the clearance channels, but counted among its encouragers the CO and the Deputy CO, people in IIA Headquarters, in other agencies, and all over the world. On the positive side also at that time was the nice gesture of the DMA award as Outstanding Career Woman" which I and nine others (probably: let's pick a round number of ten women) received in observance of 1975 as the International Women's Year (so declared by the U.N. and by the President of the U.S.). Although it seemed a little odd that women's achievements still needed a presidential decree to be specifically acknowledged as if these were something unexpected from a lesser species of mankind, the men did mean well.

Among the pleasant interruptions was e.g. also a visitor one day, who greeted me with: "Congratulations to you!"—What for?—"Your solution to the ladder problem (in the *Canadian Surveyor*) was the first correct one."—How—do you know?—"I am Mike Mogg, the editor; I should know and I wrote you a letter." Embarrassingly, as happened to me unfortunately quite often, I had not caught the name and affiliation at the introduction of the escorted visitor.

Usually, I managed to identify the visitor unobtrusively during the ensuing conversations. This time, it was not unobtrusively, but Mr. Mogg put me readily at ease. Since it took several months before an issue of the *Canadian Surveyor* in Ottawa reached my desk, my solution which I had sent off the same day I saw the problem, had not reached the editor before the deadline of the next issue. Thus, "no correct solution received" was reported there, but I had received a letter from him about that. Now, he gave me the next ladder problem. It was fun and easy to do, on the level of my Austrian high school education, and I sent it off the day after the visit. At that easy level, I wondered about the lack of correct solutions, another indication of how good my high school education had been. A later puzzle about numbers divisible by nine belonged to the requirements of my junior high school days, along with other divisibility rules, but it was entertaining to put the derivation down for the *Canadian Surveyor* in an orderly fashion. Mr. Mogg was also very kind in printing my second mnemonic verse for 31 digits of π, the first one being in my Geometry book.

"See, I have a trick The symbolic name coined by Greeks. Important to recall aptly Note too: The ratio diameter—perimeter,The sequence for pi extends limitless." Divided endlessly for pi, (*Canadian Surveyor*, Sept.1975, p.344)

Not everything was rosy and pleasant, however. On the negative side, there was the sudden close-out (not phase-out) of an advanced and timely project in my branch: the update of my widely used North American geoid chart. I had made the first geoid chart ever to cover the North American continent, in 1957, and it had been a very welcome aid for all geodetic offices concerned with pertinent projects. Upon requests by customers, I had issued an update ten years later, combining accumulated new data and a partially automated procedure. Now it seemed to be the appropriate time for another update in response to inquiries by our customers, to accommodate the fruits of the new precise surveying techniques, including satellite-derived long distances. Ray Shirley who had assisted me in the 1967 update, was taking this project over now and he planned a fully automated scheme which would permit future updates as a relatively simple rerun of the machine program. As always previously, the Coast and Geodetic Survey, now NOAA, graciously cooperated. They opened their files to Ray who collected the new data and had his automated procedures developed to the testing stage, with the help of Joan Nickless as programmer, when he was ordered to stop. At this advanced stage, the order wasted the investment of these successful efforts and disregarded the technical merits of the project. It was claimed that the project could no longer be justified before Headquarters because it had not been requested by NOAA as a requirement for their own work. Our arguments that most of our work had always been self-originated in anticipation of general need, and that a well-known product issued over almost twenty years with the prestige of our agency should be maintained at its top quality as a service to our customers, did not carry. Through the grapevine we heard that customer inquiries at Headquarters about an updated chart, with advance requests for copies, had not been refused but passed on to our agency. So the repressor was nearer to home than had been claimed. He could be traced to the same guy who years ago had proclaimed that the age of the geoid had passed and who had tried to abolish the well-known name of my "Geoid Branch". Despite the fact that contrary to his judgment and prophecy the interest in the geoid increased with more and more refinement, he still questioned our justification. This put the order to stop into perspective: this repressor did not arrive at his decision on technical grounds.

Neither did he use his (well paid for) technical judgment when faced with such a formidable task as releasing my Grenoble address on the sea level problem for publication in the IHR. Following Admiral Tison's invitation, I had turned my notes for that speech into an article right after returning from Grenoble, and expected clearance to be just a formality. "Why would you want to publish on this topic again? You already have it in the *Bulletin Geéodésique*."—"That article brought the invitation to speak in Grenoble," I explained. "People were very interested. I used a different approach there and Admiral Tison wants it for the IHR."—"Does he know of the article in the *Bulletin Geéodésique*?"—"Yes, he does, and he still wants the Grenoble presentation."—"Well, the Publication Board will meet in a couple of weeks." It didn't. I asked the omniscient secretary whether that Board had regularly scheduled meetings. No, she said, they are called together whenever someone needs it. So what's the problem? They didn't say. Was it going to be a replay, I wondered, of the clearance obstacles for my previous IHR paper, in press now, on the Kwajalein seamount? Do they object to my publishing again in the prestigious IHR, or to my publishing at all? Or is it just bureaucratic indifference, or insecure second-guessing of what a respective superior might or might not wish? Nostalgically, I remembered Mr. Kephart, the last Public Affairs Officer, who dared, at a lesser pay, to decide on factual grounds whether a new formulation of an already cleared topic needed another clearance or not. After more than two months of no Board meeting, I finally succeeded to pry the paper loose—without the Board—by referring to Mr. Kephart, the regulations and criteria for clearances, and to the unfavorable impression our agency gave in delaying Admiral Tison's simple request for so long. Years later, in 1980, a new department chief, Col. Thomas P. Baker, Jr., sent me delighted congratulations when forwarding a highly appreciative letter from Mr. P. Charlot, I.G.N. (National Geogr. Inst.), Paris, via IHR, saying that this article had inspired him in his leveling work, as a "lighthouse in the night."

My concern with the sea level and ocean geoid problems, luckily, had a strong supporter in Bob Gouker of the Advanced Technology Division who was satisfied with my monthly progress reports and provided long-term funding through renewed EAPAs (Engineering Application Project Assignments). I planned to pursue several strands of thoughts one was to continue studying the conceptual assumptions in the debate between oceanographers and geodesists about the slope of mean sea levels another considered the fact that this quarrel was based on very few and relatively short available profiles. If one had more examples, particularly longer profiles, maybe one might see better. It so happened that leveling data along the Pacific coast c South America were stored at our agency since the time our IAGS assisted the South American states with their leveling requirements. And the man involved in that assistance and familiar with these records and procedures was Mr. Jack Bray who had joined my branch some time earlier. What a unique opportunity to contribute a very long meridianal profile to the debate. I succeeded in getting Mr. Bray interested and appealed to his expertise in utilizing this unique treasure of data, not available anywhere else. By further coincidence, TC had to fill a South American request for readjustment of this material augmented with new data. Thus, Mr. Bray could advise the people working on that request in another unit, and at the same time use the material for the sea level problem. In order to produce an example comparable to Mr. Balasz' two profiles along the U.S. meridianal coasts, Mr. Bray had to collect and analyze also the tide gauge data along the coast and study the effect of currents, winds, and a number of other factors affecting these measurements. It turned into a long, painstaking study, supported by an EAPA for funding, and documented by Mr. Bray's two very useful, highly technical publications. While the sea level problem itself remained inconclusive due to the incompleteness and uncertainties of the available data at that stage, Mr. Bray made important observations and recommendations about the error analysis, the pitfalls of conventional adjustment procedures, and their avoidance when treating such long profiles.

Another strand in this problem complex was the comparison of satellite-derived marine geoid profiles with terrestrially derived ones for verification. The bathymetrically derived deflections and geoid undulations in Kwajalein were specifically mentioned as TC tasks in the "GEOS—3 Exploitation Plan" of DMA, but here I had completed the terrestrial part of the comparison and yet there were no GOS—3 data for Kwajalein in sight. (The designation GEOS—C had been changed to GEOS—3 after the launch.) We had a specially appointed GEOS—3 representative who was to be the point of contact with the other agencies, went to all the periodic meetings, wrote and circulated reports about the agenda of the meetings, spent precious travel money for meetings across the country, but despite special requests he never seemed to succeed in getting data for Kwajalein, nor saw to it that the satellite would collect data when passing over Kwajalein, nor that there would be facilities in the Kwajalein region to receive those data, not to mention the need to monitor the flow of data through the processing channels. The sudden abundance of satellite data in some areas overwhelmed these multi-agency channels, and the increasing backlog of processing made it uncertain, when or whether at all specific, processed data could be expected. We asked for a specific search for our data. When nothing could be found, we prepared and submitted a time schedule of future possibilities for the satellite to collect Kwajalein data, but also that did not lead anywhere. For another project we did receive some but not sufficient data. Colleagues who were nearer to the satellite monitoring process, wondered why DMA did not take care of Kwajalein since this was listed as a DMA task. Good question, indeed! In March 1975, I had been invited to one of these interagency meetings and had presented there my Kwajalein work as the completed "ground truth" part, in search and expectation of the promised counterpart of satellite data, and had noticed with dismay that this official DMA requirement had been "overlooked". We never did get these data. At one time, our representative was worried about justifying the various TC efforts for GEOS—3 before the TC purse string holders and arranged a briefing for them. I was allotted 3 to 5 minutes, but by carefully selecting telltale viewgraphs and weighing every sentence and every word in those few sentences, I succeeded to tie all my several pertinent efforts together into a clear picture of why and what for. It greatly helped to strengthen the support of the purse string holders for what I was doing, but it did not get me any more satellite data through the representative. At another time, this representative asked us for a listing of pending GEOS—3 papers, the satellite data required for them, and the expected completion dates. I listed several papers and promised completion "three months after receiving the data." I never received them.

Notwithstanding these circumstances, our champion of GEOS—3 made difficulties when it was time for another renewal of my EAPA. I was working then on formulas to compute marine geoid undulations directly from the bathymetry without the benefit of observed astrogeodetic deflections, since such benefit was unique to Kwajalein. My new method had been criticized as inapplicable elsewhere for this reason; and I intended to try it out in a second test case somewhere in the middle of the ocean and find some substitute for the deflections in such a case. I will tell about this work later on. Bob Gouker understood my plan and purpose, discussed with me the best formulation for the renewal and assured me that there was no problem in getting it approved. I was quite surprised upon returning from a two weeks' vacation to hear that my monthly EAPA status reports, submitted before I had left, had been held back in the department and the renewal was uncertain, because our GEOS—3 champion would not let them pass unless they were more GEOS—3 oriented. This sounded silly since the GEOS—3 aspect was clearly mentioned through the reference to the official DMA task in the Exploitation Plan. He knew very well that the expected satellite data had not come through yet; did he want me to twiddle my thumbs waiting rather than to prepare a second test case under different circumstances for a comparison elsewhere? So I spent time adding another sentence or two to the APA application, getting it back, rewriting it again and so forth, while that obviously would not make the least bit of difference to the work at hand. Contradictory to the message that this work had been cut off, was the fact that the authorizing job number for the biweekly time sheets was still valid, and that Bob Gouker had phoned me the approval and that I would get the document as soon as the paper work was straightened out: "Don't worry." I didn't, but I wondered. Through the grapevine I heard that Mr. GEOS—3 had actually gone to the office which awarded the EAPAs to protest my renewal and ask for a cutoff but had not been taken seriously in trying to stop a good project, and thus did not prevail. This little interlude and its background, although unfortunately not unique, would not have been worth mentioning here if it did not have wider implications. Apart from the fact that it had a curious, even more absurd sequel about a year later, it may serve as a point of departure to the serious general question of how to promote talented technical people up the prestige ladder in the current pyramidal bureaucratic organization. I remember an obligatory session for supervisors where the speaker from the Civil Service Commission thundered: "If you don't know how to handle people, how to organize their work, how to motivate them,…maybe you should not be a supervisor." So true; but did the good man not know that such qualifications are not usually the reasons for promotion to a supervisory position. The main reason is more likely that this is the next rung in the career ladder and there may be no other place in the organizational structure to give to that person. It could also be the other way around that a vacancy needed to be filled to uphold the structure, and someone had to fill it.

I also remember a meeting about filling out a questionnaire with one's career goals and development plans, and someone saying that people above the GS-13 rank should get out of the technical and into the administrative side of things; and that the only career goal for a branch chief could be the positions of division chief and department chief. These were the rungs of the ladder, and all had to compete for the next rung if there were a vacancy. That was not, however, in my plans. I was not interested in either of these two rungs, because they would mean leaving my projects to someone else and become instead a full-fledged paper shuffler without the authority and opportunity of leadership which these two positions had held once upon a time. Col. Jonah's prediction of the department's impending decline in the absence of a vigorous defense was visibly coming true. No defenders but only yes-sayers were chosen to preside over that decline, and—as I was told by incumbents at one time or another—they had to carry out decisions made elsewhere. My own position, by contrast, was still undiminished, full of exciting possibilities—at least for the time being—even though (or because?) I had to watch out and fight for them. Not even the salary for those two jobs, both GS-15s, was an incentive because as a researcher GS-15 I already had that salary. There was nothing to gain for me under the current circumstances but a lot to lose in my scheme of real values. Happy where I was, I felt to be at the top of the career ladder as envisioned by Civil Service years ago. All I could wish was some more freedom and resources, a couple more assistants, and maybe a chance to revitalize a general atmosphere of dedicated involvement in research by strengthening that research ladder for colleagues, helping them find satisfaction in technical work by matching their talents with the right job challenges, and opening possibilities for developing their potentials. Maybe, that questionnaire could open an opportunity at last to implement at our agency the sensible parts of the many management courses I had taken, as I had repeatedly and unsuccessfully proposed in earlier years, in order to stem the reactionary trend, manpower waste, and employee cynicism and frustration. The questions about an individual plan of developmental Objectives" lent themselves easily enough for a proposal to branch out as a consultant, to organize a commission for reviewing various aspects of manpower utilization and motivation, with special attention to scientific and technical personnel, and to study patterns of success and failure of different approaches. I added the comments "TC has spent much money to send high-paid employees to expensive management courses, but no attempt has been made yet to implement the insights gained for a possible improvement or change of problematic procedures." I realized, of course, that such an offer of utilizing my extra energies was not exactly meant in the questionnaire, but I put it down anyway to see what would happen. Sure enough, it was insisted that I write and sign a formal withdrawal from the "development plan" because my plan was "too silly to be passed on since there was no position of that sort in the structure." Obviously, they had no other ideas of their own about possible career goals and development, and just as obviously, they were not interested in an offer to contribute without asking for another chair and more salary. Neither did they know or care how to utilize the researcher ladder. I was still the only one in my agency with the title and privilege of a "research geodesist" and was considered an anomaly, although there were a number of other talented technical people around.

If the top of the pyramid has only administrative positions, then that is where the ladder leads, and more such positions have to be created. At times of a general upgrading of the whole agency, whole new management layers were created to raise the top, with the concomitant dilution of distributed authority, need of frequent briefings between layers, and alarming increase of paper work. At other times, people pushing up the ladder or displaced by reorganizations had to be taken care of by making them assistants to chiefs (even if they had no administrative talents and habitually rubbed subordinates the wrong way) or by sending them away for a year or more of training until something opened up. Some such people, good with equations, computers, or machines, but not with people and without creative leadership ability, may feel the need to overstress their new authority in an attempt to compensate for some inefficiency in the new job, unaware or unconcerned about the sorry effect and appearance they are making. That effect impinges on the subordinates but may be unnoticed or unimportant for their superiors. Thus, it may not be a case of the much-quoted "Peter Principle" (you get promoted for your competence until you reach the level of your incompetence and there you stay at your highest salary) but rather a misplacement for lack of a more suitable place in the structure. Would it not make sense to consider changing that structure (nothing sacred about a pyramid) in the interest of better utilizing available manpower?

2. APPLES AND ORANGES IN THE OCEAN

Henriksen's invitation in January 1976, to participate in another sea-level confer-ence at the University of Illinois at Urbana-Champaign, at the end of June, pro-vided a focal point toward which to organize whatever I had gleaned from the oceanographic literature about the sea-level problem. Dr. Narenda Saxena who had invited the interdisciplinary "Conference on Geodetic Measurements in the Ocean" to Urbana, asked me to join the Marine Geodesy Committee, which he chaired, and to be on the panel. Jack Bray's study of the South American coastal profile would fit there too. After the now usual difficulties to get permission to participate were mediated, the abstracts cleared and sent off, Jack and I settled down to prepare our contributions.

Since the discussions of the sea level problems in Grenoble, I had read more and more of the oceanographic classical literature and I found the explanations, discussions of the problems, and the careful formulations by seminal oceanogra-phers such as Bjerknes and Sandström, Defant, Sverdrup, and several others extremely helpful. To my ignorance as an oceanographic novice who had stuck out her neck in questioning contemporary oceanographic assumptions, it came as a revelation that these great oceanographers were on my side, or rather, my criti-cal hunch had stumbled into the right direction. I would just use appropriate quotes to underline my arguments that the slopes of the mean sea level topogra-phy as determined by oceanographers versus geodesists are not directly compara-ble since (a) the mean sea level in the open ocean is not the same as the mean water level at a tide gauge, which is subject to a number of local effects, (b) the reference surfaces against which the slopes are defined, are not the same either, and (c) of course, that the two types of observations are made at different places, miles apart.

The literature, however, did not produce for me a direct quote about the odd equatorial bulge that I had noticed at first sight in these marine profiles. And yet I still felt that here somewhere was the crux of the problem. Surprisingly none of the several authors of marine profiles had seemed to notice this, to me, striking feature. Doggedly hunting for an explanation, I found several pieces of informa-tion that suddenly merged into a coherent picture. First of all I realized that the thrust and purpose of their investigations were quite different from those of geod-esists—no wonder that the basics, adopted to achieve this purpose, were differ-ent. The oceanographers study the mechanics of ocean currents, waves, etc. (sea level topography) and the interrelationships between density, salinity, tempera-ture, and pressure in the ocean waters and the impact of meteorological and astronomic forces. The geoid as such is nowhere a primary object of interest as it is for geodesists who identify it with their basic concept of the "Figure of the Earth" and use it as the surface of departure for establishing elevations of the topography above and below. By contrast, the concept of departure for studying those oceanographic features is a theoretical standard ocean" which is by definition motionless, homogeneous throughout, of constant salinity, temperature, and atmospheric pressure, and with level isobaric surfaces. The interrelationships between changes in salinity, temperature, and water pressure at varying ocean depths were studied experimentally in the laboratory, expressed in empirical formulas, and tabulated in a number of detailed tables for practical use. Reading about the excitements and achievements of oceanographic history since the beginning of the century, I learned that the currently used tables, still the same ones as first constructed in 1910, were based on 24 water samples taken from the surface of the ocean in 1902 during the international exploration of the northern European waters. In 1908, a single water sample taken at 3000-meter depth off the coast of Portugal had been added. Later criticism that these water samples may not be representative of the whole ocean did not succeed to incorporate modifications. Similarly, the author of the tables, V. Bjerknes, was well aware (as apparently our friends in the current debate were not) that the variation in the intensity of gravity should be considered, but he had decided to use the simplification of a fixed value as good enough for the state of developments at his time, that is 1910. He did, however, add correction terms in case more precision should be desired, but these corrections have been dropped in the current tables and current procedures.

Thus, I could see the pertinence of the standard ocean for the objects of study; and the oceanographic sea level profiles under debate as deviations from that standard ocean, which had nothing to do with the geoid. These deviations were derived at particular stations, through the use of the tables, from the properties of the water column underneath each station down to a depth of 1000 meters or 4000 meters depending on the author's preference. There is no sideways connection between columns other than considering the foot of these linear columns to be at that same depth. The question whether the pressure surface down there is level or not is not relevant for a single line column; it appears with their sideways connection through the constant depth, where it falls through the cracks of unproven belief or so far unsuccessful attempts to establish such a connection. Our friends were thus correct in claiming that a refinement by introducing the latitude variation in gravity would be negligible for their type of computations. But I realized that this was so because these computations applied only to the individual linear column vertically below a station. The surface of constant pressure at the foot of that column may or may not contain the foot of another column miles away, but may pass above or below it. In that case one would say that this equipressure surface was inclined, that is, not level. The assumption that it is level ignores the difference between the real ocean (where it may not be level) and the theoretical model of the standard ocean (where it is level by definition); it also ignores the efforts of several oceanographers to grapple with this unsolved problem.

For the controversy between geodesists and oceanographers and for the puzzle of the equatorial bulges all these pieces of information seemed to mean that the respective basic ideas of the geoid and the standard ocean were conceptually different, and that the various constant parameters of the latter, since they were adopted irrespective of geographic location, pointed to an inherently spherical model. The geodesists occasionally also use a spherical model for the Figure of the Earth" (the Earth minus its topography), but only as a rough first approximation. For more refinement they prefer an ellipsoidal model ever since they have discovered, through measurements by the French Academy of Sciences in the 18th century, that the measured length of a one-degree arc in the north-south direction depended on its geographic location. Without this refinement, measurements over the globe (rather than in limited isolated places) would have shown systematic equatorial bulges. At that time also, Isaac Newton had formulated the law of universal attraction and concluded that the rotation of the Earth would flatten it at the poles. The change from the spherical to the ellipsoidal model in geodesy thus incorporated into the model both the theoretical flattening effect of the Earth's rotation and the measurable equatorial bulge. The standard ocean, by contrast, does not incorporate these features, and thus the bulges show.

Mean Sea Level and the Marine Geoid

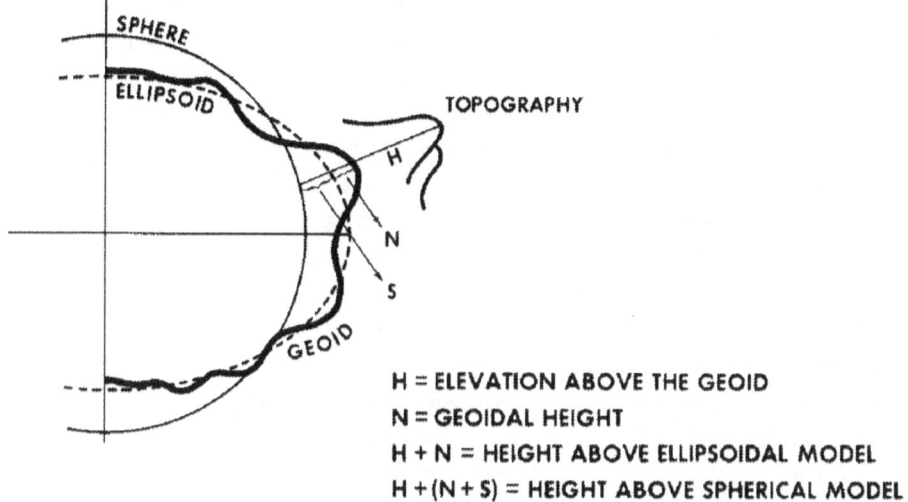

H = ELEVATION ABOVE THE GEOID
N = GEOIDAL HEIGHT
H + N = HEIGHT ABOVE ELLIPSOIDAL MODEL
H + (N + S) = HEIGHT ABOVE SPHERICAL MODEL

Figure 1 The topography and its reference surfaces.

Further refinements in geodetic capability of accurately measuring horizontal and vertical distances (which is a primary geodetic business) revealed the difference between this homogeneous, ellipsoidal model and the geoid, that is, the Figure of the Earth in the real world as molded by the effects of the existing irregular mass distribution. Thus (Fig. 1), the geodetic description of the height of a station is separated into two parts: (H) the elevation of the station, say a hill top, above sea level (the geoid) as recorded, e.g., on a topographic map; (N) the height of the geoid above (or below) the ellipsoidal model, as recorded on special geoid maps. It is important in geodesy to distinguish between these two parts like sorting apples from oranges. The leveling profiles along the U.S. coasts which E. Balacs presented at the conference in Boulder to the disgust of oceanographers constituted part (H), say oranges, the height of the mean water level at the tide gauges above or below the geoid. Part (N), the apples, would have come from e.g. my geoidal map of North America if needed, which records the geoidal height above or below the ellipsoid used in North America. That was why I had been asked whether this information could help clear up the puzzle; it could not, since the oceanographers insisted that their profiles of the sea topography (say the height of a current) were also part (1), oranges, and that they did not have nor need an ellipsoid.

Preparing now my contribution for the conference in Urbana, I had come a long way in questioning assumptions and studying the oceanographic literature; and I summarized my conclusions to the effect that the standard ocean was a spherical model; that it was roughly comparable to the spherical earth model in geodesy; that such models would produce profiles with systematic equatorial bulges; and that the oceanographic profiles were not referred to the geoid as claimed (not oranges) but to the standard ocean, and thus represented a mixed fruit basket of apples and oranges. I should have added some grapes to the oceanic fruit basket to indicate the difference between the (missing) ellipsoid and the spherical standard ocean.

It was great fun to present this paper in Urbana. It was published, as were other conference papers, in *Marine Geodesy* (vol.1, no.1, 1977, 37-59), Dr. Saxena's new *International Journal of Ocean Surveys, Mapping, and Sensing*. The people at NOAA, responsible for leveling in the United States, were pleased that my analysis proposed a solution to the puzzle other than blaming them for errors. Nonetheless, they conscientiously kept scrutinizing and refining their methods, and each little change in numerical results was eagerly interpreted in the light of the controversy. Improving the methods beyond any possible reproach was, of course, in order, but to my way of thinking in terms of concepts, a change of a few centimeters in a relatively short profile with an overall height range of less than a meter could not possibly solve a global problem. Even if numerical centimeter changes might seem to remove the debated discrepancy in slope for a local stretch, the conceptual incompatibility of the slope definitions still stands. The geodetic profiles are much too short to test the equatorial bulges. The global character of satellite-derived profiles of the marine geoid may make such testing possible but requires an accuracy of centimeters not yet available at the time. Because of the great interest and NOAA's involvement, Chovitz invited me to repeat my talk at NOAA at one of their periodic guest lecture sessions, which I did with pleasure. Also, my own department wanted to be briefed on the subject.

Jack Bray's study of the tide gauge and leveling pitfalls along the South American coast had also met with great interest in Urbana and elicited questions and discussions. He also presented it to the department, yet a repressor deputy refused to let it be issued as a Technical Report for the benefit of users. Why? I don't know. It was even better to have it published in Saxena's journal (*Marine Geodesy*, vol.1,no.2,1977, 177-197).

3. WRITE AS GHOST BUT NOT AS AUTHOR

During that time I was also busy working on a group of other papers, less absorbing than the sea level problem, but interesting in their own way. One was a translation from German of Professor Draheim's article (AVN,1971) based on his inquiry of several geodesists years ago about the status of geodesy, its achievements and expectations in the last and future twenty years, to which I had contributed at its time. It was requested for the use of the newly established Committee on Geodesy of the National Research Council, an interesting development in itself. Translating an involved technical article was an intriguing experience, since the different character of the two languages does not easily allow transmitting the nuances produced by the many little fill-in words in German that have no English counterpart.

A new experience also was to try some ghostwriting which I had never done before. The upcoming 25-year celebration of teaching geodesy at The Ohio State University took the form of a symposium with the theme "The Changing World of Geodetic Science", October 1976, and our Commanding Officer, Col. W. R. Cordova, a graduate from there, was invited to present a paper on "New Concepts in Geodetic Instrumentation", covering the past and future 25 years. I was asked to write this paper for him. After securing promises of input from a number of specialists in instrumentation, I studied the long write-up in our office newspaper, the *Topocomment*, about the background of Col. Cordova, welcoming him as the new Director when he had taken over from Col Whelan the year before. I tried to imagine what I would say addressing my university as an honored guest speaker ten years after graduation, though with his background. Although my style in speaking and writing is surely different from his, I must have hit it right, because he was delighted. Due to the long technical contributions from my specialists, the paper grew very long, which was just as well for an informative publication in the *Proceedings* (Report No. 250, Department of Geodetic Science, OSU, 1977), but I made a second, shorter version for delivery within the half hour allotted for the address. The department chief liked my humorous address and the condensed overview of the technical developments, and insisted that I give a full-blown performance to the department. The Colonel showed his pleasure by having the whole speech printed in the *Topocomment* (Oct.76) A sourpuss at DMA, however, threw cold water on the general elation when, just one week before Col. Cordova was to deliver that speech, he sent him (and us) a sharply critical note of suspiciously low caliber. Had such an attack, at this late date, really originated in that DMA office, or had it been engineered by some TC forces who had tried to gain control over the project and been rebuffed by the department chief? The contributors to the paper, each of whom was clearly more knowledgeable than the critic, were incensed and up in arms; they put their heads together as a committee and concocted a sarcastic reply: "The DMA review…gives the impression that the reviewer did not read the paper carefully.…the reviewer must be a non-technical person, unfamiliar with the subject matter and the pertinent literature…(rebuttals of several technical items)…The criterion of fifty-word sentences may apply to kindergarten or elementary school level but not to technical papers for an adult, well educated audience, quite familiar with the details of the subject matter. The reviewer is invited to read, for example, President John F. Kennedy's famous Inaugural Address, which contains several sentences of much over fifty words. President Kennedy may have reacted to the reviewer's criterion with just a smile, or maybe with a witty rejoinder of a superb 100-word sentence.…The reviewer's attention span gave out after only 19 words in a sentence of only 31 words. The 12 words overlooked specify (the answer to the criticized item)…" I am not sure whether this disrespectful rejoinder was actually passed on all the way to that DMA office, but it certainly defused the anger of the technical committee by letting off steam and most importantly, it reassured the Colonel that the paper was sound. Nothing essential was changed. Another instance to ponder whether the organizational structure should really give power to administrators over more knowledgeable researchers.

Pleased with my beginner's luck as a ghost writer, I agreed to a second such request. Obie Williams, Deputy Director at DMA, was invited to talk at the same symposium about the much debated uneasy question, "Does Geodesy Have A Future?" and he asked me to write a draft, of course, with a positive answer. Of course, what else could you say at an anniversary celebration of a school for geodesists? It naturally had to be a pep talk and I tried to make it convincing. The reason for the general uneasiness about such a question came from the fact that the ancient quest to determine the size and shape of the Earth, the major goal of geodetic ingenuity for several thousands of years, has been met and fulfilled suddenly in our own time while it had seemed only a few decades ago to be an almost endless scientific hunting ground where every question answered opened up a number of new questions and new enterprises. Now, due to the new techniques (the theme of the symposium), further inquiries about this subject were just a question of increased accuracy with improved tools and sufficient funding, no really exciting, scientific surprises. Sure, the primary geodetic enterprise, the measuring of distances and directions, was still there and it profited from the new tools, but it was considered a humble service compared to the adventurous inquiry and lofty philosophical wonder about our planet. In German geodesy the distinction is even made in the names "lower geodesy" versus "higher geodesy". And now, there was insult added to the injury: while typical geodetic services including information about the Earth's gravity field could be enhanced in scientific value by serving the scientific goals of geophysics and geodynamics, even extending them to the moon and all planets, the scientists of these fields were doing these chores themselves without realizing they were trespassing. So there was a question of delineating disciplines and defending one's turf, to assure participation in the scientific adventures of the future. The paper was a determined defense of the turf, characterizing the typically geodetic services as an ubiquitous, never ending necessity. Obie Williams liked the paper, wrote me a gracious thank-you letter, and acknowledged my name in the publication in the *Proceedings*.

A related paper, written under my own name, however, ran into totally unexpected and unreasonable difficulties. It had grown out of current discussions of whether the newly growing marine geodesy was an entirely new interdisciplinary field or part of geodesy, whether its proper name should be marine geodesy or maybe talassodesy or oceanodesy, and what its specific scope should be. Dr. Saxena, originator and editor-in-chief of the new journal *Marine Geodesy*, clearly endorsed this name by so naming the journal. He asked me for references from the literature where a definition of geodesy included the ocean floor. From my work on various historical papers I saw marine geodesy conceptually as a natural part of geodesy since ever, with the difference that only now there was the economic incentive of harvesting the resources in the ocean; and only now one could develop the techniques to do so and apply the typically geodetic services in the difficult ocean medium. This was impossible 2000 years ago, even 50 years ago. Dr. Saxena invited me to write this up as an article for his journal.

When I submitted the paper, entitled: "Marine Geodesy: A New Discipline or the Modern Realization of an Ancient Endeavor?" (*Marine Geodesy*, vol.1, no.1, 1977, 165-175) for clearance, I had to wait three weeks before I was told that it would not be cleared. Not be cleared? Why not?—Well, such type of paper was not within the agency's publication plans, and "they" had not requested it. (But ghostwriting such type of paper for a higher official was within the agency's publication plans?) There was no point in arguing with (again) a new department chief who did not seem to proffer an opinion of his own but only relayed a message. Whose message? I had a vague hunch who that might be, since such kind of reason had been tried before, but fortunately had not prevailed. Then this department chief showed common sense and a capability of making a decision: he agreed that I could treat this paper as a private effort (which it was after all) and send it out immediately without clearance, provided I took off the agency byline. This I was glad to do since my name as an author was well known and did not need the agency byline anymore; the omission would only make the agency look peculiar. Sure enough, when in due time the galley proofs arrived, the agency's name had been filled in and I had to correct it out once more. The publisher called me long distance to assure that this was not a mistake and he shook his head by telephone when told the reason. "Any other agency would be proud to have such an article written", was his comment.

Such experience made me contemplate once again the change and deterioration in managerial atmosphere compared to several years earlier when I had asked for clearance of my "Geometry" book manuscript. Not only had I been given the green light right away, but it had been accompanied by good wishes from the clearance channels, and a letter of congratulations from the CO. Now, this little paper, participating in a geodetic discussion, was refused clearance! Even on its face value, the reason given was absurd: the group of papers I had just finished (the Draheim translation, the ghostwritten papers for Col. Cordova and Mr. Williams) and my own, all dealt with aspects of the currently discussed implications of the spectacular developments in geodesy. To participate in such discussions and contribute self-originated papers was explicitly part of my job description. So "the little ogre" tried to put me in a bind: bar me from carrying out the role of expert and consultant, which I was explicitly charged to carry out by my rank. When Dr. Orlin, Executive Secretary of the NRC Geodesy Committee, had asked me about literature of interest to the Committee's planning and I had mentioned the Draheim article among other things, I had had the instinctive foresight to ask him for an official request through channels, since someone might question my doing a translation job. But it had not occurred to me to ask Dr. Saxena to request this marine geodesy article officially from DMA Headquarters, which would have worked like a charm. Don't lose time speculating what goes on in "the little ogre's" mind, just watch him for your self defense. Such a need, however, did not exactly enhance trust and respect.

What a difference between this undercover sniper who, not for the first time, abused his power of office to thwart someone else, and the agency itself. In September 1976, I received a phone call, and a few days later an official letter, from the National Civil Service League that I had been selected for their Career Service Award, as the only one from the whole Defense Department. I then learned that the NCSL was an organization of private citizens who "honor excellence in public service" by asking heads of Federal agencies to nominate outstanding career employees, and then choosing ten for their prestigious award every year. It was the first time that my agency had an awardee and they were extremely pleased. I believe, it was friendly Mr. Dewey S. Pegler, our Technical Director, who had initiated my nomination. A little while earlier I had already received letters of congratulations from all the channel steps for having been selected as the DMA nominee. Now, upon the news of the award itself, a pleased staff meeting had been called at DIA Headquarters, so I was told, to plan publicity and arrange for my escorts to the festivity at the Smithsonian Institution; and there were again congratulations from all sides. A very special, coincidental yet memorable feature at this festivity was the presence of another awardee, the former astronaut Michael Collins who had walked in space and been on the Apollo 11 flight to the Moon! He was the recipient of a "Special Achievement Award" for setting up the new impressive National Air and Space Museum of the Smithsonian Institution as its director.

4. THE BEAUTY OF THE OCEAN FLOOR

> "He had bought a large map representing the sea,
>> Without the least vestige of land;
>> And the crew were much pleased when they found it to be
>> A map they could all understand.
>
> Other maps are such shapes, with their islands and capes,
>> But we've got our brave Captain to thank
>> (So the crew would protest) and that he's bought us the best—
>> A perfect and absolute blank."
>
>> Lewis Carroll: The Hunting of the Shark

Experiencing the vastness and sameness of the open ocean surface from a ship or from the open beach, one can well empathize with Lewis Carroll's crew. And it is easy and natural to transfer this impressive sameness to the interior of the huge water masses and to the ocean bottom underneath. It is only in this century that modern technology has taught us otherwise.

In antiquity, the need to test the depth of the sea near the shore for precaution against navigational mishaps was originally met by a pole (shown in an Egyptian carving of 1500 B.C.) and later by the lead-and-line technique (mentioned by Herodotus in the 5th century B.C. as an established procedure) which was good enough for two and a half millennia. It took World War I and its need of locating enemy submarines to develop sonar methods. The technological breakthrough of echo sounding replaced the time-honored lead-and-line procedure and gave us continuous profile lines of ocean depth along the track of a moving ship, in lieu of sparse, isolated depth measurements for each of which the ship had to be stopped. This led to systematic exploring and charting of the ocean floor and a startling discovery. All of a sudden, the notion of a smooth ocean bed, smoothed by continuously accumulating debris, was shattered by the realization of great variations in depths from narrow and deep trenches to isolated seamounts and to continuous systems of mid-ocean ridges.

By mid-century, one had an over-all picture of the major ocean trenches, ridges, plateaus, and basins. In the 1960s, the National Geographic Society produced a series of beautiful, colored maps of the ocean floor, based essentially on the bathymetric studies of the Lamont Geological Observatory, artistically augmented and dramatically enhanced to give a stunning painting of the ocean bottom topography, which by far surpasses whatever the most impressive land topography has to offer. At least I was very much impressed and awe-struck.

In the 1970s when I worked on the Kwajalein project, demonstrating how bathymetric charts could be turned into highly useful geodetic information about deflections of the vertical and geoidal undulations, I read with much more than routine interest several articles about newly developed and further planned methods to collect consistent detailed depth information. Such collections were being stepped up in response to military as well as geophysical requirements. The essence of various kinds of such new instrumentation lay in the capability of side-looking sonars in multi-beam arrangements to sweep the areas on both sides of the moving ship out to more than two miles. Such blanket coverage was obviously a big step forward in compiling bathymetric charts. I wished I had had such charts for Kwajalein; or, since that was past, it would be nice to have them even now for testing how much I would have gained in accuracy. More important, I would like to have such charts for a second test area somewhere in the middle of the ocean.

At the ACSM 1974 Fall Convention where I had presented a detailed report on my Kwajalein work, there was a fascinating paper by Mr. William L. Lear of the U.S. Naval Oceanographic Office (NOO) about the application of such new techniques to small seamounts, and the computerized processing from data collection to finished contoured bathymetric charts with unbelievable detail at any scale. From the discussion I gathered that there would be "wall to wall coverage" of the ocean floor, and I asked, of course, whether I could have such charts for Kwajalein or other regions. The chairman who had given me a particularly friendly and generous introduction before my own presentation, said: "For you and your well-known work, Mrs. Fischer, I am sure they'll make them available." He was overoptimistic, however. When I sent a request to NOO for such charts in certain regions, I could not get any, not even with the help of an official liaison. It was not quite clear to me then whether "not-available" meant "not existing" or "not given out". Probably, charts for larger regions were still in the wishful planning stage, and the promise at the Conference had been some premature bragging. One day, someone from Staff brought me an unfinished sample sheet for the region of the Puerto Rican trench, to show me the beauty of its detail. Apparently, there was a discussion that we might get the job of printing such Navy charts. "Could you not use something like that in your work?" "Of course; I would love to have that." "I thought so," he said, "I can't let you have this one yet. But I could not believe it, when your front office said they had no use for something like that. So I came to you directly." (Wouldn't you have guessed!) Then, in fall 1975, I suddenly was in luck. *The Journal of Geophysical Research* (J.G.R.) carried an article on a section of the Mid-Atlantic Ridge (AR) at 260-north latitude, with a sharp reproduction of a detailed bathymetric map. It was made from a very dense survey, from narrow-beam echo-sounding profiles along tracks only a few kilometers apart, augmented by underwater stereo-photographs. Depth contours were given at 200 m intervals and at 100 m for the central part. It seemed like the answer to my prayers: a detailed central part across a conspicuous feature such as the top of the ridge, and a surrounding area—except that the latter should have been much larger for my computational requirements. But let's not be ungrateful in the face of such a windfall! I would start with what I had here, get formulas and procedures ready, see what these would lead to, and worry about supplementing the extent of the surrounding coverage later. To begin with, the map needed to be blown up considerably to allow reading and tabulating the content for the benefit of our dear pal, the electronic computer, in a grid of 5 by 5 minutes of arc and even 2 by 2 minutes in the center. Willie Nelsen helped with that and Phil Wyatt talked to the computer. Soon enough it became clear that our pal was willing to play with us, but wanted much more numbers to play with, that is, numbers way beyond the limits of the given coverage. The Mid-Atlantic Ridge is part of a huge system of underwater ridges running along the middle of the oceans, as shown so impressively on the National Geographic charts. According to the current plate theory, materials well up here from the Earth's interior, rejuvenating the crustal plates which spread out from here to either side, and are subducted into the interior again in another region. The focus of the journal article was thus on the crest of the ridge, about 2000 m under water, and particularly on the very narrow, deep rift valley that splits the crest along its axis, reaching down almost another 2000 m. A cross section would show the ridge as if it were a 5 to 6 km high, very craggy mountain with an incision at the top, and nearly symmetrical, much dissected slopes spreading out on either side for more than ten longitude degrees before reaching the deep ocean bottom. The little map, however, covered less than 1 by 1 degrees and thus included very little of the mountain slopes. For our purpose of deriving from it geodetic information at the ocean surface 2000 m and more above, the sudden cut-off of the data introduced a falsifying effect (truncation error) as if there were a quadratic prism 1 degrees wide and about 3 km high; it masked the variations on the top including the rift valley, which were further attenuated through the distance up to the ocean surface above. Would it not be nice to have the whole mountain feature, not just the rift valley, which in its narrowness was not very conspicuous among the multitude of the other crags.

I turned again to the Navy for extensions of my data, and this time they came through as helpful as ever: in spring 1976, they sent me a newly compiled bathymetric chart, still in the manuscript stage, that contained in its center the little area of the map we were already working with. The dimensions of the map sheet were the standard 6 in latitude and 100 in longitude, and thus covered a good part of the mountain feature. It was contoured at 200 m intervals and showed spot values to the meter at extreme points. Although it had not been made with the new technique, and some areas were covered less densely than others, on the whole it was a far cry from the old spot value charts I had to make do with for Kwajalein. We decided to read and tabulate its content at the same detail of the by 5 minute grid. For easier reading of the maze of contours which seemed to have no rhyme or reason, I began to outline the major contours, every 1000 m, in color, but for lack of time in the office and my increasing curiosity about the major features, I took the work home and engaged my husband to help. We were richly rewarded when out of the maze gradually a pattern emerged. The color scheme brought out an unbelievably dissected landscape of jagged mountain blocks, separated by deep canyons at right angles to the slightly NNE direction of the crest and rift valley. Blocks of the same height are larger in horizontal extent when nearer to the crest and diminish with increasing distance as if they had moved away, and were cut up in the process until dissolved. The deep ocean bottom at the sides was coming in towards the center in these deep canyons like overlong fingers. To enjoy this uniquely stark beauty of the ocean floor and share it with audiences, I made a reduced version of the map for a better focus and omitted the confusing detail within the kilometer contour intervals. From that, colored viewgraphs and slides were made for briefings, and a black and white version with appropriate shadings for the future manuscript.

Lewis Carroll's crew would never have dreamed of so much exciting variety way beneath the vast blank they saw on the ocean surface. Even less would they have cared to know of what's going on beneath the deepest ocean bottom. But scientists do care about the oddest things. Way back in the first half of this century, the Dutch geophysicist F.A. Vening Meinesz and others made gravity measurements in a submarine in an ambitious attempt to decipher these messages of what's happening underneath. And way back at the start of my own geodetic career in 1952/3, I had listened to doubts about the validity and reliability of such measurements, and had contributed a study in the Western Mediterranean Sea dispelling that doubt. Now I came full circle, happy to find a few of Vening Meinesz' gravity measurements within the area of my map sheet. Could they serve as guides in lieu of the land-based observed deflections of the vertical in Kwajalein and tell me what it's like below the ridge? Vening Meinesz thought there must be some isostatic compensation under the ridge. Somewhat to the north of our area, unfortunately not within it, there had been two scientific cruises producing two transoceanic gravimetric and bathymetric profiles, from which scientists of the Lamont Geological Observatory had deduced that there was indeed some isostatic compensation under the ridge; and they had constructed some tentative suboceanic density models to account for the discrepancy between the observed gravity and that calculated from the bathymetry. Could I use this information as a starter and modify it to fit my area? It meant that, different from the Kwajalein work, I would have to concentrate on gravity first in order to establish a reasonable density model, and derive later from that the deflections and geoidal heights. Better get busy! I asked the Navy for two more of those remarkable bathymetric sheets, the ones adjacent to mine to the east and west, which would enable me to squeeze out a little more information about the underworld from these sparse and precious Vening Meinesz messages. The additional maps would not have to be read as closely, since they were only supporting evidence to test my model constructions. With my occasional streak of perfectionism, however, I later regretted not to have taken these sheets home to read and tabulate them myself in full detail over a couple of weekends, since I did not want to impose on the time of my assistants, These side sheets contained the foot of the mountain feature, and that would have made the profiles in my future publications look more complete and prettier, not necessarily more informative.

Formulas for deriving gravity anomalies from bathymetry are related to those for deriving deflections, and I had already formulated them for Kwajalein for testing. Phil Wyatt had programmed them to produce a gravity profile across the Kwajalein atoll, and the comparison with the corresponding profile which Willie Nelsen plotted from the observed sparse values was satisfactory, except for a funny-looking sudden drop of the curve where in my expectation it should not drop. We wondered and pondered about this odd feature which I refused to accept without explanation, until we could trace it to a single ship track where the crew or the instruments or both had obviously fallen asleep on the job: the whole track which led from the deep ocean to the atoll, between islands to the lagoon and out again the other side, showed nearly constant low numbers, wiggling around—40 milligals which might be understandable in the deep ocean but not around the atoll where neighboring tracks as well as our computed values were nearer to 200 milligals. After eliminating this track, the sudden drop disappeared and the comparison was all right. I had considered following up this unexpected capability of identifying phony track results, with a consistent computation of a gravity map for the Kwajalein area to check and fill in the existent deficient one. My clock was pressing on, however, and I had not enough time left to pursue new leads, but had to choose. It was barely one more year until my mandatory retirement and I needed that time for the Mid-Atlantic Ridge.

A call for papers for the annual meeting of the ACSM in February 1977 came, as usually, several months earlier and I submitted an abstract about this study of the ridge, in August, with Phil Wyatt as coauthor. The study was interesting already now and would have a lot more to show by next February. Also Jack Bray submitted an abstract about his continuing study of the South American coastal leveling, and both abstracts were readily cleared, passed on, and accepted by ACSM for their program. The more bewildering was it that, by sudden unexplained royal decree from somewhere inside our stovepipe, these papers and also others were withdrawn, a week after we had received the formal acceptance letter from ACSM in November. Why? They didn't say, other than they wanted to limit the number of DMA contributions. But why that, particularly after they had approved them months before and ACSM had accepted them? They didn't say, or the intermediate bureaucratic links did not kmow. Col.Robert Herndon at the ACSM, called me by telephone to ask reproachfully why I would do such a thing as withdraw my paper after they had incorporated it already into the program. It would not be easy to get replacements at this advanced time. It had not been my decision, I assured him, I did not know the reason, and I myself resented the withdrawal, a thing I had never done before. Both Jack and I would be happy to present our papers if he could straighten things out with the powers that be. He tried but could not. He complained about the sudden break in the traditional support by DMA, which jeopardized their already organized and printed program. ly own resentment turned into sadly amused perception of the managerial muddle when February rolled around and, possibly on grounds of the printed program, Jack and I were given an appointment two days hence to give a dry-run to the department chief. "But don't you remember...?" No, they didn't and wanted me to brief them quickly on the details of the withdrawal order last November, so that these could be relayed up the same stairs they had come down then to begin with.

There had been other, more serious signs of managerial highhandedness, and Col. Jonah's prophesy of the Geoscience Division's demise became increasingly more likely. I only hoped the ship would hold together as long as I needed it, till my retirement next summer. The 1976 Reorganization of the department into the "Department of Surveys and Geodesy" which had taken up all the interests and energies of the administrative world for so long, primarily affected the people in the Surveys Division, uprooting most of them from their Washington homes to Cheyenne, Wyoming. Our division was apparently not involved; but the general unrest and uncertainty about future developments in the agency, and the frequent detailing of people like chess pawns away from their jobs to other units made you wonder about the farce of periodically requiring them to write down their short-and long-range career goals. Who knows what the agency will look like in a few years, and whether the current career possibilities would still exist? The people affected by the move to Chayenne with its curtailed career possibilities could not have foreseen it. In our own division, detailings of undetermined length into other units had left one of the branch chiefs all alone and wondering, some time ago. My own branch had been fairly stable so far, with one person lost to the classified area but another one gained as compensation. Now, however, details and transfers decreed from above, unconcerned about their disruptiveness, became the general way of life.

In September 1976, Phil Wyatt, my senior programmer, was suddenly transferred out of the department into Staff, leaving the Mid-Atlantic Ridge project at a rather critical stage. He had been incorporating the Navy's map sheet, as read and tabulated by Willie Nelsen, into the data collection for the electronic computer, augmenting the first data set from the little map I had found in the journal. There was some difficulty involved because the little map was bounded parallel and perpendicular to the NNE direction of the rift valley, so that the area looked like a square standing on one vertex, while the surrounding map sheet was bounded by meridians and parallels. When Phil computed the first gravity profile along the 26th parallel of the whole sheet and plotted it for inspection, I did not like it because the graph showed an odd dip that should not be there according to my own rough pre-computations. I felt sure that there must be some mixup in the data where the two maps met. The computer collection must have either gaps or overlaps along the boundary, but Phil could not find any. He did not have much time to look but, conscientious as he always was, he offered to show his successor how he had set up things and to help where and when he could. The replacement for Phil's employee slot was not a programmer, however, and some-one else took over part time, although busy with other assignments. Both he and Willie Nelsen, however, were detailed to the classified area a little while later, and that was that. What to do? I was very busy with the theoretical aspects of the project, and I urgently needed an experienced programmer to first find and cor-rect that error, then to compute the various options I was setting up for analysis and further planning, and finally to compute the results. At this stage, a new per-son dd not need to earn much about the background and intricacies of this whole project; as long as he had computer skills, I would tell him exactly what I wanted done. So I went on the lookout for a programmer-computer who might be underutilized and dissatisfied somewhere else (so many were) and could be per-suaded to work for me. The one thing I could always offer was interesting work under well organized and pleasant working conditions. Contrary to some manag-ers' viewpoint, many people don't like being underutilized, and they respond beautifully to well-planned guidance, encouragement, .and appreciation. I was lucky again this time. With the understanding support of another branch chief, Dr. Jerry Kurkowsky, one of the few people to whom I could talk about the won-ders and excitement of my project in those days, and who understood the urgency of my time predicament, I did get even two people willing to help me out: John Brady in November and Paul Berdy in December. They were both godsends.

I quickly distinguished the different capabilities and inclinations of these two men and assigned jobs accordingly. As soon as John Brady settled down to work, I heard that one of my pet little ogres" tried to lure him away but John who knew the ways of the world, refused. He got busy hunting for that error by comparing, latitude line by latitude line, the data holdings of the computer with Willie Nelsen's tabulation sheets. It was a lot of work and took a lot of stubborn patience, but eventually we were both happy and proud with his achievement of a huge clean set of data to play withs the center sheet with the finely read middle part across the rift valley contained 12,204 items, and each of the two side sheets had 8,640 items, together a little less than 30,000 items. You do learn to appreci-ate the electronic computer! In earlier times one would not even have planned to tackle a computation job with so much detailed information. John computed from it the gravity field of the region in five, fairly evenly spaced latitude profiles, with and without compensation from a number of suboceanic density models which I was designing and testing. Because of the great longitudinal extent of these profiles, John subdivided them to suit the time limits on individual research computer runs, kept track of their multitudes, and stored the information in an orderly, readily accessible fashion. I would graph the solutions for analysis and modifications. Both of us enjoyed the looks of these graphs as they grew under our eyes from our continued purposeful industry.

In the meantime, Paul Berdy helped me to squeeze the ten Vening Meinesz pendulum stations in our area for whatever clues they might hold for the sub-oceanic density conditions. It was obvious that there must be considerable compensation also in our area, since the observed values were much,much lower than the ones which John had computed from the rocks above the 6 km reference of ocean depth alone. Thus, below that depth, there must be less dense rocks than assumed as a uniform theoretical reference (particularly beneath the high crest of the ridge), so that the deficiency below would make up for the load of the ridge above, to produce the observed values. I had to construct some density model that would do that. As a starter, I applied the major simplified features of the models from the cruises in the literature to our area, and then modified them accordingly. Studying the relationship between modifications and their effect, I watched how improvements at one site might make things worse at another, and also at what distances all around a station the modifications would still have a significant effect. Since the ten key points did not happen to lie exactly on John's latitude profiles, Paul Berdy wrote a specified program to compute, at their respective sites, the effects of my model variations for further analysis and decisions. Eventually, I adopted a model that seemed to do the job, and it was incorporated into John's profiles for the whole area.

Happily ensconced in technical work, with the satisfaction of having things purr along once again, the vagaries of the current administrative world to which I also belonged, did not really disturb me so much anymore beyond a constant wonder about the difference of the two worlds in which I moved within the same working day. In an almost schizophrenic switch between them I contemplated the sterility and pettiness of the one while there were these beautiful marbles to play with in the other. I had learned long ago to put the administrative chores into perspective. There were the periodic and overlapping housekeeping reports (multiple versions of the same information), evaluations, schedules, forecasts (all to fill the files), Federalese updates and name changes with different alphabet letters, updates of work objectives and manpower needs (pro forma paper games with changes made arbitrarily along the way elsewhere), listings for higher officials (which they could have found in the files already), travel and training needs (which did not mean they were honored), nonsensical estimates of future savings (compared to what?), show business to impress someone (that at least was artistic and fun), ukase and repeated comments to new regulations that had already been decided, etc.,etc. I knew it was important and sometimes useful to be there and play ball, but I had also learned to return the ball quickly and pragmatically, with the little finger of my left hand, so to speak. Most importantly, I had adopted a sort of selective memory which made me respond to and remember the cordial conversations and cooperation of so many people, while shutting out the shenanigans of "the little ogres" and their messengers or understudies, except for a cool analysis to help me decide my next move. I would not let them infringe on my joy of life. Yet, I kept musing why office life had changed so much from the good old days in AMS when, at least within the departmunt, all seemed to play together and managers were leaders and professional colleagues, while now the kids played in different corners with distrust between them. Was it an aberration now and the normal state then, or was the present the normal state as the habitual disparagers of government workers are wont to claim, and was it an exceptional state of affairs then which I had been fortunate to witness and had mistaken for the norm?

It was good to get away from it all for a little while and recharge my batteries. My husband and I decided on a winter vacation in Yucatan to look at another culture. We had intended to do that in some winter, ever since we had been in Mexico the first time, in July 1974, on the occasion of the Panamerican Congress in Mexico City, where I had presented a paper and chaired another session. The friendly reception at the Congress had extended to several offerings such as an evening concert of classical Viennese music on the roof terrace of Emperor Maximilian's castle in the Chapultepec Park, a colorful dance group performance, and a trip to the Teotihuacan pyramids. We spent much time in the inexhaustible Museum of Anthropology, listened there to explanations of the "nueva raza" with equally cherished Spanish and Indian ancestry, compared the ethnic faces in the museum with those on today's streets, and noticed the difference between the historical violence expressed in the murals and pictures of Siqueiro, Rivera, and Orozco with the general friendliness of the people and their gentleness with the ubiquitous, smiling little children. We spent a few days in Cuernavaca and picturesque Taxco, and decided to go see the pyramids in Yucatan, but not right then in July, sometime in winter. The only thing that bothered me with these plans was the Mexican food.

In Mexico City, we had discovered a little Viennese restaurant not far from the hotel of the Congress. It was genuine Viennese cuisine, not just the famous pastries. We went there every single day for "Wiener Schnitzl", "Gulyas" (goulash), and other goodies. In Mexico City, of all places!

And now we were in Merida, Yucatan, watching the 1976/7 New Year's celebration on the plaza of the Cathedral. After an introduction to Maya art in the well-organized Merida museum which shares with Mexico City the treasured finds from the Maya excavations, we set out for Chichen Itza by bus through several exotic little villages and saw primitive thatched huts without windows, poor small houses, and large dominating churches. In Chichen Itza we set up headquarters in a delightful tropical hacienda (with good food and an idyllic swimming pool) to visit the remainders of Mayan greatness.

Although the explanations of the guides are admittedly tentative and controversial, one cannot help but be impressed and awed by what one sees even without explanations: majestic edifices with elaborate and lively sculptures and friezes, testifying to a different way of life and a different perception of its purpose and meaning, whatever it may have been. Apparently, the one site was ceremonial (with steep, square pyramids lifting small temples almost into the sky, a long wall with carved, individually expressive death masks, and the sacred water hole, the Cenote, where precious iayan artifacts have been found), and another site was probably used by the rulers and administrators (with richly decorated, multiroom one-or two-storied buildings called "palaces", and an unmistakable observatory, the means and symbol of power and control over agrarian masses). A third site, only fractionally excavated from the untouched jungle lets you understand the lure to keep digging and piecing together the as yet undeciphered messages of a long gone past. The frequently seen Chac-Mool, a striking reclining figure holding a disc on his lap, is seen by some as waiting for the heart of a human sacrifice, while others (particularly Mayan guides) explain it as a symbol of sun worship. Similarly, some say that the Cenote was used to curry favor with the rain god by ceremoniously throwing in a virgin bride for him, while others term this a misunderstood legend and say that only objects were thrown in as offerings such as were actually found there. The Mayan guides proudly identify with their great ancestors and ascribe the violence in legends and pictures (eagles and jaguars devouring human hearts) to Toltec, not Mayan, origin. The ever recurring theme of the rattlesnake with the plumed tail is interpreted as a symbol of periodic rejuvenation (from shedding and rejuvenating its skin), as is the life-giving solar disc which, in several reliefs, comes out of the rattlesnake's mouth, as does also the face of the sun god or priest. Impressive, sculptured rattlesnake heads protrude from the buildings or flank the bottom of the staircase and cast a sharp shadow which was supposedly used to tell the time of day. An interplay of light and shadow is created everywhere in the architecture with striking effects. We were also told that an acoustic effect enabled the priest at the temple atop the high pyramid to be heard everywhere on the plaza below, but we could not check that. We also were shown an example of a corbeled arch (a "false" arch) where a relatively narrow passage way is spanned by increasingly protruding layers from either side (in lieu of the capstone in a Roman "true" arch). There were more such arches in other places.

We listened to an intriguing explanation of the ball courts, which were obviously important enough to be on the ceremonial site. Ball playing is known for a fact from pictures on carved markers. Its function and fierce competition between teams was said to have been a means of settling the stations in life for the next generation. The members of the winning team, with their victory due to both, their demonstrated ability and the will of the gods, were assigned to the higher offices, and the losers would be the lowly laborers for their lifetime. Serious indeed! Compared to that, our own national fascination with ball games seems to be another ball game" altogether, or is its national appeal touching an unconscious throwback to an ancient foreboding involvement?

In Uxmal, pyramids and palaces were bigger and, if possible, even richer. We already recognized everywhere the motifs of the rain god, the plumed rattlesnake, and the recurring geometric crosshatch pattern, supposedly depicting the rattlesnake's skin pattern. A unique feature in Uxmal is the oblong rounded shape of the huge "Pyramid of the Magician". There is access on only one side, by a very steep staircase with high steps and narrow treads, barely enough for one foot sideways (no railing!)—I did not let my husband try more than the first couple of steps, but other tourists took their life into their feet and ascended. The oval shape was clearly the same as that of today's windowless peasant huts; and we learned that the original temple was on ground level and shaped like the common dwellings. It had been covered and built over repeatedly by later generations who wanted to build their own temple at the same site without destroying the earlier one. The answer lay in covering it with an increasingly higher and wider mound to lift the new temple above the old, which was filled in with soil to strengthen the construction. Careful selective digging had revealed the nested lower structures with their partly preserved sculptured and painted decorations. We had been shown already in Chichen Itza such enlarging reconstructions at the same site. This was quite different from the pre-Aztec pyramids at Teotihuacan, which have been shown to contain solid earth masses almost exclusively; and different also from the Egyptian pyramids. But excavations of Aztec pyramids at the Zocalo in Mexico City, renewed since 1978, revealed the same type of nested rebuilding. (*Science* 80, Nov.1980)

After further brief visits to other places including the ruins of walled coastal Tulum high above a steep ocean cliff, it was time to go home again to our own culture and its demands. Emerging from the depth of time as if from a spell, we felt enriched by this glimpse at the magnificent witnesses of another culture.

The Mayas stayed with me for quite a while. Today's scientific-technical ingenuity finds ways to reach the distant past as well as the deep ocean bottom and below, and the lunar and planetary realms. But there is a difference: explorations of the ocean and space now depend only on money, time, and a consensus on priorities; explorations of the past need all these and also the miracles of survival and discovery of key artifacts. It took the dryness of the desert climate and an accidental find to give us the Dead Sea Scrolls; and it took the insensitivity of the Spanish conquerors to destroy most of the irreplaceable Mayan manuscripts.

15

CATCH 22 AND A STRING OF KUDOS (1977)

1. TYING LOOSE ENDS

Thanks to John Brady and Paul Berdy, the Mid-Atlantic Ridge project progressed as planned, and by February 1977 I had decided on a density model to use as the basis for all further computations. Time was getting perceptibly shorter and needed to be parceled out carefully. Counting backwards from the fixed date of my retirement at the end of July, I allocated the last week or two for "end of business," about six weeks for clearance of the paper to give it ample time for possible red tape hurdles (that means: send it out for clearance in early June), and about a month to write it and assemble tables and illustrations (that means: start in early May). So get doubly busy now to incorporate what was needed for a rounded picture. There always could be unforeseen snags in any research.

The five latitude profiles of computed gravity anomalies with and without suboceanic compensation, looked good and in harmony with all known facts. One could see the zigzag in the graph caused by the rift valley, and I would mark that in the illustrations. Now we computed geoidal undulations and deflections of the vertical, and made contoured charts for each. As a final offering, I made versions of these charts also in the global coordinate system in which corresponding satellite-derived information would appear if detailed enough for a comparison. Clearly, no such information would be available while I was still here, but I would have completed the terrestrial part and taken it toward such a comparison as far as one could, for two very different sites: Kwajalein and the Mid-Atlantic Ridge between 230 and 290 north latitude.

Several visitors showed great interest in this unusual project and its potentialities, among them Phil Schwimmer from DMA Headquarters, visitors from Denmark, Germany, Austria, England, and others. Dr. Saxena, of course, counted on it for his *Marine Geodesy* journal, for which I also helped evaluate other manu-

scripts. Col. Knipling who had said goodbye to DMA at a very nice farewell luncheon, June 1976, to which I had been invited, was teaching at the Naval
Postgraduate School in Monterey, California, and asked for a set of all my papers
for the School where geodesy courses would be added to the Oceanography
Department. The Navy updated its "American Practical Navigator (Bowditch)"
and Mr. Ernest B. Brown of the Hydrographic Center who was responsible for
this tremendous job, asked me to help with the appendix on geodesy. (When the
monumental Volume I of nearly 1400 pages was finally published in fall 1977 in
hard cover with gold lettering, he sent to me at my home a personal copy with
my name in gold on the upper right corner of the cover. Among the long list of
consultants, I was listed as the one for geodesy.) Mr. C. Boucher, France (*Bull.
Géodésique*) visited and mentioned among other things the idea of an IAG Working Group for the history of geodesy. Would I be interested? Yes, of course; and I
gave him reprints of my historical papers. Considering my approaching retirement, however, I did not think I would be able to participate. Wrong!—Now in
1980, such a Working Group has actually been established with J. Levallois as
chairman, and I have accepted his invitation to be a contributing member.

In April 1977, there was a three-day in-house government course on "Leadership and Women," given by the Civil Service Commission within the Federal
Women's Program. There had been quite some propaganda to make us attend,
and so I went to see what they had to offer. Soon I felt out of place, but I also recognized the seriousness of these women, and I reproached myself for feeling
bored and uninvolved. Why really? I decided to stay and use the time to observe
them and to analyze why I was bored, why my experiences were so different from
theirs. For one thing, the title was a misnomer. Although some of the handouts
were literature references for the two topics, I was disappointed that the course
itself did not deal with leadership *per se* in any serious way beyond a questionnaire
exercise with triple-choice answers how to lead a group discussion. The actual
content of the three days was a rather primitive tackling of course objectives as
they were later correctly described in the TOPOCOMENTS: "to learn how to
become more effective at work; how to plan and organize; and the major psychological and social barriers facing women in their efforts to be successful in management." There were enumerations and complaints about sex stereotypes and
sexist language, and there was an exercise for subgroups to produce related drawings or cutouts. Was this an attempt at releasing feelings of frustration? A simple
cutout showed the man lazily watching TV while the woman cooked and
cleaned, with children around her. One large stunning drawing, made in the few
view of himself minutes by an obviously very talented woman, showed Man's

view of himself as the center of the universe with submissive women all around him: wives, playmates, secretaries, etc.; there was no love lost on him for sure. Then there were these silly (yet often used) two-minute role-playing exercises supposed to teach assertiveness, business courtesy, and supervisory fairness. The interesting thing for me was the serious involvement of these kids (all so much younger than I) in these kindergarten-style activities, signaling their need for basic reassurance, which I thought they should have been given at home when they were little.

The reason for my aloofness was the fact that my own generation of European refugees was way ahead of this generation. We had already been raised as "liberated" persons, standing our ground in school, university, and profession as a matter of course. In the crucible of the emigration with the loss of our means and security, we had deeply appreciated the good formal education and professional experience which enabled us to start again from scratch in a new country. We had no time to worry about traditional sex roles and assertiveness training, because we needed to make money to live, and our husbands were our companions in filling that need as best we could. It did not occur to us to "avoid female success" as discussed in the course's literature as a woman's dilemma, because we needed all the success we could achieve to rebuild our lives. Choosing between home and job was a luxury for the well to do. Sexist slurs could not take more time than a fleeting amused smile, but we did teach our children, daughters and sons alike, the value of a good education in earning a living in an enjoyable activity as independent persons.

In fact, my first experience of a sexist slur in this country, still amuses me. Among my early ways to make a little more money, was a small job with Professor Vassily Leontieff at Harvard University in 1942, correcting and grading the homework of his students in Mathematical Economics, until one of these proud Harvard boys somehow found out and complained that a woman was grading their papers. What a sacrilege! Professor Leontieff was very apologetic about having to discontinue my job. Would the women in this course have reacted with a feeling of humiliation and a negative self-image? For me it was a good story to tell about the Harvard scene. I did the same grading job at the math department of the M.I.T. for Professor Norbert Wiener, whb had a whole corner of his office stacked with blue books: "If you don't do these, I am afraid, they won't get done." Apparently, the M.I.T. students did not mind or did not know. Neither did Brown and Nichols, the Preparatory School in Cambridge for the future Harvard boys, see anything inappropriate in my teaching there. But my little daughter picked up a sex role viewpoint from her little playmates on the street:

"Mom, I wish you too would stay in bed until mid morning, take a bath at noon, and then go out to play bridge."

2. AGU FELLOW AND OTHER KUDOS

Coming back from this course on "Leadership and Women" and the reminiscences it had aroused, I found an interesting questionnaire waiting to be filled out, which the President of the Geodesy Section of the AGU, then Obie Williams of DMA, had circulated. It asked for reactions to proposed steps to be taken to enhance the standing of geodesy within and outside of the AGU, and invited related comments. Theoretically, the proposed steps seemed rather obvious, but practically they were inapplicable to DMA employees under the current policy of discouraging and undercutting their professional standing. More symposia? (Would DIAI let us go?) More active cooperation with groups such as ACSM,...(DMA just withdrew Jack's and my and other papers there over our protest, and without giving any reasons.) Unrestricted attendance of scientists at international meetings? (DMA takes them off the travel schedule even if their presence is needed and expected for their papers and offices, and sends instead administrators who can neither contribute nor report back what has been discussed there.) Geodetic journals? (DMA tries to prevent clearances with delays and subterfuges.) If anyone could do anything about such deterioration, which clearly contradicted the official DOD and DA Instructions, it would be Obie Williams with his two hats of a high official in AGU as well as in DMA. So instead of a bland response, as apparently all other AGU members at DMA sent in (I was told later that my response was the only one critical of DMA procedures), I told it "like it is" and documented that with concrete examples.

One day in mid-May, Chovitz was on the telephone: "Congratulations!" What for? "Haven't you heard yet? Maybe, I called too early. You should have received a letter from the AGU that you were elected an AGU Fellow. So I am the first one to congratulate you." I was floored from the sheer unexpectedness of that honor. When the official letter arrived, it turned out to be an even greater honor than I had thought, since it was limited to not more than three percent total of the entire membership at election time. There was a very nice and festive "Honors Banquet" for the awardees, and, as usually, a very nice write-up in the *Topocomments*.

At the AGU business meeting, the gist of the questionnaires had been discussed, but Obie Williams wanted to discuss mine privately at DMA since it was specifically directed at difficulties at DMA. So I wandered over to Headquarters one morning in the following week, armed with a bundle of documentary support for my comments. I stayed for a couple of hours for a pleasant and mutually informative discussion. One of the useful revelations concerned a peculiarly restrictive regulation for GEOS-3 papers that had used Obie Williams' name but misinterpreted his directive. Other recent instructions about publications were obviously written by people unfamiliar with the ways scientific journals function. One, e.g., claimed the right to "select and approve the place and method of publication" for the department chief, blissfully unaware that it is the journal which accepts or refuses or asks modifications of submitted papers. Another directive required that requests for publication approval by DMA be accompanied by "appropriate justification for the publication and the estimated total cost of page charges" (Unaware that the costs are only known after a specific journal accepts it, prints it, and sends the final version in a page proof). DMA and not the author would be the one to officially submit a DMA approved article to the publisher for a particular volume and date (unaware that the journal decides the date depending on its backlog) and assumes reproduction rights (which cuts out copyrighted journals such as the prestigious J. G. R.). Another "guidance" required that prior to initiating a paper, the request for approval must include the need for it (who would decide that?), the estimated man-hours to prepare it (how long does a worthwhile research take that has not even been started?), the estimated sale price and the number of copies for initial printing (both obviously not applicable to journal articles and thus a confusion with entirely different matters). The upshot was that Obie Williams asked me to help him straighten out this silly, uninformed mess by writing a draft for a sensible DMA Instruction. This was easy to do since I had written suggestions in green and in blue for various officials for years on this topic. All I had to do was quote the genuine DoD and DIA Instructions in my files, which were quite supportive of professionalism and recognized its "obvious value for the success of agency programs". A Memorandum from the Assistant Secretary of Defense, 16 February 1977, said: "The opportunity for employee development afforded by participation in professional programs is a significant factor in creating the kind of working atmosphere that helps attract and retain competent personnel." My draft, dated June 13, 1977, referenced the good instructions and cancelled the obnoxious "local" guidances. Whether it was used, I don't know. I have not seen any original publication in the J.G.R. or *Marine Geodesy* or *Bull. Géodésique* after I left, authored by anyone from my agency, other than an agenda listing of a symposium.

On 6 June 1977, I submitted the Mid-Atlantic Ridge (MAR) paper for clearance and sent along the security release obtained on 8 June. It was a very substantial paper and it gave all of us involved a good feeling of achievement after such prolonged intensive work. I had written awards for John Brady and Paul Berdy in appreciation of their assistance (both were, of course, also mentioned in the paper); also for Jack Bray in recognition of his own painstaking research. (His second South American sea level paper would be distributed at the XIII PAIGH Conference in Quito, Ecuador, in August 1977, and he presented it also to ACSM in February 1978.) Ray Shirley had been recognized the year before, and I recommended him now, in an official memorandum to the department chief, as the temporary successor to my position until the whole division was to be dissolved as planned. The reorganization plan, which did not contain a place for the Geosciences Division, had come as an unpleasant surprise to the division chief who had not recognized the increasingly stronger handwriting on the wall during the last couple of years, and had not even been directly apprised of the plan while it was in the making. The directive to restrict the number of high grades in the reorganization raised problems of what to do with the people who had them. Since I was not competing due to my leaving in a few weeks, I was asked to serve on the personnel panel to evaluate the applicants for the few high-grade slots. These were educational days. While I knew approximately how these panels worked, it was still another thing to see it firsthand. The adding up of points assigned to the various qualification types such as education, experience, etc., looked objective enough, but I had heard from comments of previous panel members that if the objective outcome did not please the expectations in a specific case, the vacancy was revoked and shelved for another time with a modified job description. It also happened that people serving satisfactorily, even with an award, in an "acting" position for a prolonged time, were surprised to find that the panel for the final appointment had omitted them from the necessary "best qualified list". Someone who had served on many panels for lower grade positions commented that, while ways to circumvent the intent of the procedure existed and were used, the system in general was still better than the previous complete arbitrariness. The interesting thing for me in the panel where I served in these last weeks, were some of the comments and musings of the members. One comment referred to one of the high-grade slots, which would not be missed if abolished; other desks could easily pick up the little business without any hardship. But that would lose us a slot where to park now an employee who had a high grade. Another comment gave vent to some regrets of having been "the idiot who gave high grades to certain people" who should not have them.—Why did you do that?—Well, at the time it looked like the thing to do. Another queried: did you find out whether the Colonel likes the way this ranking here shapes up, or will we get it back? Also, I had a chance to chat with several people while I waited for the meeting to begin, and was surprised to find that some had definite plans for an early retirement two to three years hence, which would afford them a cushion on which to base another enterprise which was already in the making.

3. CATCH 22

On June 15, the assistant to the department chief returned my MAR paper: it would not be forwarded for clearance because it was not complete, because it did not contain GEOS-3 work. At first I could not believe my ears. This man here knew very well that no GEOS-3 data had come in for my project. And even if they did come in tomorrow by some miracle, there was no time left for me to work with them since I had to leave in July. Besides, had not he himself written a guidance saying that all GEOS-3 papers would be classified CONFIDENTIAL with few special exceptions? So if I still had a chance to add GEOB-3 work, then the paper would be classified and thus unpublishable? I wondered whether this was another of those games to try and suppress my publication. A "Catch 22"? I had better make it quite clear right now that I had no time left to play any games and would insist on clearance. I explained that while this paper was part of the over-all GEOS-3 project, it stood on its own feet as a well rounded, substantial contribution, already long enough (39 pages) to exclude any further topic even if that were available. I questioned the legal justification of refusing clearance for an unclassified research paper in the face of the "freedom of information" instruction, and mentioned that I would seek the help of a senator if necessary. So what are you going to do next? Mr. Assistant said he did not know the answer but would inquire and let me know in a couple of days. I almost pitied him: an otherwise bright person, willing to disconnect his own thinking cap in favor of someone else's illogical and illegal bidding. Now I knew at least, from where the wind blew.

Of course, I did not wait a couple of days for his answer, which could easily have dragged out to a week or a month or more. I resubmitted the paper immediately to the department chief himself, requesting clearance as soon as possible or I would seek the help of a senator since time was pressing. "Oh come on, you know you don't need a senator with us!" And he claimed that he had talked already to the Colonel about it and had informally inquired at DMA Headquarters from where the idea had come to turn the paper back as incomplete. So my hunch had been correct! To save face, the department chief said that the paper had been misinterpreted as the final report for the over-all GEOS-3 project, and as such it was indeed incomplete. But if I were still to write a final report (which I was doing anyway for my Staff Engineer to close out this project which he had funded), then this MAR paper was clearly seen as a phase and would be released. And he promised to help speed it up. It took until 8 July, however, before the paper left the building for DMA Headquarters, although the Clearance Board at TC had indeed released it within a couple of days. It would have taken even longer, had I not kept track of its whereabouts. It was stuck at an office that had to write a formal cover letter for the journey to DMA. When this did not happen for a whole week despite repeated inquiries by Mrs. Leilani Jensen, our always-helpful department secretary, I asked the department chief to go with me to the rescue. It turned out that a minor misspelling in the cover letter still waited to be corrected! I insisted on waiting in that office until I received the corrected cover letter, the signature of the office chief, and the appropriate official envelope, and I took it to the mailroom myself for the next shuttle to DMA Headquarters. Everyone was very nice and helpful: it is just one of those things, we cannot let a misspelling go out of our office; wish you good luck now. And greetings from the chief to my husband whom he remembered from the AMS days. Dr. Macomber at Headquarters had called me the day before to tell me that he had received a letter from Dr. Saxena of the *Marine Geodesy* asking for my MAR paper for his journal. (I had learned my lesson to take no more chances this time, and had alerted Dr. Saxena.) Dr. Macomber said he would help speed up the clearance at DMA. Thank you very much!

Another ten days went by and I had found out who held the paper up. A penny for your guess! After this roadblock, an additional week at least would be needed for clearance in the Pentagon, enough to make it almost too late. Maybe that was the plan? It was clear to me that if I did not get the paper released while I was still here and in in-house contact with all these people, it would be much harder from the outside. I called Obie Williams for help. Could he pry the paper loose and dispense with the trip to the Pentagon, which was legally not even required for a technical research paper?

The following day, July 20, at 800 a.m., the Public Affairs Officer at DMA called me: it had been decided to dispense with the Pentagon trip; the paper had been released at DMA; and I was free to mail it out to wherever I wanted to, even before the clearance copy got back to me. "That's wonderful, thank you." At 8:30, a friend from Staff at Headquarters called me happily with the same message. "Thank you very, very much." At 9:30, Obie Williams called to ask whether I had received the message and everything was all right. "Yes, thanks to you. You certainly made my day!" And off the paper went. ("The Effect of the Mid-Atlantic Ridge in Terms of Gravity Anomalies, Geoid Undulations, and Deflections of the Vertical", *Marine Geodesy*, vol. 2, no. 3, 1979, 215-237) Mr. Hold-up had refused to come to the phone when I had called him the day before to ask his help and to say goodbye. And he had complained to my department chief that I had dared to call. Why shouldn't I? One can even call the President of the United States, and one can talk to God. Or were we prisoners held incommunicado? One cannot say that I had now made an enemy, because he had been that already with respect to all my papers since he was with us. Why? I really don't know. I had greeted this colleague from another agency warmly when he joined us, hoping naively that he would reinforce professionalism in our scientific aspirations. It did not turn out that way. Some observers suggested: professional jealousy and/or osmotic absorption of a certain poison spread underhandedly. Whatever his personal hang-up was, the ensuing behavior points out one of the serious problems in giving power of office to the wrong people in the higher bureaucracy.

4. THE FINAL REPORT

In the meantime, I had been writing the Final Report for the Staff Engineer to close out the project. Because of the hullabaloo around the clearance of the MAR paper, I decided to make it into a comprehensive, 11-page record of wider interest, reviewing the objective of the assignment, its four phases, and their respective results. For the benefit of Mr. Hold-up and his messengers, I delighted in showing the papers he had tried to suppress as the tangible, well-received products of the first three phases, useful in their own right even without Phase IV (the comparison with GEOS-3 results), while Phase IV had been made impossible by the agency's managerial non-support and failure to provide the data for the very job they had assigned and expected to be done. Furthermore, a guidance on publications was in force classifying practically everything connected with GEOS-3; and I continued that this "was corroborated by my own experience when my bathymetric paper for Grenoble: "Deflections and Geoidal Heights Across a Seamount—A Ground Truth Check for GEOS-3 Altimetry," had the innocent part: 'A Ground Truth Check for GEOS-3 Altimetry,' deleted in the clearance process. The lesson was clear: (a) Don't spend time on planning GEOS-3 papers for open publication; you won't get them cleared. (b) Don't even allude to GEOS-3 in any other paper, or you won't get it cleared either. The further intimation through channels that requests for clearance will be turned back unless the paper contains work on GEOS-3, sets the interesting scenario of a Catch 22, considering the lack of GEOS-3 data, the lack of manpower, and the lack of time prior to the date of my retirement."

Having set the record straight in an official document which I planned to distribute to certain people, I relaxed in the happy feeling of all's well that ends well, gratitude to all the good people who had helped me to come out on top once again, and affection for my agency that had given me so many precious years. Why not tell them? I added another paragraph to this Final Report: "Acknowledgmentl Since this is the last Report I'll ever submit to DMATC, I may be permitted the unconventional step of concluding it on a personal note. I joined the Geodesy Department of the Army Map Service in mid-career, at a time when to my surprise as an outsider the size and shape of the Earth was not adequately known. I was privileged to witness and significantly contribute to the glorious decade of geodetic development when this millennia-old problem was solved, first through the historic super long triangulation arcs, then confirmed and refined by satellite geodesy. While searching successfully for new solutions to my agency's requirements, I also had a chance to do a variety of intriguing research studies as byproducts such as applying my newly derived world datum to the topic of the ice-age or to a re-evaluation of the classic determination of the lunar parallax made half a century earlier. And I became involved eventually in the newest geodetic frontier, marine geodesy, with major contributions to the problem of the sea level slopes and the topic of...the utilization of bathymetry, which conclude a long list of timely geodetic research well received in the international geodetic community. Reviewing a rich experience of creative geodetic activities, I acknowledge with gratitude the opportunity given me at the world-renowned Army Map Service, later the Defense Mapping Agency Topographic Center, for a happy and fascinating second career."

On my birthday, I took the day off from the office, and my husband and I drove out to Harper's Ferry for a private celebration. It was a glorious sunny day, and we reminisced on our long happy years together despite harsh life experiences, and we hoped and planned for a happy retirement together. The children were grown and far away, our daughter in California and our son in Massachusetts, but both had agreed to come along for a shared family experience in exotic lands unknown to each of us: a month's trip to several countries in the Far East.

Before that, however, there would be still the farewell party at the office and cordial goodbyes there. And that is where you, the readers, came in many pages ago. If you reread that Introduction now, it will conclude the story—but not quite. There still were some leftovers from work on rigorous formulas, accuracies, etc., for which there had been no more time in the office. I managed to finish that up at home and utilize it in a paper presented to the ACSM in February 1978. ("The Usefulness of Bathymetric Charts in the Computation of Detailed Gravity Anomalies, Deflections of the Vertical, and Geoidal Undulations Due To Specific Oceanic Features," *Proceedings* ACSM, 38th Ann. Meeting 1978, Washington, D. C.) It was a pleasure to be able to do things like a grownup person without the rigmarole of government clearance charades, and also to make up for the enforced withdrawal from ACSM the year before. 5. POSTSCRIPT KUDOS

Some postscripts to my story were written by other people. While we were traveling in the Far East, the PAIGH Consultation of Cartography convening on the other side of the world, in Quito, Ecuador, remembered me kindly with official Resolution No. 9, entitled "Appreciation to Dr. Irene Fischer" containing a beautifully worded "*voto de aplauso*" on the occasion of my retirement. There also was a lovely retirement write-up in the *Topocomments* by my assistant, Ray Shirley. For a short while, I kept in occasional contact with my office to follow the reorganization effects on those I knew, but I soon lost track of the new office names and musical chairs. That closed the chapter, I thought,—but to my surprise, not entirely yet.

In April 1978, I had a surprise phone call from the new Director of DMA, Lt. General A. Martin, who wanted to be the first one to congratulate me on having been selected as the first "Federal Retiree of the Year" by the National Association of Retired Federal Employees (NARFE); DMA had submitted my name and back-up record, and was proud to have the winner. It turned out that this was a completely new award and I had been chosen as the first recipient from ten finalists in a government-wide competition. There were several more congratulatory phone calls from DMA, from President J. F. McClelland of NARFE and others, and many letters including one from the Under Secretary of Defense for Research and Engineering, one from U.S. Senator Paul S. Sarbanes, one from the Mayor of Takoma Park, etc. This award, with my picture as the cover girl of a NARFE magazine issue, and with the news in newspapers and on the radio, had more publicity than all my other high awards and medals taken together. There was a big ceremony for all finalists at the Mayflower Hotel, each introduced by someone from his/her agency. Several high officials of DMA had come, General Martin gave an informative speech and Rep. Gladys Spellman made the presentation of my award. I agreed to be a guest speaker at a local NARFE chapter at some later time to explain some of the mysteries of my profession, called geodesy. My beautiful viewgraphs of my *Basic Geodesy* came in very handy then.

And that was not the last word either. In February 1979, I received two mystifying letters from the President and the Home Secretary respectively, of the National Academy of Engineering, with congratulations at having been elected a member, and with some literature about the Academy and a press release about all the new national and foreign electees, including the citation for each. Mine read: "Pioneering in geoid studies for application to defense..." How would they know? At first sight, none of the many names rang a bell. My husband and I studied the literature and tried to figure things out. But then came letters of congratulations and the puzzle fell into place. My former bosses, the Corps of Engineers to which the Army Map Service belonged before it was turned into the new Defense Mapping Agency, had remembered me. Several people of the Corps were members and they claimed me as one of their own. Major General T.J. Hayes wrote: "Dear Irene: The National Academy of Engineering (NAE) assures me the 'cat is out of the bag' and it is proper for me to be the first to congratulate you on your election to the NAE. I do so, with great pleasure! This is, as you may know, the highest national honor that can be bestowed on one in our profession....I know of no one who better exemplifies the best in meeting such challenges and pioneering new techniques than you..." The Chief of Engineers, Lt. General J. W. Morris, wrote: "...While this is a great "personal achievement for you, I am especially proud of what your accomplishment reflects upon the Corps of Engineers, which you have served so well. We all bask in the honor you have achieved. For that, I thank you."

EPILOGUE: The 1978 Civil Service Reform

On May 2, 1978, President Carter outlined his proposal for a civil service reform in an address to the National Press Club and a message to Congress. There followed a multitude of hearings in Congress where the "Civil Service Reform Act of 1978" was formally introduced into the Senate as S. 2640 and into the House of Representatives as H.R. 11280. Like many other ring-seaters with firsthand, long-time insight into the workings, weaknesses and strength of the nearly 100 year old merit system of the Federal Career Civil Service, I could not help visualizing the effects of some proposal points in terms of concrete circumstances, individual people, and help or harm to agency missions; and, although I had retired from government service by then and thus was not affected personally, I reacted with putting my two cents' worth of reservations into letters to senators and representatives involved in the debate. I also had the transcripts of the many witnesses before the Senate and House committees mailed to me. Inasmuch as this Reform Act did go into effect, the criticisms voiced then are still pertinent now, and they give a good view into the nature of the Reform and its pro-management philosophy.

The witnesses before the committees fell clearly into two camps: managers who single-mindedly and glibly hailed the enormous increase in their power, and employee organizations (professional as well as unions) who pointed to the gross discrepancies between the sure-winner claims of needed improvements and the proposed legislation which actually subverted these claims into abolishing employee rights; it actually eliminated the merit system and subtly resurrected the old spoil system which had existed a hundred years ago before the Civil Service Act of 1883; this would open the doors to ominous politization and manipulation without any serious protection against abuse. The language of these objectors was strong and the evidence factual and clears they had done their legal homework. It foreboded a long time of unrest and battles. Out of the great number of illuminating detail, I will mention here only a few of the major points made by the witnesses.

(1) The President's Task Force for working out this proposal (June—Dec. 1977) consisted entirely of personnel officials from federal agencies, the Civil Service Commission, and the Office of Management and Budget. It was charged that the anti-employee bias of these people whose entire federal career was devoted to contesting employee appeals, complaints and grievances, was already apparent from the disparaging language in the preface of their report. No federal union or other employee organization was appointed to the Task Force. The hearings now were the first opportunity for employee groups to present their side of the case. These facts explained the resentment against stacked cards and the exasperation over an uphill fight.

(2) There was general pleased concurrence with the proposal to split the Civil Service Commission, which played the roles of rule maker, prosecutor, judge and jury in appeal cases, and assign its conflicting management and appellate functions to two separate offices, an Office of Personnel Management and a Merit System Protection Board with its Special Counsel. But it was felt that the latter could not really carry out its proclaimed independent watchdog mission because its chairman was placed on a lower Executive Level than the director of the Personnel Management Office, and the authority of the Special Counsel was much too limited to really protect "whistle blowers' as advertised. There was no real independence from the Personnel Management Office whose director was a political appointee of the President and omnipotent in controlling the whole system at the pleasure of the White House, without the bipartisan checks and controls of the present system. The Special Counsel also serves at the will of the President and can be removed without causes thus not only could serious allegations of impropriety by federal officials go untouched if that is the desire of the White House, but the Special Counsel could even be directed to use his power against employees who are out of political favor.

(3) In the many legal discussions of the bill's reduction of employee rights when appealing an adverse personnel action, some aspects were pointed out as particularly ominous. E.g., the agency's decision to dismiss or discipline an employee need no longer be based solely on the charges listed in the letter of notification to the employee, but other failures during the preceding year may also be considered. Thus an employee would have to defend himself against unknown charges. He is not guaranteed a hearing. If he is allowed a hearing, the burden of proof is shifted to him and made virtually impossible: he must prove that (a) the agency's procedure contained an error that substantially impaired his right or (b) there was prohibited discriminations or (c) the decision was arbitrary or capricious. In other words, the agency can fire a person for any reason except for arbi-

trary and capricious ones, and that person is no longer innocent until proven guilty, but is guilty until he cab prove that the agency made a procedural error or acted arbitrarily or capriciously.

The Administration spokesman claimed that this reduction of rights was necessary because a dismissal of incompetent employees was now virtually impossible. But this was declared an outright misleading myth using incorrect arbitrary numbers instead of official ones that tell a different story. The further Administration claim that better private sector management principles were being introduced was countered with pointing to the omission of also introducing the parallel commitment of the private sector to full employee rights, including collective bargaining, which was not even mentioned in the federal bill. (4) Grave concern about the danger of politization revolved particularly around the establishment of the "Senior Executive Service" (SES) under the proclaimed need for more flexibility and responsiveness. Attention was drawn to two previous very similar attempts at gaining political control of the bureaucracy under President Hoover in 1955 ("Senior Civil Service"), and under President Nixon in 1972 ("Federal Executive Service System"), both of which had been rebuffed. According to the *Federal Times* (March 20, 1978) a proponent of the Nixon proposal had even congratulated President Carter on that similarity. The revival of these previous unsuccessful attempts could create a pool of "yes men" from the supergrade with the incentive of unusually high annual bonuses (up to twenty percent of salary) if they please their political bosses! and the federal retirement fund would be raided irresponsibly by including these awards in the calculation for higher retirement benefits without making the normal retirement (and life insurance) deductions, without the normally matching government contributions, and also allowing a preferential retirement calculation (two and a half percent per award year instead of the normal two percent). In other words, no money for these extra goodies for the top-paid SES would be provided by the government, but it would be taken from the normal retirement contributions of and for the rank and file of current and future retirees.

(5) Under the new "Merit Pay System" an interesting division of the work force is established, by disallowing for the group of GS-13 to GS-15 positions with managerial or supervisory responsibilities the normally periodic within-grade increases and comparability adjustments, and replacing these by varying increases tied to performance appraisal, "within available funds". Since these funds are determined for each agency by the Office of Personnel ianagement before the beginning of the fiscal year (considering the amount of increases which would have occurred for this group under the old system continued for the other

employees), it follows that higher raises for some must be offset by smaller raises for others. I could well imagine that under such a system, a consistently good employee could not be highly rewarded annually because of the rising resentment of the others; thus one would probably let the highest award rotate, defeating the intent. Had I myself not been told by an envious administrator that "with all your medals, it is time for someone else to get recognition!" And there was not even any money involved nor was there a set limit of available medals. Even he could have gotten one, had he accomplished anything appropriate to earn it. This award system with its built-in demeaning expectation of apple-polishing could easily be used for personal favoritism and political manipulation. The impact would not be limited to the group affected, but be felt throughout the work force.

Attention was drawn also to the fact that cost savings were to be an item in the performance appraisal, and supervisors might be tempted to withhold requests for needed funds or deny eserved romotions in order to advance their own chancesfor a merit pay raise.

Although the immediate purpose of these hearings was the evaluation and possible improvement of detailed items of the poposed bill, there were also some sober voices of more general concern. Professional groups advocated that merit recognition through the proposed high bonuses and opportunities at the SES level for executives should be available also to successful individual specialists, thus creating a dual career ladder up to equivalent top levels. As I know from my own case, such dual career ladder had been established by the Civil Service Commission years ago when I was given the "Research Geodesist" title with the promise of personal merit promotions. This promise had been sabotaged. Here was a feeble voice from a professional organization for revitalizing this concept. It was pointed out also that it is not an organization by itself that establishes proper function but the motivations, ethics, and character of leadership. A good manager can motivate, reward, discipline, and achieve under the present structure with little problem. (So true!) A bad manager cannot do well even given the flexibilities of unquestioned authority. (So true again!) So why destroy a system that over-all has worked well? If it has flaws, they should and can be corrected without extensive legislation or excessive costs. There is no need for wholesale changing of the present system with far-reaching consequences, merely for the sake of change. Reminiscing, I saw the procession of managers through the years, inspiring leaders and devious underminers, all under the same system! What for, indeed, do we need such an elephant in a china shop, trampling the good things along with the bad? As all observers of drawn-out wholesale government reorganizations well know and I described repeatedly, they create—not the proclaimed greater efficiency, motivation, and trust in government—but the opposite: widespread uncertainty, restlessness, distrust, and cynicism.

Nonetheless, the Reform Act of 1978 was enacted, the old Civil Service Act of 1883 summarily abolished, and agencies were given until October 1981 to figure out the details and establish the new system. I have saved a few newspaper clippings from that time. In answer to the qiestion "what for exactly?", it was recalled there that President Carter's 1976 election campaign promise had been concerned with government reorganization to eliminate duplication in federal agencies, not with personnel policy. But the White Houses needed a bill to be able to point to a domestic policy success. Considering the public mood of attacking bureaucrats as an easy whipping boy for government failures, a civil service reform looked easy to get, and a spectacular enactment would receive historic attention, while the turmoil of the implementation could be left to those affected who have to grapple with it in the future.

Evidence from the private sector was cited that the dependence of merit pay on performance appraisals made colleagues into competitors, undermined trust, and changed intrinsic interest in the task itself to watching why an anticipated merit pay increase had or had not been forthcoming. Even monkeys were shown to lose their previous enthusiasm in solving puzzles when rewards for success were made the point of the game (Experiments by Professor Harry Harlowe, University of Wisconsin). Besides, not all people are motivated by money above their basic needs (Government spent a lot of money for management courses to teach us to understand the hierarchic levels of motivation) and the best of civil servants are known to be hard working idealists interested in and motivated by quality and service. There were indications that many employees felt uncomfortable with the idea of competing against their peers for pay raises divided from a common salary pot. Some figured they would be better off anyway if they could stay outside this special pay system. But to be or not to be under it, will depend on each agency's interpretation of the rather vague definitions of a manager and a supervisor. It could include nearly everybody, even some who neber thought of themselves as part of management; and the group affected now, the President indicated, could later be extended by Congress to other managers and supervisors.

But if that group is meant to eventually include nearly everybody, then the interesting current acknowledgment of the difference between professional and managerial functions will be lost by the wayside. Yet, if the 1978 Reform is not just a publicity gimmick and a means to legalize political manipulation as charged by some critics, if the rhetoric about restoring public confidence in the vast majority of dedicated federal employees and just dealing with bureaucratic frustration were to be taken at face value, then this difference of functions might have been a useful point of departure for finding the reasons for that frustration as well as a way to cut down on waste. Reviewing my twenty-five years' immersion into the daily life of a vibrant research office (described in these pages as a testimony to the ideal and inspiring circumstances prevailing during the first two thirds of the time and the increasing frustration later on) clearly connects the deterioration of atmosphere with the increase of management layers and the traditional, but mischief-inviting allocation of authority. Let me elaborate a bit.

The typical researcher and the typical manager (convenient abstract notions for discussion purposes) follow different guidelines. One is work-oriented, the other people-oriented! the one wants to find correct objective answers and produce objective results, and the other wants to make other people work by organizing, planning, setting objectives, monitoring, supervising, writing progress and performance reports, accounting for funds, time (of others), and manpower, etc. In real life, however, a competent researcher does most of these same functions as an inherent part of his bid for success of his or his group's research and he resents the interference of administrators. On the other side of the fence, the full-time administrator whb is not usually himself an active researcher anymore, measures his own success not by an objective work output, but by the approval of his superiors. He is a link in the chain of a necessary two-way communication between the research leaders and the agency headquarters, where gvernment policy is translated into authorized agency missions. While this communication is recognized as an all-important function, frustration and resentment revolve around the lengthening of that chain, where Parkinson's Law created multiple management layers with a corresponding increase in paperwork, wasteful multiple briefings, meetings, reports, etc., in an attempt to make up for the loss of message up and down the lengthening chain and to appear informed before the higher links about what is going on in the workshops.

Under the best of circumstances, an administrator in a key position along that chain may also be a competent researcher in his own right and thus not only be able to command respect all around in his work-oriented judgment, but may also be a leader who inspires and motivates his research group and authoritatively presents their efforts to agency heads and the outside. I am obviously thinking of the leaders to whom my story has given warm tribute. Under the worst of circumstances, an inefficient full-time administrator may not be interested or able to discuss research on its objective merits, but sees it only in terms of what it may mean in the eyes of some management superior. Compensating for his own inability or insecurity, he may want to exercise his organizational power by interfering with actual work, piling up busy work (sometimes initiated at the top of government and passed down the line), or controlling the flow of messages through delaying, stopping, garbling, or detouring if personal politicking is his game (does any reader recognize the references to my real life stories?); but he will avoid the risk of taking an official stand or making a decision by hiding behind management group responsibility and the anonymity of a stovepipe system. The management network becomes a slippery entity, a politicking and powerful end in itself, duplicating functions to ensure growth and importance, while the researchers might have thought naively that their work for the agency mission was the *raison d'etre* of the agency in the first place.

Is there a rational solution to preventing the wasteful growth of top-heavy management? Cutting the agency's budget has not had more than a transient and insignificant effect in the past. It tended to ause the dismissal of lower-paid workers and shortages in paper and pencils, but left the high-paid management layers sacrosanct as a rule. But cutting the career-incentive to leave productive professional work for overhead" administrative work might have an effect. Some talented professionals are loath to leave, but have no other opportunity for advancement. But since the professionals and not the administrators do the actual work on the agency's mission, their career ladder should be strengthened to be more prestigious, more lucrative, and endowed with more power and authority than that of full-time administrators whose role, after all, is supportive and not primary. Both career ladders should lead to equivalent top levels so that first-rate, top-level administrators would still get top-level pay.

To utilize managerial talents of professionals, they should be invited to accept additional (part-time I administrative chores without having to give up their primary interest, with the incentive of greater opportunities of influence and research due to a larger subordinate work force; such additional duties would carry an additional proportionate pay at the lesser administrative pay scale, which would cease with the end of that extra duty. Maybe, a scheme of rotation for these chores and opportunities might be set up if desired. It stands to reason that people primarily interested in productive work would not tolerate busy work and would not need multiple formal briefings and reports to understand what is going on. They would function more efficiently since they are inherently trained to distinguish important from less important things, look for the significance of efforts, and use their time expeditiously. The often heard cliché that professionals are narrow specialists not willing or capable to manage is a transparent self-serving label put out by administrators to defend their vested interests. I have heard comments of disgust by professionals who have watched administrative fumbling without being allowed to do the obvious about it. Of my own experience, I only refer to my unsuccessful attempts of many years to be allowed to streamline the publication clearance channels, according to the actual government regulations, common sense, and successful precedents.

Managing tasks requiring full-time attention could still profit from work-oriented managerial talents of professionals by arranging a time-limited leave of absence from the basic position, again with an additional pay at the lesser scale. The real incentive would be the opportunity and variety to serve from a different vantage point. The same would apply to the new "Senior Executive Service"(SES) serving at the pleasure of the White House, similar to the invitation of experts from academia or industry to serve in a cabinet position. But the service should always be time-limited with automatic return to the basic position, to permit this person's continued technical involvement and his maintaining professional standing among his peers. To attract high-quality people who are interested primarily in the project and an opportunity to serve and gain insight, no absurdly high monetary bribes are needed. Conversely, dangling money as incentive, you get people interested primarily in money.

A double career ladder which allocates greater authority, power, and pay to the primary (professional) functions than the supportive (administrative)ones, would do wonders in deflating the lure of currently greener administrative pastures. A realistic solution of cutting government waste by cutting the now all too powerful and sprawling multi-layered management complex down to size, however, does not have much of a chance to even be considered under the present circumstances. But while the idea may be utopian, it is not wholly absurd. Is it?.

LIST OF PUBLICATIONS BY IRENE K. FISCHER

1. "The Deflection of the Vertical in the Western and Central Mediterranean area." *Bull. Géod.*, no. 34, Paris, December 1954.

2. "A New Determination of the Figure of the Earth from Arcs." Chovitz and Fischer, *Transactions*, Am. Geophys. Union, Vol. 37, No. 5, October 1956.

3. Item 2 reprinted in *Revista Cartográfica*, no. 6, Buenos Aires, Argentina, 1957.

4. "Ellipsoidische Parameter der Erdfigur 1800—1950," Review of the book by George Strasser, *Surveying and Mapping*, vol. XVIII, no. 4, Oct-Dec. 1958.

5. "On the World Geodestic System," Proceedings of the Seminar on Military Geodesy, Air Force Cambridge Research Center, Bedford, Mass., Dec. 1958.

6. "A Tentative World Datum from Geoidal Heights based on the Hough Ellipsoid and the Columbus Geoid," *J. Geophys. Res.*, Vol 64, No. 1, Jan. 1959.

7. "The Impact of the Ice Age on the Present form of the Geoid," J. Geophys. Res., Vol. 64, No. 1, Jan. 1959.

8. "Map of Geoidal Contours," unsigned, *The Military Engineer*, Vol. 51, No. 343, Sept.-Oct. 1959.

9. "The Hough Ellipsoid," *Bull. Géod.*, No. 54, Paris, December 1959.

10. "The Influence of the Distant Topography on the Deflection of the Vertical," B. Chovitz and I. Fischer, *Bull. Géod.*, NHo. 54, Paris, Dec. 1959.

11. "A Map of Geoidal Contours in North America," *Bull. Géod.*, 57, Sept. 1960.

12. "An Astrogeodetic World Datum From Geoidal Heights Based on the Flattening f = 1/298.3, *J. Geophys. Res.*, Vol. 65, No. 7, July 1960.

13. "The Present Extent of the Astrogeodetic Geoid and the Geodetic World Datum Derived from It," *Bull. Géod.*, No. 61, Sept. 1961.

14. "Die gegenwärtige Ausdehnung des astronomisch-geodätisch bestimmten Geoids und das davon abgeleitete geodätische Weltdatum, Deutsche Geodätische Kommission bei der Bayerischen Akademie der Wissenschaften, no. 32, MÜnchen 1962, translated into German from the *Bull. Géod.*, No. 61, by L. Kolb.

15. "The Parallax of the Moon in Terms of a World Geodetic System," *The Astron. J.*, Vol. 67, No. 6, August 1962.

16. "Recent Determinations of the Earth's Shape," The Fifth National Surveying Teachers Conference, *Proceedings* 1962.

17. "Comments on: Comparison and Combination of Satellite with other Results," North Holland Publishing Co., *The Use of Artificial Satellites for Geodesy*, Amsterdam 1963.

18. "The Distance to the Moon," *Bull. Philosophical Soc. of Washington*, vol. 16, No. 2, Dec. 1963.

19. "The Distance to the Moon. A Recomputation of Geometric and Dynamic Determinations in Terms of a World Geodetic System," *Bull. Géod.*, No. 71, March 1964.

20. "Triangulation to the Moon," unsigned, *The Military Engineer*, No. 374, Nov.-Dec. 1964.

21. "From the International Ellipsoid to the Mercury Datum," Geodetic Objective Symposium, Cameron Station, Alexandria, Va., *Proceedings*, Oct. 1954.

22. *Geometry*, Allyn and Bacon, Inc., Boston, 1965, reprinted 1967.

23. "The Mercury Datum," unsigned, *The Military Engineer*, no. 375, Jan.-Feb. 1965.

24. "How Far Is It From Here to There?" *The Mathematics Teacher*, vo. LVIII, no. 2, Feb. 1965.

25. "A Study of the Geoid in South America," Translated into Spanish by the Panamerican Institute of Geography and History (PAIGH) for Publication: Un Estudio del Geoide en Sud America, *Revista Cartográfica*, No. 14, 1965.

26. "Gravimetric Interpolation of Deflections of the Vertical by Electronic Computer," *Bull. Géod.*, No. 81, Sept. 1966.

27. "Choosing a South American Reference Datum," unsigned, *The Military Engineer*, No. 379, Sept.-Oct. 1966.

28. "A Revision of the Geoid Map of North America, *J. Geophys. Res.*, Vol. 71, No. 20, Oct. 1966.

29. "Slopes and Curvatures of the Geoid from Gravity Anomalies by Electronic Computer, *J. Geophys. Res.*, Vol. 71, No. 20, Oct. 1966.

30. "A Geoid Profile in North America from a Combination of Astrogeodetic and Gravimetric Data," *J. Geophys. Res.*, Vol. 71, No. 20, Oct. 1966.

31. "The Shape and Size of the Earth," *The Mathematics Teacher*, vol. LX, no. 5, May 1967.

32. "Un Estudio del Geoide en Sud America," PAIGH Publicatión No. 293 de la Comisión de Cartographia, Buenos Aires, 1966.

33. "Deviations of the Geoid from an Equilibrium Figure, *Österr. Zeitschr. für Vermessungswesen*, Sonderheft 25, Vienna, 1967.

34. "A Preliminary Geoid Chart of Australia," *The Australian Surveyor*, Dec. 1967.

35. "Do We Need a New I.A.G. Ellispoid? *Allgemeine Vermessungs-Nachrichten* 74. Jahrgang, Heft 8, August 1967.

36. Gravimetric Interpolation of Astrogeodetic Deflections by Electronic Computer, *Revista Carográfica*, No. 16, Buenos Aires, 1967.

37. "New Pieces in the Picture Puzzle of an Astrogeodetic Geoid Map of the World." *Bull. Géod.*, No. 88, Paris, June 1968.

38. "Investigations Concerning the Astrogeodetic Determination of the Geoid on a Common Datum, Combined with Gravimetric and Satellite Interpolations," Secretary's Report of Sec. V, IAG, Lucerne 1967, *Travaux of IAG*, Paris 1968.

39. "From Pythagoras to a Modification of the Mercury Datum (Fischer 1968), Annual Convention of Canadian Inst. of Surveying, Edmonton, Canada, 1968.

40. "Geoid Determinations," *Proceedings of the 1968 Army Science Conference*, West Point, 1968.

41. "Modification of the Mercury Datum," *The Military Engineer*, May-June 1969.

42. "The Geoid in South America Referred to Various Reference Systems," *Revista Carográfica*, No. 18, Buenos Aires, 1969 (English and Spanish).

43. "The Uses and Limitations of Satellite Geodesy in Specific, Real-life Geodetic Problems. *Proceedings*, Annual DoD Geodetic, Cartographic and Target Materials Conference, Cameron Sta., Alexandria, Va., 1969. (see also PAIGH 1973).

44. "La evolución de la geodesia desde Pitágoras hasta la última modificación del "Datum Mercury," *Bol de información* Núm. 6, Serivico Geografico del Ejercito, Madrid, 2° trimestre 1969, translated into Spanish by Capitán de Artilleria, Geodesta, D. José Antonio Puerta Navarro.

45. "The Development of the South American Datum 1969, Techn. Papers from 30th Annual Meeting, Am. Congr. *Surveying and Mapping*, March 1970.

46. "The Development of the South American Datum 1969," *Tech. Papers from 30th Annual Meeting*, Am. Congr. *Surveying and Mapping*, March 1970.

47. "Constructing a Geodetic Datum that Fits a Continent," *Proceedings*, Fourth South African National Survey Conference, Durban, Natal, South Africa, July 1970.

48. "Specific Practical Applications of Satellite Geodesy," *Surveying and Mapping*, Dec. 1970.

49. "The Geoid in South America Referred to Various Reference Systems," *PAIGH Publicación* No. 325.1 de la Comisión de Cartografía, Buenos Aires, 1970, 1-39.

50. "El Geoide Sudamericano Referido a Varios Sistemas de Referencia," *PAIGH Publicación* No. 325.2 de la Comisión de Cartografía, Buenos Aires, 1970, 41-81.

51. Über die Schwierigkeiten in Einfachen Problemen, Gezeigt an der Berechnung der Monddistanz," *Festschrift zum 70. Geburtstag von Professor Dr. Karl Ledersteger*, November 1970, Österreichischer Verein für Vermessungswesen, Vienna, Austria, 1971, 77-84.

52. El geoide Sudamericano Referido a Various Sistemas de Referenceia," Instituto Geografico Militar de Chile, *Boletin Informativo*, Tercer Trimestre 1970.

53. "Constructing a Geodetic Datum that fits a Continent, Submitted by the Government of the United States of America, United Nations Economic and Social Council, Sixth U.N. Regional Cartographic Conference for Asia and the Far East, Tehran, 24 Oct.-7 Nov. 1970, Item 8, 1-11, Ann. 1, 1-8.

54. *Basic Geodesy, An Invitation into the Mysteries of Geodetic Concepts* (with colored slides, viewgraphs, and narrator tape), U.S. Army Topographic Command, Washington, D.C., 1971.

55. "Contribution to H. Draheim's article 'Die Geodäsie ist die Wissenschaft von der Ausmessung und Abbildung der Erdoberfläche,'" *Allgemeine Vermessungs-Nachrichten*, vol. 78, no. 7, Karlsruhe, July 1971.

56. "A Refined Procedure for Computing Geodetic Distances from HIRAN Observations, Surveying and Mapping, March 1972, 53-60.

57. "Recent Geoid Studies," Secretary's Report of Section V, IAG, Moscow, 1971, *Travaux de l'Association Internationale de Geodesie*, Paris, 1972.

58. "Interpolation of Deflections of the Vertical," Chairman's Report of Special Study Group No. V-29, IAG, Moscow 1971, *Travaux de l'Association Internationale de Geodesie*, Paris, 1972.

59. "A Layman's Search for Understanding," *Judaism*, Vol. 21, no. 1, 1972, 134-150.

Technical Reports, U.S. Army Map Service, No.s 60-69:
60. TR-13: The Deflection of the Vertical in the Western and Central Mediterranean.
61. TR-19: A New Determination of the Figure of the Earth from Arcs.
62. TR-22: A Tentative World Datum from Geoidal Heights.
63. TR-27: U.S. Army World Geodetic System, 1959, Part I: Methods.
64. TR-28: U.S. Army World Geodetic System, 1959, Part II: Results.
65. TR-34: The Parallax of the moon in Terms of a World Geodetic System.
66. TR-51: Conversion Graphs for an Astrogeodetic World Datum.
67. TR-62: Geoid Charts of North and Central America.
68. TR-67: A Modification of the Mercury Datum, Fischer 1968 (Geodetic Mem., No. 1624).
69. TM-5-241-35: Latitude Functions, Fischer 1960 Ellipsoid.

70. "Extended Range Missile Study, First Interim Report, U.S. Army Topographic Command, Washington, D.C., Dec. 1971.

71. "Astrogeodetic Geoid Charts of North America." *Survey and Mapping*, June 1972.

72. *Basic Geodesy, Student Pamphlet*, U.S. Army Engineer School, Fort Belvoir, Virginia, March 1972.

73. Extended Range Missile Study, Threshold Report, Defense Mapping Agency Topographic Center, Washington, D.C., July 1972.

74. "The Role of Africa in the History of Geodetic Concepts," *Festschrift for the 70th Birthday Celebration of Acad. Prof. Dr. V. K. Hristov*, Bulgarian Academy of Sciences, Sofia, 1972/3.

75. "The Basic Framework of the South American Datum 1969," *PAIGH, Revista Cartigráfica*, 1973.

76. "Uses and Limitations of Satellite Geodesy in Specific Real-Life Situations," *PAIGH, Revista Cartigráfica*, 1973. (see also items 43 and 48).

77. *Geodesia Basica*, Sociedad Cartigráfica, de Colombia, Boletin XV, 1972, XVI, 1973, XVII, 1973, translated into Spanish by Ing. Rodolfo Llinas Rivera.

78. "South American Datum," *The Military Engineer*, No. 424, March-April 1973.

79. "The Basic Framework of the South American Datum of 1969," VI Brazilian Congress on Cartography, Rio de Janeiro, Brazil, July 1973.

80. "Deflections at Sea," IAG International Symposium "The Earth's Gravitational Field and Secular Variations in Position," Sydney, Australia, November 1973.

81. "Interim Report of Special Study Group V-29, IAG, International Symposium, "The Earth's Gravitational Field and Secular Variations in Position," Sydney, Australia, November 1973.

82. *The Geoid—What's That?* Defense Mapping School, Ft. Belvoir, Va., 1973.

83. *Geodesie Fondamentale*, Translated into French by the Canadian Topographical Survey Directorate, Ottawa, 1973.

84. "Is the Astrogeodetic Approach in Geodesy Obsolete?" GEOP IV Research Conference, Boulder, Colorado, August 1973, *Survey and Mapping*, June 1974.

85. "The Determination of the Geoid for National, Continental, and Global Purposes," *Memorias* del I Congreso Panamericano, III Nacional, Fotogrametria, Foto-interpretacion, Geodesia, Mexico, 1974.

86. "Deflections at Sea," Journal of Geophysical Research, Vol. 79, No. 14, May 1974.

87. "Deflections of the Vertical from Bathymetric Data," International Symposium on Applications of Marine Geodesy, Columbus, Ohio, June 1974, *Proceedings, Marine Technology Society*, 1974.

88. "A Detailed Deflection Chart for an Oceanic Region from Bathymetry," *Proceedings American Congress on Surveying and Mapping*, September 1974.

89. "The Role of the Geoid in Datum Definitions," International Symposium on Redefinition of North American Geodetic Networks," Fredericton, New Brunswick, Canada, May 1974, The *Candadian Surveyor*, Dec. 1974.

90. "A Continental Datum for Mapping and Engineering in South America, Commission V of the XIVth Congress of the Federation Internationale Géodésique (FIG), 1974, *Surveying and Mapping*, December 1974.

91. "Deflections of the Vertical and Geoid Undulations Across the Kwajalein Test Range," For the use of the Ballistic Missle Defense Systems Command, Huntsville, Alabama. Defense Mapping Agency Topographic Center, Washington, D.C., Nov. 1974.

92. "Esta Obsoleta la Geodesia Astronomica?" Translated into Spanish by P. Compos V. *Boletin Informativo*, III Trimestre, Instituto Geográfico Militar, Santiago de Chile, 1974.

93. "La Structura Basica del Datum Sudamericano de 1969," Translated into Spanish by the *Boletin Informativo*, Instituto Geografico Militar, III Trimestre and IV Trimestre, Santiago de Chile, 1974.

94. "The Figure of the Earth—Changes in Concepts," *Geophysical Surveys*, Vol. 2, No. 1, 1975, 3-54.

95. "Does Mean Sea Level Slope Up or Down Toward North?" Bulletin Géodésique, No. 115, March 1975.

96. "Another Look at Eratosthenes' and Posidonius' Determination of the Earth's Circumference," *Quarterly Journal of the Royal Astronomical Society*, Vol. 16, No. 2, June 1975, 152-167.

97. "A Mnemonic Verse for π," *The Canadian Surveyor*, Vol 29, No. 3, Sept. 1975, 344.

98. Report of Special Study Group V-29, IAG, "Astrogravimetric and Astrogeodetic Methods for Determinating the Shape of the Geoid," Grenoble, France, August 1975. *Travaux de l'Association Internationale de Géodésie*, Paris, 1976.

99. "Deflections and Geoidal Heights Across a Seamount, International Symposium on Marine and Coastal Geodesy, International Union of Geodesy and Geophysics, Grenoble, France, August 1975, *The International Hydrographic Review*, Vol. LIII, No. 1, January 1976.

100. "Does Mean Sea Levl Slope Up or Down Toward North?—A Response to M.G. Arur and Ivan I. Mueller's Comments on my Paper of that Title, *Bulletin Géodésique*, Vol. 50, No. 1, March 1976.

101. "On the Mystery of Mean Sea Level Slopes," *International Hydrographic Review*, Vol LIII, No. 2, July 1976.

102. "An Inquiry Concerning the Present Status of Geodesy, by H. Draheim," Allgemeine Vermessungs-Nachricten, Vol. 78, No. 7, 1971, translated for the Use of the Committee on Geodesy, National Research Council, May 1976.

103. "Marine Sea Level and the Marine Geoid—An Analysis of Concepts," Conference on Geodetic Measurements in the Ocean, University of Illinois, June 1976, Marine Geodesy, Vol. 1, No. 1, 1977, 37-59.

104. "New Concepts in Geodetic Instrumentation, Ghostwritten for Col. W. R. Cordova's Presentation to Ohio State University Symposium on the The Changing World of Geodetic Science, October 1976, with technical input from A. Carlson, E. Cyran, W. Doxey, I. Fischer, j. Kurkowski, T. Robinson, G. Schie-

bel, F. Varnum. Proceedings of the Symposium, Report No. 250, 1977; also in *TOPOCOMMENTS*, 29 October 1976.

105. "Does Geodesy Have a Future? Ghostwritten for O. W. Willaims, Headquarters, Defense Mapping Agency, for presentation to Ohio State University Symposium on the The Changing World of Geodetic Science, October 1976, *Proceedings*, Report No. 250, 1977.

106. "Marine Geodesy: A New Discipline orthe Modern Realization of an Ancient Endeavor?" *Marine Geodesy*, Vol. 1, No. 2, 1977, 165-75.

107. "Author's Reply to Comments by W. D. Forrester on my paper 'Mean Sea Level and the Marine Geoid—An Analysis of Concepts,'" *Marine Geodesy*, Vol. 1, No. 4, 1978, 391-92.

108. "Geoid Undulations, Final Report on Technical Improvement Programs (TIP), Assignment No. 301, Defense Mapping Agency, Topographic Center, July 1977.

109. "The Usefulness of Bathymetric Charts in the Computation of Detailed Gravity Anomalies, Deflections of the Vertical, and Geoid Undulations Due to Specific Oceanic Features, *Proceedings*, American Congress on Surveying and Mapping, 38th Annual Meeting, 1978, 242-51.

110. "The Effect of the Mid-Atlantic Ridge in Terms of Gravity Anomalies, Geoidal Undulations and Deflectins of the Vertical," *Marine Geodesy*, Vol. 2, No. 3, 1979.

111. "Geodesy," *McGraw-Hill Encyclopedia of Science and Technology*, 5th edition, Vol. 6, 1982, 166-71.

112. "At the Dawn of Geodesy," *Bull Géodésique*, No. 55, June 1981, 132-42.

113. "Ropes and Knots," *International Hydrographic Review*, LIX (1), 1982, 58.

114. "The Map as Talisman—Different Ideas in Early Mapping," *Surveying and Mapping*, June 1982, 127-37.

115. "The Role of the Sea and the Seafarers in Early Geodesy," International Hydrographic Review, LX (1), January 1983, 107-18.

116. "Geodetic Triangulation and the Figure of the Earth," *History of Cartography*, Vol. 4, University of Chicago Press [1984].

117. "Geodesy: Historical Introduction," *The Encyclopedia of Solid Earth Geophysics*, ed. David E. James, *Encyclopedia of Earth Sciences*, Van Nostrand Reinhold Comp., New York, 1989, 447-53.

118. "Abraham's Aliyah to Mount Moriah (the Akidah, Gen.: 22:6-8)," *Midrashim of the Saul Bendit Institute*, Fall 1992.

119. "Memories of AMS," (invited by and for the Association of Mapping Seniors), AMS Life, vol. 13, No. 2, March 1992, 7-9.

120. "A Tribute to Brigadier Guy Bomford," Bombford Symposium at the School of Military Survey, Newbury, Berkshire, 5 December 1996.

121. "A Tribute to Dr. John A. O'Keefe," AGU Spring Meeting, 1997.

122. "The Size of the Earth," Encyclopedia of Events, People and Phenomena," *Science of the Earth*, ed. Gregory A. Good, New York and London: Garland Publishers, 1998, 188-191.

SELECTED REFERENCES

Bomford, Guy. 1952. *Geodesy*. Oxford: Clarendon Press.

Bursa, Milan. *Gravity field and dynamics of the Earth* [translated from the Czech by Jaroslav Tauer]. Berlin and New York : Springer-Verlag, c1993.

Deetz, Charles Henry and Oscar S. Adams. 1945. *Elements of Map Projection*. 5th edition. Washington, D.C.: Government Printing Office.

Dille, John. 1958. "The Missile-era to Chart the Earth", *Life Magazine*, May 12.

Hayford, John F. 1917. *Hypsometry. Precise leveling in the United States, 1900-1903, with a readjustment of the level net and resulting elevations*. Washington, D.C.: Government Printing Office.

Hayford, John F. and William Bowie. 1912. *Geodesy. The effect of topography and isostatic compensation upon the intensity of gravity*. Washington, D.C.: Coast & Geodetic Survey.

Heiskanen, Weikko A. 1967. *Physical Geodesy*. San Francisco: W.H. Freeman.

Heiskanen, Weikko. A. and H. Moritz. 1964. *Methods in Physical Geodesy*. Columbus: Ohio State University, Research Foundation. ["Prepared for Air Force Cambridge Research Laboratories, Office of Aerospace Research, United Air Force."]

Heiskanen, Weikko, A. and F.A. Vening-Meinesz 1958. *The Earth and its Gravity Field*. New York: McGraw Hill.

Helmert, Friedrich Robert. 1898. *Beiträge zur Theorie des Reversionspendels*. Postdam: B.G. Teubner

Hosmer, George L. 1930. *Geodesy, including astronomical observations, gravity measurements, and method of least squares*. 2nd edition. London: Chapman and Hall.

Jordan, Wilhelm, Eggert [und] Max Kneissl . 1950. *Handbuch der Vermessung-skunde.* Stuttgart, J.B. Metzler.

O'Keefe, John A., A. Eckles, and R. K. Squires. 1959. "Vanguard Measurenents Give Pear-Shaped Component of Earth's Figure." *Science*, vol.129, 565-566.

Strasser, George. *Ellipsoidische Parameter der Erdfigur 1800-1950.*

Vening-Meinesz, F.A. 1941. *LinkTables for regional and local isostatic reduction (Airy system) for gravity values T=20 km, 30 km and 40 km, R=O km, 29.05 km, 58.10 km, 116.20 km, 174.30 km and 232.40 km.* Publication of the Netherlands Geodetic Commission. Uitgegeven door de Rijkscommissie voor Geodesie. Delft: D. Waltman.

Vening-Meinesz, F.A. 1932. *Gravity Expeditions at Sea. Publication of the Nether-lands Geodetic Commission.* Delft: D. Waltman.

Index

978-0-595-36399-5
0-595-36399-7